YALE AGRARIAN STUDIES

The Great Meadow

FARMERS AND THE LAND IN COLONIAL CONCORD

BRIAN DONAHUE

YALE UNIVERSITY PRESS   NEW HAVEN & LONDON

Published with assistance from the Annie Burr Lewis Fund.

Designed by James J. Johnson and set in Adobe Caslon type by Tseng Information Systems, Inc.

Printed in the United States of America.

The Library of Congress has cataloged the hardcover edition as follows:

Donahue, Brian, 1955–
The Great Meadow : farmers and the land in colonial Concord / Brian Donahue.
p. cm.
Includes bibliographical references and index.
ISBN 0-300-09751-4 (alk. paper)

1. Agriculture—Massachusetts—Concord Region—History.   2. Agricultural ecology—Massachusetts—
Concord Region—History.   3. Meadows—Massachusetts—Concord Region—History.   4. Concord (Mass.)—
History.   5. Concord (Mass.)—Social life and customs—To 1775.   I. Title.
S451.M4 D66   2004
630'.9744'4—dc22
2003026037

ISBN: 978-0-300-12369-2 (pbk.)

A catalogue record for this book is available from the British Library.

10   9   8   7   6   5   4   3   2   1

For my parents, Thomas and Esther Donahue

Contents

Illustrations

PLATES

A Note on the Illustrations

Most of the photographs used in this book were taken by Herbert Gleason at the beginning of the twentieth century. Although Gleason's Concord had seen dramatic changes since colonial times, some aspects of the landscape looked very much as they might have in 1750. Still, readers will notice such anachronisms as wire fences, purebred cows, large dairy barns, and legions of invading white pines. The Gleason photographs are from the Concord Free Public Library collection and are used by permission. The contemporary photographs were taken by Jerry Howard and are used by permission.

The maps were produced from a comprehensive GIS (Geographical Information Systems) map that constitutes the major research for this book. The base layers showing hydrology, surficial geology, and modern roadways and town boundaries were either digitized from USGS topographical quadrangles or taken from the Town of Concord's WebGIS site. The Native Landscape map in chapter 2 is an interpretive map, made by modifying the surficial geology layer.

The land division maps that underlie chapters 4 and 5 were not copied from original plats of Concord, because none exist. Instead, they were constructed from lists of First- and Second-Division land grants in the Concord Town Records. These lists were compiled sporadically from the 1650s to the 1670s and can be found in multiple versions for some proprietors and not at all for others. Boundaries on these maps are approximate, and the placement of some parcels tentative. Nevertheless, they are reasonably accurate as to pattern—that is, the great majority of parcels appear in their correct location relative to known Concord landmarks and to adjacent parcels.

The East Quarter landownership maps that underlie chapters 6 and 7 were drawn from deeds and probate, working back from modern assessors' maps. While some boundaries and dates had to be estimated, on the whole these maps are quite accurate. They were constructed from reasonably complete primary sources, leaving only small gaps to be filled by guesswork.

A Note on the Illustrations

The 1749 land use map featured in chapter 7 is somewhat different. Information on land use was drawn from deeds, probated estates, tax valuations, and field observations. These data were available at or near 1749 for some properties, but for others only at other dates or not at all. Therefore, this is an *interpretive* map in which patterns discovered at several points were extended in a logical fashion across the entire landscape. It is consistent with all the available evidence but goes beyond a simple visual display of that evidence. Like a piece of historical writing, it combines evidence and argument.

Preface

I began the work that led to this book in 1980. I was living in Concord, at Fairhaven Bay on the Sudbury River, in the arcadian interstices of the suburbs. I worked in the nearby town of Weston as a farmer and woodcutter, but my attention had shifted to the more famous place where I lived. Canoeing on the river, walking in Walden Woods, I grew curious about the environmental history of Concord—a story of changes in the land that preceded suburbanization and then continued alongside it, within it. Why had the land gone back to forest since Henry Thoreau's time, but even more interesting, why had it been so excessively cleared during his time? The answers might be embedded in the quiet woods and sleepy river, if I could only find ways to uncover and elucidate them. I didn't want to hear again the timeworn New England tale of rocky hill farms succumbing to hard economic reality because as an ambitious young farmer I wasn't buying it—the land seemed kindly and responsive to me. I wanted answers that went to the sandy soil and swamp muck, the chestnut stumps and huckleberries. Walking in the woods one day, I decided to return to school and try my hand at history.

It soon became clear that mid-nineteenth-century Concord farmers hadn't simply fallen victim to economic progress—they had cut the ground from beneath their own feet. Tax valuation data confirmed what Thoreau had charged: even as the forest was falling to nothing, the town was filling with ragged, depleted pastures. At the same time, a controversy erupted over flooding along the Concord River, which spoiled the town's greatest natural gift, its hay meadows. These ecological disruptions appeared to be the consequence of two centuries spent clearing forest for farmland and wearing it out, a habit of extensive farming born in the colonial era when acres were many and yeomen were few. Even during the nineteenth century, when New England farmers were supposedly improving their husbandry, it seemed they were still frontiersmen at heart, dependent as ever on virgin land—which they went right on *unimproving* until it was used up and gone. In 1850, they were suffering the consequences of a long reign of extraction that millions of other nineteenth-century American farmers were just inaugurating in the vast interior of the continent. That was the conclusion I reached in 1982, as I completed my interrupted undergraduate career.

But there was a problem with the idea that nineteenth-century environmental degradation in Concord was simply the culmination of bad old colonial habits. The timing was wrong. Timing is important to historians — it's the cleanest swing we get at causality. New England historians had shown that by the end of the colonial period our earliest communities were already packed full of farms. Because of steady population growth they were straining at the seams, their soil worn out, their crop yields falling, their children departing for the frontier. I agreed with those who argued that this described Concord between 1750 and 1775. The town had been doubling in population every fifty years up to that point; then, for a century, it all but stopped growing. But there was something odd: why was Concord still full of trees? In the final decades of the eighteenth century, forest covered 30 or 40 percent of the town. If healthy young Concord husbandmen were so desperate for farmland, why not just clear more? Why didn't they convert the last of the woods until almost a century later, before the eyes of Henry Thoreau? Nearly all of it did prove to be farmable, or at least worth a try. Something must have been holding them back, and then that something must have changed.

Clearly, I couldn't understand what went wrong in nineteenth-century Concord without a better grasp of how farming had worked in colonial times. But whatever had been going on back then, especially during the first three or four generations of settlement, was a black box. A few dozen Puritan families went in one end and a couple hundred Yankee farmers came out the other, but as to how they got the place settled with working farms, I didn't have a clue. The first comprehensive tax valuation — which recorded, for each farmer and for the town as a whole, the number of acres plowed, planted in orchard, growing meadow hay or English hay, in pasture, in woods, unimproved; the number of cows, sheep, and swine kept; the bushels of grain harvested, tons of hay mowed, barrels of cider pressed — wasn't taken until 1749. Anyway, such figures by themselves couldn't tell how the diverse landscape of Concord had actually been farmed. I wanted to know not just how many acres of tillage or pasture there were, but *where* they were. What *kind* of land was being plowed, what was being mowed, what grazed? Did each kind of land use stay put year after year, or did some of them move around? *Was* land being worn out? The way to find out was to map a few farms. The only way to do that, as far as I could tell, was to take modern assessors' maps, run the deeds back to the first English owners in 1635, and see what these farms looked like as the generations passed.

This took longer than I expected. For one thing, I was distracted for more than a decade by the demands of running a nonprofit community farm in Weston. That endeavor is described in a book called *Reclaiming the Commons*, which is a companion piece to this one, at least in terms of my own experience. For another, until desktop GIS (Geographical Information Systems) mapping came along in the late 1980s, I found it impossible to create the three-dimensional maps I needed: deep maps covering several thousand acres over three and a half centuries, made up of five- and ten-acre blocks that were owned for a few decades at a time. Once these maps were completed I could slice the database horizontally and display landholdings for any year, or slice it vertically and follow the career of any family. I could see how farms were laid out over soil types. Using information contained in deeds, probated estates, and tax valuations, I could determine how most of these parcels were used at particular dates. Once I put enough tiny pieces together, larger, sometimes unexpected patterns were revealed.

Having grasped the method, I then mapped all the original land grants in Concord, some forty-five square miles divided among fifty or so proprietors into nearly a thousand pieces, in order to reconstruct the layout and division of the town between 1635 and 1673. With the starting point nailed, I had a reasonably complete picture of how land was settled and farmed in colonial Concord, and a story fit to be told. This is that story.

My contention is that colonial agriculture in Concord was an ecologically sustainable adaptation of English mixed husbandry to a new, challenging environment. Mixed husbandry combines livestock and tillage into a single integrated system whereby, among other things, the stock fertilize the crops by recycling or transferring nutrients. As far as I can tell, farming in colonial Concord was not extensive. It did not rely on fresh land, or on long fallows, as its primary source of continued fertility. Instead, it rested upon *husbandry*—on the careful balance and integration of diverse elements across a varied and difficult landscape. Once established, these elements—plowland, orchard, meadow, pasture, and woodland—were for the most part conserved. In particular, this mixed husbandry system relied on manure, derived from meadow hay, to maintain the productivity of the tilled fields. The native hay meadows lay at the heart of the system and were in many ways its most stable, intensively managed component. That is why this book is called *The Great Meadow*.

This finding surprised me. It did not sit well with much I had read about colonial agriculture, and it certainly did not match what I had found for the nineteenth century. But it became less surprising when I examined the history of English agriculture and rural society and considered what kind of people these Concord settlers were. Even if most of the first generation of immigrants had not themselves been farmers in England, their agriculture derived from an ancient tradition of ecological adaptation to a very similar environment. Even if they always had an eye out for profitable commodities they could export, the mixed husbandry upon which they depended for their daily survival and prosperity was deeply embedded in the expectation of long-term family and community life in a well-known place. It was thus bound by a set of ecological and cultural constraints that guarded against unbalanced exploitation of land. It was not designed for efficient production of a single commodity—its first object was stable, diversified production; that is, security. It was meant to provide what its practitioners called a comfortable subsistence. The version that became established in Concord was not without flaws, but it was fundamentally sound. Given small improvements that were within the grasp of preindustrial technology, the colonial husbandry system could have gone on supporting the community for a long, long time. That was, after all, its history over the previous thousand years in England.

Tightly coiled within agrarian life in Concord, however, were two powerful drives for expansion, destabilizing forces that were making themselves felt by the end of the colonial era. The first was as old as agrarian life itself, the second was something new. The old, cyclical disturbing factor was demographic growth. In European agrarian society—as in many human societies—the available land had been repeatedly filled to the ecological limit, forcing painful adjustments. This crisis recurred every few centuries, sometimes ending in disaster, but also sometimes driving innovation. Population growth was spectacularly rapid in New England, so that by the fifth generation the ancient drama was playing out again in old towns like Concord: they were full. Concord had not worn out its old land, and it had not yet run out of potentially farmable new land, but it had reached the limits of certain critical resources in its complex, interdependent system of mixed husbandry. It had become difficult to form any more new farms that included every kind of land required to provide a comfortable subsistence. More and more children had to leave for frontier towns, or come down in the world. How would Concord respond to this increasing demographic pressure? The complex relation between family size and ecological limits was one great tension in Concord's system of husbandry.

The new force in the world was capitalism. Concord was settled as part of a great wave of early modern European expansion that was largely driven by the rise of a revolutionary new economic system of market relations. Farmers were hardly outside this transformation—from the small dairies of East Anglia to the tobacco plantations of Virginia, they were at the heart of it. By specializing in a few profitable commodities for trade, farmers everywhere (at least, those with sufficient land and capital) could seek to increase their prosperity. If successful, in exchange they could purchase goods produced more cheaply elsewhere, and more of them. But once engaged with the market, both as producers and consumers, farmers might be tempted to rapidly exploit and deplete natural resources and the land itself—especially where these were plentiful and labor scarce, as in America. Those who did not might be unable to compete with those who did.

New Englanders certainly participated in this expanding Atlantic economy and its new ways of thinking and behaving, but historians have long debated how soon and how much. Some have argued that the Puritan settlers were thoroughly market-driven from the start. Others have argued that during the colonial period, towns like Concord remained largely isolated from outside markets for both geographical and cultural reasons. But few doubt that by the middle of the eighteenth century the Atlantic economy was gaining influence, even far inland. As generations passed and Concord filled with farms, a second great tension arose: that between the possibilities and requirements of market production and consumption and the ecological and cultural limits of Concord's existing system of husbandry.

These tensions in New England's colonial society have been the subject of a great deal of scholarly discourse regarding the pace of economic growth, distribution of wealth, environmental change, family and gender relations, religious and communal order and discord, and political conflict within the British Empire. What I hope to contribute in this book is a clearer picture of how the activity at the heart of agrarian life, farming, actually went on. How had English husbandry been adapted to New England's soil? and how well was it holding up after five generations? Documentary sources alone, whether descriptions in diaries and travelers' accounts or hard data in tax records, estates, and account books, are not equal to this task. I have tried to bring two new perspectives to bear on the problem: that of the landscape ecologist and that of the farmer.

The ecological approach I have already described. It is embodied in the maps. I have tried to discover how Concord husbandmen made use of the components of their environment, how the elements of their agroecological system were deployed over different kinds of soils. I have done my best to reconstruct how these elements functioned as a simplified ecosystem, the productive role of each, the flow of energy, nutrients, water, livestock, and human labor among them. Above all, I have followed the nutrients because I take the maintenance of soil fertility to be the central problem of sustainable husbandry. But in saying that I have taken an ecological approach, I do not mean to imply that I have paid much attention to the fate of the many *un*husbanded creatures of Concord. What became of native species inhabiting the forests, wetlands, and waters as the English agroecosystem was put into place is an interesting and important question, but it has been pushed to the margins of this study. Without minimizing the losses that accompanied that ecological transformation, it is a concern here only as it affected the sustainability of the system of husbandry that Concord's human inhabitants came to depend upon. Neither have I paid adequate attention to women, in spite of their central role in the internal functioning of this human ecological system, the immense labor (and the great joy and sorrow) of bearing, raising, feeding, and clothing their families. That is because women played only a small recorded part in the subsystem I have investigated, the ownership and working of land. The boundaries of ecosystems are famously elastic, and for this study I have drawn mine around the male world of colonial husbandry, leaving the women within the houses and gardens and the wildlife beyond the pasture wall.

I have already alluded to my own career as a practicing farmer, mainly as a distraction. It turned out to be more than that: it turned out to be another kind of scholarship. When I began this work I had been farming for only five years, and my experience was limited mostly to growing vegetables. As I finish writing *The Great Meadow* I have another twenty years of farming ground into my skin—and my head. Our community farm has given me a chance to practice all sorts of husbandry across a wide range of soils, on a small commercial scale. We have grown fruits and vegetables on twenty-five acres scattered among several fields—some the plowland of colonial farmers and all the better for three hundred years of cultivation. We have planted apple orchards and pressed our share of cider. We have made maple syrup—something my colonial predecessors here could not do because they had no sugar maples to tap. We have logged hundreds of acres of forest for both timber and fuel, some of it regrown farmland, some of it ancient woods. We've used oak and pine not only to heat houses but to build barns, sheds, tables, apple boxes, and the beds of trucks and wagons. For many years my wife, Faith, and I kept sheep on a dozen pastures and helped make hay on several farms in the neighborhood. Now we raise only a small flock of freezer lambs each year on a single pasture—which is far too dry because it is right behind the barn. The location may be convenient, but the home field was placed close to the barn not for ease in pasturing the stock, but for ease in carting the dung. Where we live, good plowland is often poor for grass, and good grassland is difficult to plow.

I don't recall ever reading this simple fact in a book about New England's agricultural history, amidst all the worn clichés about fields full of stones; yet that history is all but incomprehensible without it. It is a thing you learn by watching the arable home field scorch in an August drought, while those rockbound hillsides stay green. If we hadn't been herding sheep in several places at once, I'm not sure I would have noticed this. Of course there is danger in reading your own experience backwards into the past. Colonial farmers had different crops and tools, different economies, different ideas and ambitions. Like any other scholarly implement, personal farm experience must be employed with restraint. Still, I think it can yield insights that might be difficult or impossible to obtain otherwise.

In the spring of 1980, a few months before I began my research on Concord, we planted an apple orchard in Weston. It was on the site of an overgrown orchard across the road from our cornfield, sloping down to a pond. It looked perfect for apples, and I assumed that those who had planted the previous orchard knew what they were doing. They didn't, and neither did we—although we dug the holes two feet deep and carefully backfilled with compost. In my previous book I recounted how our new orchard failed to thrive, stressing the difficulty of overcoming apple scab fungus. Once we went organic, we gave that orchard up—today, we simply mow beneath trees that seldom bear. In that book, I went to great lengths to explain why colonial farmers could grow apples without fungicides, but we can't seem to.

But I didn't tell the whole story. The truth is, some of the apples we planted were scab-resistant varieties, but those didn't thrive either. Worse, we ordered about twenty extra trees and planted them for friends who also wanted a small orchard, on another hillside a mile away. Sometimes we got over to spray those trees after spraying our own for scab, sometimes we didn't. But while the trees we coddled with agrochemicals scarcely grew, the ones we neglected shot up like weeds. While ours were barren, theirs were bearing fruit—scabby, misshapen fruit, to be sure, but the trees themselves were flourishing. It about drove me nuts.

After puzzling over this for a decade or two I think I get it. We planted our orchard on the wrong kind of soil. It was the edge of a plain of glacial outwash, an ice-contact slope of coarse sand and gravel, fine for peaches but too droughty for apples. In dry years, the young trees withered and did not grow. By contrast, our friends' hill was composed of compact glacial till, full of rocks, hard to dig—but it held water. We once grazed sheep on a pasture across the street from it and could depend on the grass even in the driest summer. Those boulder-clay till soils may be tough to plow, but they can be perfect for grass and trees. Then I looked at my maps of the first generations of farmers in Concord, feeling their way onto this new land, and didn't I see them steadily moving their orchards away from sandy outwash and onto stony till? The significance of that would probably have escaped me were it not for the vivid memory of how treacherously easy it had been to sink all those holes through loose sand and gravel. Now, though it costs me more blisters from the crowbar and spade, I know to set all my future apples among stones.

I believe there are two conflicting traditions in the relation between Americans and land. The first, which has dominated U.S. history at most times and places and continues to rule today, has been to use land ruthlessly and efficiently for immediate gain, in spite of long-term consequences. This is the market tradition, and it has brought us Americans (our society, not our farmers) great wealth at great cost and great risk, for we are still fattening on the exhaustion of several kinds of natural and human capital. Yet a second, subservient tradition has been to cultivate nature with more understanding, skill, and restraint and to care for the places where we live as though we meant for our children to live here, too. This current is also very strong and deep, though often submerged—it is not simply an invention of urban romantics but has long been a cherished, though beleaguered, tenet of agrarian life. I believe what is compelling in our agricultural history is not that our interaction with the land has always been one way or the other, all either farming as a business or farming as a way of life; but that these two drives have coexisted, in complex and changing tension, within the same people.

As a community farmer and forest steward I have worked to promote the second tradition, struggling against what I take to be a deepening cultural and ecological crisis of industrial logging and farming. In scholarly work, one is obliged to be a bit more circumspect. I hope you find this book is suitably nuanced. The odd thing is, I set out expecting to find that my forebears on this soil and in these woods were primarily agents of the dominant tradition of market exploitation, perhaps in spite of themselves. What I found instead—and what I think is most significant in the story I have to tell—is that here we have an unusual interlude in American agrarian history in which the tradition of sustainable husbandry was, for several generations at least, more powerful than the extractive drive. Most of us do not want to live like colonial New England yeomen and goodwives again, for more reasons than one. But perhaps we can still learn something from them about how to live in this place, or in any place.

Acknowledgments

In twenty years of research and writing one acquires debts. For help with computers and mapping I'd like to thank Brad Botkin, Kathy Dalton, Ryan Arp, Eric Pierson, and Ted Chapin. Thanks to my colleagues in the National Park Service, Larry Gall, Nancy Nelson, Dan Datillio, Lou Sideris, Terrie Wallace, Doug Sabin, and Barbara Yocum. This work would have been impossible without Leslie Wilson, Marcia Moss, Joyce Woodman, and Barbara Powell of the Concord Free Public Library and its Special Collections. Also helpful in Concord over the years have been Dan Monahan, Joe Wheeler, and Tedd Osgood, among many others. Thanks also to David Wood, Rob Tarule, Jack MacLean, Shirley Blancke, and David Foster for helping to clear up various historical and ecological puzzles. Thanks again to my editors, Jean Thomson Black and Lawrence Kenney. The research was supported by a fellowship from the National Endowment for the Humanities, an independent federal agency; and by a Marver and Sheva Bernstein Faculty Fellowship from Brandeis University.

The manuscript was strengthened by the comments of anonymous readers and of Sam Bass Warner, John Demos, Richard Judd, and Daniel Vickers. I owe more than I can say to a pair of patient mentors throughout my undergraduate and graduate careers, and still today, Donald Worster and David Hackett Fischer. I owe most of all to Faith, Liam, and Maggie, and to my parents and their enduring love of that most exacting kind of mixed husbandry, passionate engagement and critical thinking.

The Great Meadow

CHAPTER I

Introduction

EVERY SUMMER a million visitors come to Concord, Massachusetts, to listen to the echo of the "shot heard round the world." They walk over the rebuilt arch of the North Bridge and read Ralph Waldo Emerson's famous poem, inscribed in a stone monument. Here these daughters and sons of liberty stand where the farmers stood and pay homage to Concord. This remarkable New England town was present not only at the birth of a great democratic republic, but at the birth a few generations later of a distinctive American literary voice and (some would say) of the environmental movement. Lewis Mumford once ranked Concord among a handful of places of greatness in the history of civilization.

But there was an earlier Concord, a place not yet great—though already drawn to the word. The small, dark stream that creeps seaward beneath the bridge was once known as the Great River. Around a bend downstream the valley opens into a broad floodplain of marsh grasses fringed by bottomland forest: the Great Meadow. Today it is the Great Meadows National Wildlife Refuge, dedicated largely to ducks. There is an impoundment for migrating waterfowl, an observation tower, a trail loop for walkers and joggers. On the shore nearby sits the town's wastewater treatment plant. A perfectly fine meadow, you might say, but what's so great about it?

For two centuries, the Great Meadow lay at the heart of husbandry in Concord. It was the largest of the town's many meadows, providing the hay that drove the system of agriculture by which the inhabitants lived. Agrarian life in Concord, as in many other colonial New England towns, revolved around hay meadows and their management and, when necessary, their defense. Long before the Minutemen, and long after them into the days of Emerson and Thoreau, Concordians lived by a complex system of mixed husbandry that was intimately adapted to a diverse local landscape, at the center of which rested the meadows.

Concord was the first inland settlement in Massachusetts Bay colony. English settlers were attracted by the native grass that grew on the Great Meadow. That grass grew here because of what had gone before—other Concords of even greater drama and duration, communities of people and nature, legacies of rock and ice. All cultures inhabiting the place that came to be Concord have been deeply influenced by what went before, whether they acknowledged it or not. Before the English went the Indians. Native American foragers and gardeners lived for thousands of years alongside these meadows and named the place Musketaquid: Grass-ground River or River of Meadows.[1] They not only lived here and named the place, they did much to make the place. Without them, Concord would have been different. Without them, in 1635 there might have been no Great Meadow.

Concord is home to a human presence very nearly as old as the earth itself—that is, as the earth that washed out of the retreating glacier. Before the Indians there was ice a mile thick, a clean start. The glacier left behind a raw mix of soil materials; sandy lands, rocky lands, and moist lands. The lay of those lands, the flow of water through them, the growth of forests and meadows upon them, their long cultivation by human inhabitants—all went to form a place with a particular range of ecological opportunities and limits. Different cultures made different things of those possibilities, but what each did was conditioned both by the land's bedrock nature and by history. The land's nature includes its history. This book is about what a few dozen square miles of land surrounding the Great Meadow were made of, how that land came to be the way it was, and what the English husbandmen who arrived in 1635 made of it. It is about how one human community shaped, and was shaped by, a place to which they gave the harmonious name Concord.

But the Great Meadow itself is not the best place to begin this story. No one lived at the meadow during the era of mixed husbandry. Its mowing lots belonged to yeomen who lived in the nearby village, and later to their descendants who lived anywhere from the village to the borders of the town, three or four miles distant. To discover how Concord was settled and how its system of husbandry took shape, the place to begin is the village. So let us leave our car and walk half a mile south from the bridge to the village square by the meetinghouse, and then east along the oldest street with its grand sugar maples and pleasant New England homes. What is now Lexington Street was first known simply as the "way to the Bay" because between Concord and the tidewater settlements there were no other towns. The steep slope north of the street, later to be celebrated as Revolutionary Ridge, was just the Hill. Buried beneath these handsome Georgian houses lie the first dugouts and hovels that followed the town's incorporation by the General Court on September 2, 1635, which declared that "there shall be a plantation at Musquetaquid and . . . there shall be six miles of land square to belong to it" (fig. 1.1). As we walk, let us make an initial attempt to read the land.

Past Emerson's house, past the Alcotts' Orchard House, a few doors past the Wayside, near the east end of the village, stands the modest Grapevine Cottage. Here a man named Ephraim Bull bred the Concord grape from native stock in 1840, an achievement

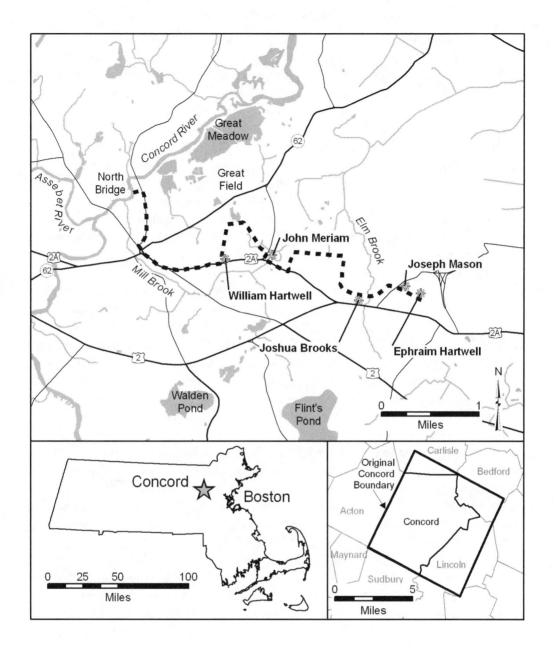

FIGURE I.I.

A walk through Concord. Three maps showing the location of Concord, the original boundary of Concord overlaid on modern town boundaries, and the route of the described walk through the east part of Concord and along the Battle Road trail.

commemorated by a small plaque set into the yard fence. Even the historical marker looks historic, as if Bull had placed it there himself in 1841. Two centuries before Bull this was the homestead of William Hartwell, whose seventeenth-century home may yet reside within the structure we see today. The Concord homestead of this pioneer was very similar to those of his neighbors along the way. Why did these hardy, determined folk choose to settle here, in this place? As the first inland settlers in Massachusetts Bay colony, Concordians could more or less pick their spot, so their choice should tell us something. Were they wise or misled in deciding where to undertake their "wilderness work"?

The immediate advantages of planting dugouts and then raising frame dwellings here under the Hill are easy to see. The long, low, steep ridge afforded a sunny southern exposure and shelter from the cold northwest wind. That they could readily dig into this sandy hillside the Puritans may have credited to the divine hand of Providence, as they did so many things. Most of the credit should go to the glacier, which, had they known about it, they would have deemed a mere surficial expression of God's will, readying the place for them. In terms of geomorphology, the Hill is a kame delta formed of sand and gravel dumped against the flanks of the shrinking continental ice sheet some thirteen thousand years ago, as meltwater swept debris from a canyon in the ice that once stood before us into a great lake that stretched for miles behind us. We stand at what was once the bottom of that body of water, looking up a slope of coarse sediment that rises to the height of the lake's former surface. Such sandy outwash soils are very common in Concord, covering about half the town.

But the glacier left behind more than one kind of soil. The Concord proprietors would have had a much harder time digging in had they settled on packed, stony glacial till, deposited beneath the weight of the ice itself and never inundated by water and sediment. But Providence decreed that most of the stiffer uplands composed of till would be located at the far corners of the square of land that became Concord. In 1635, those rocky outskirts were covered with denser forests than the sandy pine plains flanking the river meadows where the plantation was settled. It was going to be a few generations before many descendants of Concord's founders got seriously into glacial till.

Across the Bay Road from the first houses, just behind us, the English settlers found a long, narrow stretch of native grass they christened Town Meadow and, here farther upstream, Bridge Meadow (fig. 1.2). The soil beneath these meadows had been deposited at the bottom of glacial Lake Concord (so named by geologists more than ten thousand years after it drained away). The fine silt loam of this ancient lake bottom underlies thousands of acres stretching through the center of Concord and off to the northeast. It is fertile ground, but much of it was wet and prone to flooding as the Puritans found it. Was this wetness a liability? Today, our instinct says yes, certainly—without extraordinary efforts such land appears agriculturally worthless. Emerson's improving neighbors agreed, and during the 1840s laid in tile drains. This progressive, nineteenth-century judgment has left the impression that colonial farmers did not do enough to improve the land. But to William Hartwell and his neighbors, this wetland was a great blessing in

FIGURE 1.2.
Bridge Meadow, looking south from the Bay Road near William Hartwell's house site.
Herbert Gleason, 1904.

its "natural" state. The coarse native hay growing there provided the fuel that drove the colonial husbandry system and was for many years its most critical resource.

Through the meadow flowed the Mill Brook. At the west end of the Hill where the brook turned more northwardly stood a weir, built by the Indians to catch alewives running upstream to their spring spawning grounds. This weir was purchased by the English to make a pond to drive their gristmill—another central requirement of an English village. By thus turning a permeable weir into a solid milldam the English made the first of many ecological changes in Concord. It was the initial step in the long, slow transformation of local waterways into a more tractable servant of an English agrarian order, and also the beginning of the end for the fish. Or was it? Dams built across the Concord River itself during the eighteenth century caused controversy because they harmed fish runs, and we know that the fish were all but completely cut off by larger dams during the industrial era. But was this small milldam, together with the other alterations that were made to the Mill Brook during the era of meadow husbandry, so bad for migrating fish? More broadly, at what point did the *changes* in the land undoubtedly imposed by the English amount to ecological *degradation*? Since nature itself is not an unchanging pond but a moving stream and has not run free of human interference since the first day the alewives came up, there can be no simple answer. We can hazard one kind of answer by limiting the question: did the system of husbandry put in place by the first proprietors and their descendants undermine its own ecological foundations, or could it be sustained?

William Hartwell's First Division holdings began with his narrow nine-acre house-lot that stretched from the Mill Brook across the road up to the top of the Hill behind his dwelling. But along with that he owned a dozen other small lots scattered widely among the general fields and meadows throughout the east part of town. He also had rights, as did his neighbors, to cut wood and graze livestock in the undivided commons, which covered the remaining three-quarters of the town's land. What was the purpose of distributing land in such a dispersed fashion? Why was so much of this commons system given up by the time William Hartwell died in 1690? What were the central features of the New England system of husbandry that evolved from it? To find answers to these questions we need to walk from Hartwell's village houselot to the eastern boundary of Concord, crossing the intervening lowlands to reach the heavily wooded uplands three miles away. There near Rocky Meadow, William Hartwell's grandson Samuel was among the first to establish a farm—but not until some sixty years after Concord was founded. This walk will introduce us to the families who will inhabit this book—the Hartwells, Brookses, and Meriams—and to the sand plains, rocky hills, and wet meadows they labored so diligently to improve.

First, we must make our way behind the Hill to the Great Field, as William Hartwell must have often done to reach his planting ground. But this is not an easy task today; and indeed it may not have been all that easy for Hartwell or for his oxen pulling a creaking cartload of manure. The hill behind the village houses is steep, and the back side of the ridge where the land slopes down toward the sandy plain beyond is steep again and hemmed with kettlehole bogs left behind by melting blocks of ice. Concord husbandmen drove their teams to the Great Field by cartways and bridleways that passed through the "vaults and vallies" of the Hill or went the long way around to the east. Let us tramp straight up and over the ridge, as Hartwell's son Samuel may have done when sent home to fetch a mislaid dung fork, and see what we find.

After slipping between the modern houses on Alcott Road and Independence Drive, we drop down the back side of the Hill and emerge from the dark woods into the bright sunshine of a cultivated field. The land here behind the Hill is almost flat, an outwash plain laid down during a lower stage of Lake Concord when the ice front had receded northward. The soil is not quite so gravelly and droughty as the Hill behind us. Still, it is coarse and sandy enough to make us wonder how reliable a corn crop it could have provided, especially in a dry summer such as this. Today, aluminum irrigation pipes with sprinklers on tall risers are being laid through the dark green rows of sweet corn, and the nutrients for this heavy-feeding crop have no doubt been supplied by timely doses of fertilizer. Before the arrival of the English, this was a Native cornfield—part of a shifting "forest fallow" rotation between planting ground and pitch pine forest. But why did the Native women favor such sterile-looking soils for their gardens? Were the husbandmen who followed them misled when they adopted the very same field to grow their own bread grains of English rye and Indian maize? Was this land hopelessly marginal, and did they soon wear it out? Marginal or not, much of it was kept in tillage by their descendants for centuries, and some is being plowed to this day. We stand upon one of the most ancient pieces of arable ground in Concord, and here it is still growing corn.

The English named their largest expanse of planting ground the Cranefield, after a Bedfordshire village where several of them were born. William Hartwell was among those who came to Concord from that open field country in the English Midlands, bringing traditions of common husbandry with him. He owned several tillage lots in the Cranefield along a lane that ran north toward the river, not far from this spot. His plowlands lay intermixed with those of his neighbors. A stretch of sandy plain nearly two miles long and a half mile wide was brought into cultivation during the first few decades of Concord's existence: the Great Field. Here a few dozen husbandmen owned at least forty separate planting lots, ranging in size from one acre to twenty-five. Not every acre in the Great Field was put to the plow, as here and there across the plain were steep eskers and deep kettles that couldn't easily be improved; but every acre was enclosed within a single common fence to keep wandering livestock out of the crops. Thus these immigrants turned what had been a scattering of Native gardens (many no doubt on their way back to forest) into something resembling an English open field, with its arable lands, intervening wastes, and ways of passing with carts and teams. In spite of the decline of the commons system in Concord, the Great Field survived through the entire colonial period, and much of it remained divided into small, intermixed tillage lots well into the nineteenth century.[2]

Three centuries ago, we could have looked down the gentle slope before us across half a mile of open field and seen all the way to the Great Meadow, stretching along the river. The Great Meadow continued downstream more than three miles east and north to the Billerica line and was divided by its proprietors into at least forty parcels. As its name implies this was the largest and most important meadow in Concord, a place where many of the town's leading citizens owned substantial mowing lots. William Hartwell, a man of middling means, owned only four acres in the Great Meadow—and his lot lay nearly three miles from his door, at the lower end of the meadow. Altogether, the Great Meadow provided more than four hundred acres of native hay to winter the cattle. At least, it did following those summers when the river was polite enough to stay out of the meadow during the hay mowing season. For more than two centuries, taming this "great and peevish" river (which appears meek and mild today) remained the central improvement for which Concordians most diligently labored and fervently prayed. Nature was not the only obstacle they faced—they also had to deal with neighboring towns, who had other designs upon the river. A host of issues of ecological adaptation and conflicting access to resources were fought out upon the Great Meadow, all of them lying hidden today beneath a quiet lagoon and a wild marsh. Concord's leading citizens no longer venture there with oxcarts and ditching knives, but with cell phones and binoculars.

Here among the village houselots, the Great Field, and the Great Meadow, we have seen (at least in imagination) how Concord was initially laid out. The combination of the glacial origins of the landscape and the uses to which Native people put it furnished the materials with which the English newcomers set out to construct a familiar, tightly knit common field system of mixed husbandry. By this First Division the proprietors of Concord distributed only one-quarter of their nearly thirty thousand acres; the rest remained as commons until a generation later, when it, too, was divided. How differently was this Second Division land settled and farmed? Let us head east, toward the outskirts of Concord, to look for more clues.

An ancient lane leads southeast back to the Bay Road at Meriam's Corner. Across Bedford Road from the end of the lane stands the old Meriam house, built about 1705 (fig. 1.3). A second, even older house apparently once stood at the corner, and on the afternoon of our walk archaeologists are painstakingly searching for its shadow on the subsoil, scraping a shallow trench across the parched yard toward Lexington Street. The Meriam house and much of the property for several miles along the road to the east are now part of Minute Man National Historical Park. It was at Meriam's Corner that the running battle resumed on April 19, 1775, as the Regulars tramped back to Boston. The old lane down which we have just ambled was the route no doubt followed by many of the militiamen, hastening to reach the corner and fire on the British flank.

Meriam's Corner marks the eastern boundary of the original village, and the beginning of a new phase of land distribution and settlement in Concord—the Second Division. Concord's first generation of settlers did own a few private parcels of upland and meadow lying far from the village, but by and large they made their *homes* within a mile of the meetinghouse. The Second Division began in 1653 as Concord's second generation of husbandmen sought landholdings of their own. During the next few decades the proprietors of Concord divided almost all of their remaining common land among themselves. This sweeping act of privatization enabled each family to pass substantial parcels of land directly to their children, to settle and use as they saw fit—although, as we shall see, they remained so tightly enmeshed with neighbors and kin that one should not imagine that their use of land became completely private. During the second generation of farm making in Concord it was largely the second mile from the village center that was settled. Typically, one son—often the youngest—remained in the village and inherited the dispersed holdings that had been worked by his father, after older brothers had dispersed to more consolidated Second Division lands.

John Meriam was of the second generation. The small collection of meadows and up-

FIGURE 1.3.

Joseph Meriam House, built ca. 1705, from the Bay Road. Probable site of John Meriam
House, built ca. 1663, vanished ca. 1805, foreground. The trail begins. Jerry Howard.

lands that John and his descendants accumulated here at the corner formed the nucleus
of a swarm of Meriam family farms for the next five generations, well into the nineteenth century. As fast as the Meriams gathered land, they divided it again among their
offspring. Often there were three or four Meriam households at the corner, their dwellings and shops within easy hailing distance of one another, their nearby home fields and
meadows lying intermixed. Many of these Meriam brothers and cousins lived as much by
blacksmithing and locksmithing as by husbandry. The Meriams furnish an example of
the development of a small neighborhood of tightly interlocked family farms and trades,
one of the typical patterns of the colonial period. To get a look at other opportunities afforded by the Second Division, we need to continue our walk east, away from the center
of Concord.

The Park Service has built a trail from Meriam's Corner east to Fiske Hill in Lexington, sometimes closely paralleling (or even running upon) the old Battle Road where
the king's army marched, sometimes winding through back lots to follow the route that
might have been taken by the militiamen as they raced ahead to engage the Regulars
at the next bend in the road. The trail crosses a branch of the Mill Brook, circles north
through a hayfield and woodlot, and emerges on top of a rise about half a mile east of
Meriam's Corner and a quarter mile north of the Bay Road. Here we may as well have
stepped into another world. The suburbs have disappeared. Before us, plantings of young

{ 9 }

FIGURE I.4.
Looking north from the Bay Road. Brickiln Field, lower left. Overgrown wetland in Island
Meadow, center. Brickiln Island, top right. Field and meadow divisions lie much
as they did in the seventeenth century (see plate 5B). Jerry Howard.

corn, winter squash, and pumpkins run down along an ancient lane to the rear of a white
farmhouse and barn. The earth between the crop rows is dark and freshly cultivated and
looks moist and cool in the summer sun. If it weren't for the cars passing silently on the
road, out of our hearing with the light northwest breeze over our shoulder, we would not
be sure into what century we had strayed.

We are standing in the midst of what three and a half centuries ago was the Brickiln
Field (fig. 1.4). Where the brick kiln itself stood we don't exactly know, but there is a band
of clay in this field, so it would not have been far. William Hartwell owned a small parcel
of plowland here, but in time it was sold to a neighbor. The Brickiln Field was managed
as a general tillage field until late in the seventeenth century, and so was excluded from
settlement by the second generation. Not until the third generation would the grandsons
of several original proprietors establish houselots along the Bay Road within the field and
begin piecing together irregular farms that ran from the road back through these rolling
uplands and meadows. But working against each generation's efforts to consolidate their

holdings was the countervailing force of fragmentation arising from partible inheritance. We will take a close look at what it took for an aspiring young husbandmen to assemble a working landholding in colonial Concord and then to provide in his own turn for each of the large family he had sired to help work a yeoman's farm.

Behind us, beyond a low stone wall, another cornfield runs down the slope to the north toward a distant hayfield in the meadows, where a green John Deere tractor is circling and baling hay in the early afternoon haze—a sight now grown almost as rare in Concord as the stalwart forkers and stolid oxen that preceded it. Rising over the faint chug of the machinery comes the periodic whack of the kicker, reaching us while the tumbling bale is already high in flight toward the trailing wagon. Beyond the hayfield we can see the encroaching red maples and purple loosestrife in the swamps along the Mill Brook, in what was once called the Dam Meadow. Why a meadow needed a dam is another question we will explore. To the east, the lane runs down to another stretch of brushy wetland called Island Meadow, crosses the brook, and rises again to cultivated ground on Brickiln Island: but that low gravelly mound probably never boasted a brick kiln, and it is not an island. Other than that, it is accurately named. It no doubt came to be Brickiln Island because it lay beyond the Brickiln Field, and it was called an island because it was completely surrounded by low-lying meadows. Simply to understand how this landscape was named obliges us to adopt the low, meadow-eye view of Concord husbandmen.

To the southeast rises a bench of sandy outwash, swept against the hills beyond by meltwater when the ice front stood where we stand now, and Lake Concord had not yet formed. We head that way, crossing the intervening meadow. This upland was Second Division granted to Thomas Brooks, a merchant from London, and was settled by three of his sons. Where the trail cuts behind the old Brooks farmhouses on the north side of the Bay Road it passes through young woods. Here and there among the slender ash, oak, elm, and maple trees are attenuated apples, reaching desperately for a little sunlight, slowly failing to find it. We are walking through an orchard abandoned some decades ago—but for how long did this orchard stand before that? Surely these old trees are not colonial, although most colonial farms did have cider orchards. Apple trees have occasionally been known to live as long as two centuries, but too much change has taken place here for many colonial apples to have survived. This stretch of road boasted several estates during the nineteenth century and a commercial orchard that lasted through the first half of the twentieth century. Although they may not be colonial themselves, these old trees nevertheless recall the long history of orchards in this landscape and their important role in colonial husbandry. What these people drank may tell as much as what they ate. Why did these confirmed English beer drinkers turn to apple cider? The answer, as we shall see, is ecological.

The trail passes close behind the Job Brooks house, recently restored by the Park Service. Job was a grandson of Joshua Brooks, the man who built one of the first houses here on his father's substantial Second Division grants about 1660. For six generations Joshua

Brooks and his descendants also ran a tannery, meeting a critical need in the local economy, and until the nineteenth century they prospered and their farms spread over the surrounding uplands and meadows. The tanyard was in the meadow beside Elm Brook. Tanning combined the bark of black oaks from the rocky hills above with the hides of red cattle from the grassy meadows below. Even today, the brook runs tea-colored with the tannin of fallen leaves, and we can imagine that it once imparted a certain leatheriness to all those who lived alongside it. This tanyard reminds one of how important wood was in Concord, and not just for timber and fuel. The forest soaked into the look, the feel, and even the flavor of this world—into its shoe leather, cider barrels, huckleberry puddings, and the plain color of its everyday clothes.

But the tannery also reminds one that cattle were ubiquitous in this economy, providing not only milk and meat but also manure and locomotion, along with everything from leather to candles. The husbandmen who lived here were obliged to supply the great bulk of their domestic needs directly from local resources, and this required a diverse, integrated approach to the land. It also demanded of their wives and daughters an equally broad range of skills and even greater stamina, to transform those diverse raw materials into food, clothes, and other household goods. That close relationship with the local landscape was paramount in shaping their lives. It was reflected in the household economy, and more completely in the dense exchange economy of the community as a whole. At the same time, these families were engaged with the emerging Atlantic market to obtain cash to meet some of their needs—most notably to acquire land itself. To this end, too, cattle were a crucial means of accumulating wealth. The cattle lead us down both the market highways and the local cartways of this economy and always bring us back to the ecological bottom of the matter, which was the grass growing in the meadows.

Beyond the Brooks tannery the path crosses Elm Brook and the Tanyard Meadow, a vibrant magenta on this July afternoon (fig. 1.5). This meadow was the uppermost in a long string of meadow lots that ran along Elm Brook for almost four miles northeast into the present town of Bedford, and in a second string curling westward along a branch of the Mill Brook three miles to the Concord millpond. We began by the Mill Brook at William Hartwell's houselot and crossed it at John Meriam's home meadow. Now we are crossing it again—or are we? On some old maps this was called Mill Brook, on others, Elm Brook. In fact, it was both. The water parted just below where we stand and flowed two ways, north and west, into two rivers, the Shawsheen and the Concord. No brook would naturally behave in that fashion. The Brooks brothers had a hand in this diversion, which later generations nevertheless judged to be natural. The changes they engineered and the legal wrangle that ensued will give us deeper insight—by about two feet—into the intricate work of improving these meadows for the employment of husbandry.

The Brooks family was endowed with extensive meadows, lying primarily in the upper end of Elm Brook Meadow stretching from here north to Brickiln Island. Ample hay allowed them to keep large herds of cattle and ensured them a modest prosperity—a competency, or comfortable subsistence, as they expressed it. The key to it all lies be-

FIGURE 1.5.
Crossing Elm Brook Meadow, near the site of the Brooks tannery. The uplands are now overgrown with hardwood forest, while the meadow lots, stretching away to the north, have been invaded by cattails, loosestrife, shrubs, and trees. Jerry Howard.

neath us in this wet meadow ground. But it seems hard to believe today, as we cross the marsh on a long boardwalk. Even in this dry summer the Tanyard Meadow is surely too soft underfoot to support oxen and hay carts. It is too choked with brush—led by purple loosestrife, the unstoppable wetland scourge of the twentieth century—to feed cattle. A remarkable change has come over these meadows, all but erasing them from memory. As we have walked along, we have seen meadows overgrown with red maple, buckthorn, alder, loosestrife, cattails, and sedges. Most of these are native species, so this change has not been simply a matter of foreign ecological invasion. The change has been as much cultural as ecological—we value this wetland now as habitat for wildlife and for its ability to regulate streamflow and recharge aquifers. And so, by law, we now "protect" what we once "improved."

All along our walk we have glimpsed shallow ditches choked with silt, muck, and a little nearly stagnant water. Getting water to drain from the sprawling wetland that covered the low spots in this ancient glacial lake bottom was clearly no casual matter. Having arrived here in Elm Brook Meadow from the uplands, the water had no great ambition to move on. Yet a little archival digging reveals that within a few decades of the arrival of the English in Concord, the flanks of these wandering brooks were lined with dozens of native hay mowing lots. How much of this meadowland was natural, and how much was a work of cultural transformation? Was the resulting system of husbandry that relied on these meadows sustainable, or were there ecological flaws that ultimately undermined the land's productivity, leading to its abandonment? How comfortably did this meadow hay husbandry suffice once Concord itself filled to the brim and had to be drained of excess farmers?

Beyond Elm Brook rise the glacial till uplands of what is now Lincoln, a town formed in 1754 from the rocky corners of Concord, Lexington, and Weston. Lincoln is our local version of a hill town.[3] This upland marks the beginning of the third mile from the Concord meetinghouse, running eastward from here to the old Cambridge line. We have reached a section of Concord not much settled until the third generation, at the very end of the seventeenth century. Here some of the grandsons of the original English settlers came to try their own hand at improving farms and achieving a comfortable subsistence, in the established manner of their forefathers. How well did this generation and *their* offspring fare on this hard ground? We need to press on, up the hill to the east.

There is a dramatic change as we pass under the trees on Elm Brook Hill. The temperature drops a good ten degrees. In spite of the summer drought, the ground underfoot is dark and moist with a lush undergrowth of blueberries, viburnum, and ferns. As we cross the stone wall at the meadow's edge we see the slope ahead is strewn with angular stones—a classic glacial till. But along with the granite boulders, this ground holds water—much more the sandy terraces we have traveled so far. And the trees show it: despite the seemingly thin, stony soil this is an excellent hardwood stand, dominated by straight, vigorous red oaks and red maples. White pines two feet in diameter shoot up through the hardwood canopy, and even a few unusually tall pitch pines are scattered about. Given all the large stones still on the surface, this is obviously land that has never been plowed. Yet it certainly can grow trees.

We pass up the slope, quartering northeast to keep to a gentle grade, admiring the verdant woods. Surely forest is what nature intended for this rocky land, an ecosystem that could flourish undisturbed for centuries, renewing itself for thousands of years. And perhaps it could. But then, we encounter another reminder of the dramatic changes this land has in fact undergone. The trail reaches the stone wall that marks the north line of the parcel, and growing there against the wall is a magnificent white oak tree, well over a century old. Its great lower limbs reach out toward us in that sturdy, knotted, grandfatherly manner of white oaks. Of white oaks, that is, that have grown in the open. And so we realize that we have been walking through the former pasture into which the branches of this oak once freely spread, before the forest grew back up around it (fig. 1.6). Cows loitered here in its shade, it may be, in colonial times.[4] This land may have grown grass for as long as two centuries—but not anymore. A pine forest most likely sprang up here more than a century ago, as was often the case with abandoned pasture. When that blew down in the hurricane of 1938 (or was cut down in the early part of the twentieth century), these hardwoods followed. Here, as in so much of New England, where forest once fell to agriculture, it has risen again.

Are we looking at the implacable verdict of the market on land too marginal to be profitably farmed? This is the standard conclusion, if not the starting assumption of many accounts of New England history. Were Concord farmers pressing onto impossibly stingy soils when they reached these stony hills—just as, a century later, their progeny would fling themselves against the still more forbidding slopes of Maine, New Hampshire, and Vermont and briefly drive back the forest there before giving it up as a bad job? As we continue up the slope beyond the oak tree we pass another wall, this one disappearing into the woods to the north. This double-thick wall contains many smaller stones, and heaped against it at intervals are piles of granite shards and cobbles. Some husbandman (if not generations of husbandmen) industriously sledded innumerable loads of stones to the lower edge of this small hilltop field. Year after year they removed rocks in order to mow the field for hay or perhaps even sometimes plow it for grain, and when the wall had absorbed all it could hold, they simply dumped the rest. How richly was that monumental labor rewarded?

The trail reaches the top of the hill and rejoins the Battle Road. The upland land-scape, now mostly reforested, is honeycombed by stone walls on every side.[5] We pass the Bloody Curve, where some of the heaviest fighting took place on the road from Concord to Lexington, the militiamen finding cover among the many trees that stood in these pastures, orchards, and woodlots. Just beyond the curve was the small farmstead of Joseph Mason, a currier of leather who lived here at the time of the battle, a poor man. Before Mason, this rocky outpost was the homestead of a cooper named Timothy Cook; before that of a weaver named Ebenezer Brooks, who gave the place up and moved to Grafton in the 1740s. For all the success of the Brooks family at carving out farms in Concord

FIGURE 1.6.
Pasture oak and stone walls, young pine woods advancing in background.
Herbert Gleason, 1899.

FIGURE I.7.
The centuries meet at the Hartwell Tavern: eighteenth-century building, late nineteenth-century sugar maples, late twentieth-century visitors, interpreters, and rebuilt stone walls. Jerry Howard.

and Lincoln, in every generation some of their progeny looked beyond Concord for new homes. Before Ebenezer Brooks, one Joseph Wheat farmed here for a few years. Where he went, I cannot tell. Wheat bought the place in 1706 from Benjamin Whittemore, who had lived here for only a short while before thinking better of it and moving to another homestead half a mile farther east. Whittemore had bought fifteen acres of partly improved upland with a house in 1692 from a man named Moses Whitney, who appears to have been the first to settle here among these stones. This litany of broken ownership and hasty removal—at least six families in less than a century—seems to confirm our suspicion that this glacial till upland was not the sort of soil to hold sons and daughters to its bosom for long generations.

And then, a few more steps down the road, we come to the Hartwell Tavern (fig. 1.7). Ephraim Hartwell built this house, now maintained by the Park Service, on land given him by his father, Samuel, in 1733. Samuel Hartwell had homesteaded here in the 1690s and built the next house down. That ancestral hearth is now represented only by a bolted timber frame surrounding the original central chimney—the house itself burned down in the 1960s. With the tavern and the cellar hole nearby, we have arrived at our destination: Samuel was the grandson of William Hartwell, at whose homestead in Concord center our walk began. Samuel Hartwell put together this farm from his grandfather's First

Division mowing lots in Rocky Meadow and his adjoining Second Division upland, and from lots purchased of neighboring landowners. Ephraim Hartwell carried on his father Samuel's habits of accumulation and improvement: two of Ephraim's sons inherited the pair of Hartwell houses in their turn, and three more generations of Hartwells inhabited this land and throve until the end of the nineteenth century. The two parallel Hartwell lines racked up six generations here on this rockpile. They were solid, prosperous yeomen, most of whom came to style themselves gentlemen. Their stocks of everything from cattle to cider were ample, and their grain yields were better than those of many of their contemporaries who worked longer-established farms nearer the center of Concord. By 1775, Ephraim Hartwell was a moderately wealthy man—as was his neighbor down the hill, the tanner Joshua Brooks. By the standards of their day, the Hartwells were a successful farm family, flourishing here in the middle of what strikes us now as a woeful collection of cold bogs and mossy boulders.

How did they do it? The tavern Ephraim Hartwell ran on the main road from Concord to the bay surely crowned his success. But that alone explains neither the good showing of his farm nor the similar performance of his neighbor Nathaniel Whittemore to the east. Beginning about the same time as Samuel Hartwell, Benjamin Whittemore and his sons also developed two thriving farms on what appears today to be mostly submarginal ground. How? We might postulate that these yeomen succeeded for a time by extensive means, by constantly clearing fresh land and enjoying a brief run of limited fertility until the soil was exhausted. But this hypothesis does not withstand scrutiny. It can't explain this: when Samuel Hartwell established his farm in 1694 he purchased eighteen acres of land already "for the most part sowed in" from his neighbor Richard Rice, half a mile down the road near the Cambridge line. A century later, in 1793, Ephraim Hartwell divided the same field between his sons Samuel and John, and it was still being tilled. This was the Hartwell's plowland, and their plowland it remained. This field was referred to repeatedly as plowland in deeds, surveys, and estate inventories, and no other parcel the Hartwells owned ever was. In fact, it is still being tilled today. As far as one can tell, it has seldom been out of cultivation for three centuries. True, this is an unusual patch of sandy loam, a flat, workable soil amidst these rocky hills, but the colonial system of mixed husbandry kept it consistently productive. Pastures and meadows were similarly steadfast throughout the colonial period.

Something is going on here. Let us pull up a chair at the Hartwell Tavern and consider what we have seen. We have passed over a wonderfully variable, convoluted landscape of gravel knolls, sandy plains, boulder clays, silt bottoms, and spagnum bogs, left in a jumble by a departing glacier. If this is all marginal land, it is certainly marginal in an entertaining variety of ways. We have seen the ready mixing of native and exotic plant species across this edaphic substrate, a slow ecological dance continuing today. We have found hints that this land was not in a state of perfect equilibrium before the English settlers arrived but had always undergone natural change and had long been husbanded by its Native inhabitants. In terms of our primary purpose, which was to uncover clues to the early development of the town of Concord, we have seen ample evidence of a young agrarian community being driven to expand by a powerful demographic impulse—no surprise there. We have seen the hazy outline of how land was distributed to accommodate that growth, and something of the way farms were settled by the first few generations as they spread from the center to these rocky outskirts.

Above all, we have glimpsed the startlingly deliberate manner in which different parts of this diverse landscape were assigned specific roles and crafted into a complex, integrated farming system. We have seen evidence of many farms that succeeded and remained ecologically stable for generations. This raises some interesting questions, because history has not been kind to farming in New England. Colonial husbandry in particular has generally been portrayed as a makeshift affair, held back by marginal soils and lack of markets. Yeomen in towns like Concord have long been regarded at best as extensive farmers who showed great industry but little enduring regard for their land, and at worst as rank exploiters who exhausted the soil and degraded the environment through ignorance and greed. Joseph Mason and Ebenezer Brooks, on the hardscrabble homestead next door, seem to fit this model: they gave up and left, or stayed and stayed poor. Our landlord here at the tavern, Ephraim Hartwell, does not fit the mold at all.

For two centuries following the birth of the Republic, almost nothing complimentary was written about colonial agriculture. Colonial husbandmen throughout America were cast as crude, extensive farmers because land was cheap and plentiful and labor scarce and dear. They were charged with choosing to deplete soil fertility and clear new land rather than care intensively for what they already had in cultivation. Beginning with the likes of the anonymous (and non-American) author of *American Husbandry* in 1775, European gentlemen accused these yeomen of mistreating their livestock, wasting their manure, and taking an unkempt, slovenly approach to tilling their soil. Nineteenth-century New England leading men, bent on progress and improvement, passed the same harsh judgment on their forefathers—following the lead of Timothy Dwight, president of Yale. In his *Travels in New England and New York*, published in 1821, Dwight famously wrote, "The principal defects in our husbandry, so far as I am able to judge, are a deficiency in the quantity of labor necessary to prepare the ground for seed, insufficient manuring, the want of a good rotation of crops, and slovenliness in cleaning the ground."[6]

In the twentieth century, economic historians, taking the indictment of these early critics as their primary evidence, likewise convicted colonial agriculture of gross neglect. Like Dwight, these historians ascribed extensive colonial farm methods to pioneer conditions—subsistence farming with few market incentives for intensification—and at least excused it as rational, lightening the sentence: time off for good economic behavior. New England farmers, they said, occupied some of the most remote, marginal land and so were among the last to improve. Once urban markets emerged in the second quarter of the nineteenth century, New England agriculture finally began to show grudging signs of progress—but then quickly sank beneath competition from farms on better soils to the west, and so farmland returned to forest. The colonial period was pictured as an anomaly—an era during which land poorly suited to the plow was nevertheless cultivated because inland New England was both physically isolated from markets and culturally bound by a Puritan ethic that endorsed austere living. This view was framed by Percy Bidwell in the early part of the twentieth century and still guides popular understanding of the history of land use in New England.[7]

Beginning in the 1970s, a group of social and community historians put a different spin on this colonial New England detachment from commercial farming. Instead of denigrating it, many of them pointed to its communitarian virtues. These historians, including Kenneth Lockridge, Richard Bushman, James Henretta, Christopher Clark, and others, saw the yeomen who settled colonial New England towns as either deliberately, or at least effectively, withholding themselves from the socially corrosive influence of emerging world capitalism. Because of their religious convictions and the closely knit communities in which they had settled, they forged economic relations based more on reciprocity among neighbors and kin than on simple market calculus—a "moral economy." Their villages resembled the traditional peasant communities of Europe. Over time, however, this isolation from the wider commercializing world broke down, and Puritan villagers became calculating Yankee farmers. This transition was explained as the result, in part, of environmental shortcomings in colonial agrarianism: expanding population and extensive farming. These forces undercut the stability of older communities, particularly the ability of fathers to pass on workable landholdings to their numerous sons. This brought stress and change to colonial society and eventually helped pull farmers into increased production for the market and a very different set of economic and social aspirations. One of the best-wrought studies of this difficult transition is Robert Gross's *The Minutemen and Their World,* concerning Concord itself.[8]

But another group of economic and social historians has vigorously disputed the idea that New England husbandmen withheld themselves from the market, or that their economy was isolated and stagnant, or that there was ever any deep conflict between Puritanism and capitalism at all. Winifred Rothenberg demonstrated that the New England rural economy was aligned with wider markets by the late eighteenth century. Others argued that New England farm families engaged in a broad range of surplus production from early in the eighteenth century, in order to satisfy rising demand for imported consumer goods. That much could perhaps have been reconciled with the idea of a society moving from a local world tightly bound by family and community toward a wider world of individual commercial striving and market exchange. Still other historians, including Daniel Vickers, William Martin, Stephen Innes, and Gloria Main, have pushed further, insisting that a driving ethic of industry and enterprise was central to Puritan New England from the start. They have portrayed the ability of these yeomen to not only wrest a living from the unpromising New England soil, but to build a thriving economy of diversified family farms linked with a prosperous maritime trade as a small miracle. Innes took the moral economy of the social historians and slyly rechristened it "moral capitalism."[9]

As that phrase indicates, these historians do not argue that New England farmers were completely free of restraints upon their acquisitiveness, imposed by their family, community, and religious obligations. Neither do they claim that colonial farmers in New England were producing exclusively or even mainly for export, which they obviously weren't. What they do claim is that market production was the prime mover in the story, and that the colonial economy was *always* either growing or straining to grow through diverse, broad-based enterprise. With its expanding population and its entrepreneurial drive pressing the limits of the available land by the late eighteenth century, New England, in this view, was perfectly poised to spearhead the astounding economic growth of America in the nineteenth century. The breakthrough was initiated by a progression into more intensive, thoroughly commercial agriculture within New England itself, Rothenberg argues. Then it culminated in a big way with a surging diaspora of Yankee farmers and loggers across the fertile farmlands and forests of New York and the upper Midwest and with the powerful sluicing of pent up Yankee capital, labor, and talent into the wheels of the industrial revolution at home. Far from an anachronism that had to be dragged reluctantly into the modern world, colonial New England husbandry is seen by these historians as the cradle of the "distinctive productive drive" of American capitalism.[10]

But what was the impact of this vigorously expanding colonial husbandry on the environment? The transforming of "raw land"—in truth, former Indian homeland—into family farms and the wringing of production from every available natural resource may have been an engine of economic growth in colonial New England, but were these "improvements" sustainable? On this question environmental historians, led by William Cronon and Carolyn Merchant, have weighed in. Environmental historians have agreed that the English settlers were committed to the market (especially by contrast to the Native Americans who preceded them), but they have emphasized the ecological damage that resulted from this revolution to a "world of fields and fences." They have argued that the profit motive drove these newcomers to commoditize and exploit natural resources as rapidly as possible, rather than conserve them as the Indians had done—clearing forest, depleting wildlife, exhausting soils, eroding hillsides, fouling streams. In this view, the extensive approach to the land that was once seen as the mark of crude subsistence agriculture is ascribed precisely to the market itself. The economic growth of New England may have been an impressive achievement, but it was founded not only on family labor and a strong work ethic but also on the expropriation of land from the Indians, the forced labor of slaves (since New England's maritime economy depended so crucially on trade with the sugar plantations of the West Indies), and the headlong consumption of nature. Such growth could never have continued had it not seized upon and expended a whole continent, indeed an entire world, of natural capital.[11]

The environmental view thus interprets New England history in a way that is almost diametrically opposed to the progressive economic thinking that has dominated the past two centuries. Native Americans are no longer dismissed for failing to improve the land. They are credited with maintaining ecologically sustainable ways of life over millennia thanks to their deep familiarity and spiritual closeness with the environment, to a cultural ecology that was imbedded in intact natural systems, and to the small population and limited material wants that marked their premarket economy. The English suffer by comparison. Colonial planters, in New England and elsewhere in America, are accused of clumsily imposing inappropriate European traditions of land use upon an unwelcoming environment, rather than learning sufficiently from their natural surroundings and Native mentors. They are charged with commoditizing nature and treating it as a mere storehouse of resources, to be mined as rapidly as possible for wealth. From these unbridled commercial drives came environmental degradation, beginning immediately upon the arrival of the colonists and intensifying ever since. Where economists see slow but steady progress from wilderness toward civilization, environmentalists see a long slide from ecological harmony toward alienation and abuse.

These contrasting views of the pace and strength of the development of a market economy frame a rich matrix of possible explanations of why New England farming degraded the environment. But are we sure that *colonial* New England farmers degraded the environment? Did an extensive system of husbandry undercut its own ecological base? Larger questions about the relation between capitalism and nature cannot be resolved by a detailed study of farming in a single New England town, but neither can they be resolved without such studies. What does this walk through Concord suggest? There were definite environmental tensions here by the end of the colonial period—shortages of land, declining grain yields—that need to be carefully examined and explained. Were they caused by exploitation driven by market forces, or was the problem the same one that had periodically dogged this agrarian system in England, time out of mind: demographic pressure leading to an imbalance between arable fields and the grasslands that manured them? In order to begin to answer any of these questions, we need a clearer picture of how colonial husbandry actually worked, on the ground. That is the main business of this book.

We need to entertain the possibility that colonial farms were not extensive and exhausting. Almost lost beneath layers of more recent change, we have seen traces of an impressive ecological transformation that took place in the century that followed English settlement. What if we were looking at a largely *successful* adaptation of traditional, diverse English mixed husbandry to a new set of environmental and social opportunities and limits in New England—what would that do to our thinking about the relation between the development of market capitalism and nature in America? The answer would still depend, of course, upon just how market-driven we took Concord's system of husbandry to be. Once established, was this system able to supply a wide range of goods and to support a fair-sized community from the local environment, in a balanced, sustainable way, or not? If so, what social, economic, and ecological limits constrained it, and how long could the chafing of those bounds be tolerated?

Looking at the long success of the Hartwell farm, we are driven to consider some interesting propositions. Perhaps this stony upland was not as irredeemably marginal for *all* agricultural purposes as might appear to us today. Perhaps these were skillful farmers who deeply understood their land. Perhaps they managed to husband this intricately walled and ditched landscape and to make all its diverse elements work together *without* constant recourse to fresh land, supplying a wide range of food and other materials for their wives and daughters to transform into the stuff of everyday life. Maybe these people knew what they were doing. If that should be true, we would want to understand the circumstances that enabled them to succeed.

This book will describe in the greatest possible detail how land was settled and how mixed husbandry came to operate in one town, Concord. It will explain what both the native landscape and the English husbandmen brought to this encounter, the ecological challenges that had to be overcome, and the working landscape that resulted. The book will then consider whether this system of husbandry was sustainable or whether it caused environmental degradation. Avoiding degradation or achieving sustainability are very difficult standards to define, and they are not necessarily the same standard. Loss of biodiversity and a decline in the integrity of the long-established native systems that preceded the English might well be degradation of a kind, and there might be ways to measure it. The losses attending this sweeping ecological transformation do concern me, especially as it continues to this day on a global scale and at a galloping pace—nearly half the planet's primary production is now turned in some way to supporting humans, by one estimate.[12] But that is not the central concern here. This study is confined mainly to a more narrow, human standard of sustainability: could Concord's system of husbandry, once established, continue to deliver the desired level of natural products and ecological services to its human inhabitants more or less indefinitely, or did it undermine itself? Was ecological capital being depleted in ways that were felt by the townspeople? The book will conclude by examining the ecological limits and tensions that had been reached by the end of the colonial period and suggest ways in which the people of Concord would respond to them.

Having seen the broad outlines of the task before us, we need to make our way again over the countryside we have just explored, this time slowly and carefully. We have taken our first excursion during summer, seeing the land at the height of its growth. Now we must retrace our steps in winter, as it were, stripping the tangle of vegetation back to structure, so that every stone wall and every ditch is revealed in that clear, low light that casts long shadows. We are going to cut away the undergrowth and dig down to the soil that underlies and incorporates history, "standing stoutly to our labours every one that can lift a hawe to strike it into the Earth," to paraphrase the wonderful Edward Johnson.[13] We will see how these farms were laid out over the tailings of the glacier, see how the land grew and changed with time. Dusty documents in hand, we will do our best to read this land—wet meadow, sandy plain, rocky upland, Native gardeners, English husbandmen, and all.

CHAPTER 2

Musketaquid: The Native Ecological System

There is an incessant influx of novelty into the world.
—HENRY DAVID THOREAU, *Walden*

FIRST, there was the land. But the land itself is in motion—its stones and trees are never at rest. Concord in 1635 was not a blank slate waiting for Englishmen to write upon it, but a work in progress. The soils of Concord were formed of ancient rock, but their crumbled remains had been recast comparatively recently by the glacier. The pattern in which these materials were laid—boulder clay dropped by the ice, sand and gravel deposited by meltwater, and silt that settled at the bottom of glacial lakes—proved fundamental to the shape of both natural ecosystems and human cultures in Concord. The shifting mosaic of forests and meadows that covered these soils comprised species that were not fixed in their places, but had learned during the past two million years of the Pleistocene to cope with sweeping changes in climate imposed by periodic glacial expansions and interglacial contractions. They knew how to move, and they were—and are—in motion.

During the present interglacial period we call the Holocene, natural systems have reconstituted themselves more or less continuously in response to warming and cooling climate, to the latest arrival of migrating species, and to disturbances such as wind and fire. In addition, from the moment the land emerged natural ecosystems have responded to people: human beings have been shaping nature in Concord since the melting of the most recent glacial ice. The power to remake the world belonged to Native as well as European inhabitants, although the direction and degree of human influence have varied markedly over time. In Concord, there was no land before history. Nature has included people since the dimly remembered days when the rocks were still wet.

This chapter reviews the long-inhabited place called Musketaquid that greeted the

English settlers, who renamed it Concord. In 1635, three ancient, ongoing ecological stories met: nonhuman natural forces, Native American foraging and horticultural practices, and the husbandry traditions of English villagers. The first two had of course merged long before the first Europeans arrived—not because the Indians blended invisibly into the landscape, but because they, too, changed the way it looked. But if there had been no Native Americans there would obviously still have been nature in Concord. Nonhuman nature was all there had been "in Concord" for eons of the earth's prior existence, after all. It is worthwhile to review what soils and trees brought to the story, before addressing the ways in which the Indians and the English modified those natural elements to give us the only Concord whose history we can actually investigate. That story began as the ice departed.

Ice and Earth

It was not always dry land where we dwell.
—HENRY DAVID THOREAU, *Walden*

So first, before the land, there was the ice. In Concord, the land as we know it was left behind by a great thaw. Land and life resumed together between 13,500 and 13,000 years ago with the wasting of the Wisconsin glaciation that had covered the region for a large part of the previous 100,000 years. This was only the latest in what are now thought to be glacial cycles paced by small perturbations in the earth's orbit around the sun and its wobble and tilt on its axis, and amplified by global systems such as atmospheric carbon dioxide levels and deep ocean currents. Glacial expansions have occurred regularly during the past 2½ million years, the Quaternary Period (which includes the Pleistocene plus the 11,500-year Holocene tacked on the end). At first, these cold periods came and went in a rhythm governed mainly by the 41,000-year cycle of the earth's changing tilt. But within the past million years the pattern has shifted to longer, more pronounced pulses of glaciation punctuated by brief warm contractions, closely matching the 100,000-year cycle of elongation in the earth's eccentric orbit. Cold and colder conditions have been the norm, and interglacial respites like the present have been short-lived. At the start of each glacial cycle the climate has cooled and the great continental ice sheet has gradually expanded, sometimes burying Concord under a mile of ice. These longer cold periods have lasted on the order of 90,000 years. By contrast, warmer interglacials during which the ice retreats to the poles and mountain peaks have usually lasted little more than 10,000 years.[1]

We are presently living in what may well be a similarly brief interglacial period we have named the Holocene Epoch. It might be more honest (and more humble) to work on the assumption that we still live in the Pleistocene. To us, our time appears to come *after* the Ice Age, but to the future it will probably appear to have fallen during a brief mild spell in the midst of it. Setting potentially calamitous anthropogenic disruptions aside, the end of our interglacial summer would very likely be upon us. The climate appears to have passed its warmest period several thousand years back, and a few centuries ago we were given a foretaste of what may lie in store, aptly named the Little Ice Age. This several-hundred-year cold spell reached its nadir just when English Puritans had landed and were trying to discover how to survive in New England. The Little Ice Age was perhaps both a reminder and a portent. If we think metaphorically of the one-hundred-thousand-year glacial cycle as a single year, then the Concord that human beings have known, with its comely rivers and ponds and its diverse, attractive landscape of meadows, hardwoods, and conifers—in brief, its natural *concord*—is a kindly world that rushes through spring, summer, and autumn in only about six or eight weeks. The rest has been ten months of all too easy sledding.

The glacial cycle is crucial to the environmental history of Concord in two ways. The advance and retreat of the glacier both created the underlying structure of the landscape and conditioned the movements of the plants and animals upon it. As the ice periodically made and unmade the land at northern latitudes over the past two million years, it also pushed and pulled vegetation up and down the continent thousands of miles before it, along with the wildlife that roamed the land, and more recently the human beings who followed the game and then remained to shape the land. The glacial cycle thus lies behind the composition and the dynamics of Concord's forest. Perhaps most fundamental of all, the last glacier left in its wake a fresh assortment of soils that deeply influenced the patterns by which both the Indians and the English inhabited and worked the land. Ice grinding down from the north and then melting back scraped, mashed, and dumped the very stuff of Concord. To begin at the bottom and work up, we start with what the glacier left behind: the bare rubble of earth.

In 1792, a Harvard student named William Jones wrote a "Topographical Description of the Town of Concord" for the newly founded Massachusetts Historical Society—the first of many term papers on Concord. "The soil is various;" Jones reported, "consisting of rocky, sandy, and moist land; but it is in general fertile."[2] Jones was a native son of Concord. This tripartite division of his hometown into rocky, sandy, and moist soils was not offhand—it reflected the way Concord's yeomen had long perceived and used their land. These three categories in turn derived mainly from the land's differing glacial origins: till, outwash, and lake bottom. The soils to which the dying glacier gave birth can be broadly segregated first into two classes: those born directly of the ice, the glacial tills; and those born of the meltwaters that flowed from the ice, the glaciofluvial and glaciolacustrine deposits. The tills are Jones's rocky land. The waterborne deposits can be further subdivided into the coarser upland sands and gravels laid down in streambeds and around the shores of postglacial lakes and the finer lowland silts and sands that settled to the bottom in the still waters nearer the middle of those lakes. These became the sandy lands and the moist lands. The Concord Soils Table (table 2.1) shows the breakdown of these soils (together with a smaller amount of more recent alluvial and swamp deposits), while the Surficial Geology map (fig. 2.1) shows their distribution across Concord.

TABLE 2.1   Concord Soils and Their Origins

Spodosols	
Rocky Land	*Percent*
Glacial Tills	25
Sandy Land	
Glacial Outwash—Fine to Coarse Sands and Gravels	35
Glacial Lake Bottom—Well-drained Fine Sands and Silts	15
	50
Histosols	
Moist Land	
Glacial Lake Bottom—Water-logged Mucks	10
Swamp Peats	15
	25

Glacial till covers approximately 25 percent of the Concord landscape, primarily in upland areas. Till was formed by material scraped from the bedrock and carried along at the bottom of the glacier as it moved south. When the ice stopped advancing and began to melt, the underlying material was packed in place by the weight of the ice above it. Till is unsorted. It consists of particles ranging in size from minute clay to large stones, all rolled together—in fact, till containing a high proportion of clay is aptly termed boulder clay, capturing this jumbled fruitcake texture. Sometimes, the compressed lodgment till from the bottom of the glacier is overlain by a thinner layer of material, called ablation till, that was riding on top of the ice and that settled back to earth when the ice was all melted. Ablation till is looser than lodgment till and tends to have had its finer particles washed out as the ice melted.[3]

If the ice had simply evaporated, subliming into thin air, virtually the entire Concord landscape would have been left covered with nothing but till. But of course the ice melted, so in many places the land was exposed to torrents of liberated water, which deposited silt, sand, and gravel along meltwater streams and beneath postglacial lakes. Hence, till soils remain exposed only in those parts of the landscape where they were not buried by waterborne sediments—primarily the most elevated spots. Because of this, till soils are usually sloping, either mantled over the tops of hills in the underlying bedrock or built up into drumlins—long mounds that formed beneath the ice, composed almost entirely of till.[4]

Glacial till contributed something indelible to the character of New England: a wealth of stones. Stone walls in Concord are found primarily (though not exclusively) in areas of till. Because of the embedded stones, these would seem to be difficult soils to improve for agriculture—especially for the plow. And so they were, and are. Once cleared of the worst stones, however, some glacial tills make excellent farm soils for certain pur-

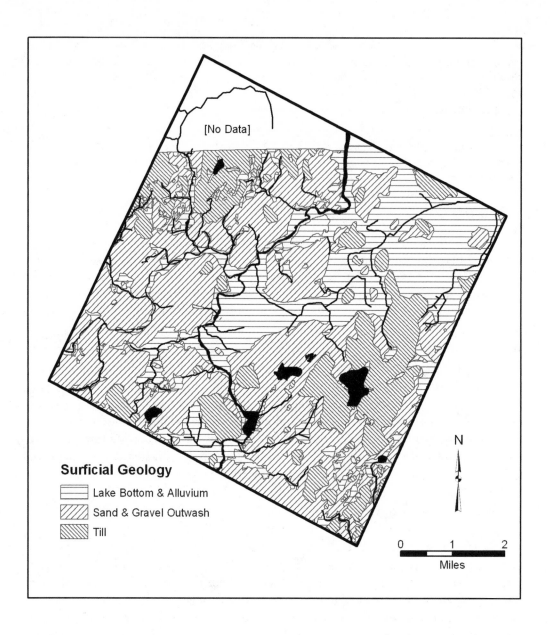

[No Data]

Surficial Geology

Lake Bottom & Alluvium

Sand & Gravel Outwash

Till

N

0   1   2
Miles

FIGURE 2.1.
Surficial geology of Concord aggregated into rocky till, sandy outwash, and moist alluvium.
The top corner is blank because the geology has not been mapped by USGS.
After Koteff, Nelson, and Hansen.

poses. Where the till lies thick and a hard-packed pan in the subsoil retards drainage, one may find first-rate soils for pastures and orchards. There water moves downslope near the surface during even a dry summer, providing reliable pasture when grass on more arable, well-drained soils has grown parched. On the other hand, where the till layer is thin and permeable and bedrock outcroppings abound, farming of any kind is hard going. Concord's tills divide about equally between these two categories.[5]

The rocky land came from the ice; the sandy and moist lands came from the waters. Which way did those waters run? The Concord region drains toward the north and northeast, down the Concord and Shawsheen rivers to the Merrimack. But during the thaw, drainage to the north was blocked by the glacier itself. Large meltwater lakes were ponded against the ice as it pulled back, the water rising until it was able to escape through spillways to the south and east into the Charles River drainage. Capes and islands of ice were stranded behind the ice front, releasing summer streams that ran downslope into the lakes. Icebergs sometimes became half-buried in sediments sweeping into the waters around them. These trapped blocks of ice slowly melted away after the lakes had been drained, leaving only their impression in the land behind. One such kettlehole is now called Walden Pond.

The meltwaters running from the ice into the lakes and then overflowing again through spillways toward the sea laid down beds of sand and gravel in their courses. Where these waters poured into glacial lakes, great deltas of sand, gravel, and cobbles were built up. Two glacial lakes were important in creating the Concord landscape: Lake Sudbury and Lake Concord. Lake Sudbury lay mostly south of Concord, reforming in several stages along the Sudbury River valley as the ice withdrew. During the final stage of Lake Sudbury, the ice front stood about a mile south of Concord Center. A great mass of sand and gravel was carried into the lake, covering the area from the hills of Lincoln all the way across the future Sudbury River bed, leaving only Fair Haven Hill, Mount Misery, Emerson's Cliffs, and a few other isolated peaks of bedrock and till sticking up through the rubble. This coarse soil was the foundation of the future Walden Woods.

This giant sheet of debris, together with a large block of ice wedged between the hills just below where Fairhaven Bay now lies, formed a dam that kept Lake Sudbury from extending farther north as the ice continued to retreat. A second lake, known to geologists as Lake Concord, filled the lowlands stretching northeast from Concord Center to Bedford Center. More deltas were built up by meltwaters at the margins of this lake. The steep slope of coarse sand and gravel that backs Concord village was deposited with its flanks to the ice when Lake Concord was at a high stage. The sandy plains of the Great Field behind the hill to the north were laid down later with the lake at a lower stage. Finally, the Shawsheen River valley opened as the glacier pulled farther north and Lake Concord drained away. Some time after that, the water cut a channel through the hills below Fairhaven Bay, draining Lake Sudbury across the bed of Lake Concord and down the Shawsheen as well. Eventually the ice withdrew far enough to allow the Concord River to flow north, restoring the now-familiar interglacial watercourses of the country.[6]

Deposits laid down by meltwater streams and built out into deltas at lake margins underlie the most widespread soils in present-day Concord, covering some 35 percent of the town. They gave rise to the sandy pine plains that abound in the early records of the Town. Although these soils all consist of well-drained sands and gravels, there is wide variation among them. In places like Walden, the deposits are so coarse and droughty that they are virtually worthless for agriculture and were only occasionally tried and then quickly abandoned. Elsewhere, outwash soils composed of finer sands and silts provide prime agricultural lands, some of which were favored by the Indians and continue to be farmed to this day. These gradations often lie together cheek by jowl. Knolls of coarse gravel and pockets of fine silt are sometimes marbled across the same field.

During those stirring late-Pleistocene summers the meltwaters ran milky with silt and clay, which were carried out to the calmer waters in the middle of the lakes. These smaller particles settled into the lowest portions of the landscape, forming glacial lake bottom deposits. Lake Sudbury's bottom soils are found along the river valley as far north as Fairhaven, just inside the Concord bounds. Bottom soils of Lake Concord extend from Nut Meadow across the Sudbury River northeast through Concord center, up Mill Brook and down Elm Brook into Bedford. In some areas, lake bottom deposits were later overlain as well by several feet of windblown loess consisting of fine sand and silt. These lake bottom soils cover about 25 percent of Concord. With a drop of only a foot or two in elevation, they grade from the best of the sandy land into the more challenging moist land. Those that are dry enough to cultivate bring the sandy lands to about half of the surface of the town.

The remaining 15 percent of soils in Concord are mucks and peats of more recent alluvial origin. These wet soils, together with the lowest-lying lake bottom soils, total about one-quarter of the landscape. Moist soils formed under anaerobic conditions that left them rich in organic matter but too wet to plow. However, they supported a suite of wetland vegetation that had a special role to play in the cultural ecology of both the Native and English inhabitants of Concord.

Once the ice had departed, the creation of soil began. The debris laid down by the melting glacier was not yet soil, properly speaking. Soil scientists refer to this ground-up rock as soil "parent material." In other words, the glacier was but the *grand*father (or grandmother, if you prefer) of the soils of Concord. True soil is formed from mineral parent material by life; by the growth and decay of vegetation and soil organisms. Soil sometimes contains more biomass than the vegetation growing upon it. But just how much organic matter accumulates in any soil is determined by the complex interplay of minerals, vegetation, and climate over time.

All of the upland soils in Concord, that is, about 75 percent of the town, are classified as *spodosols*. These include both the rocky lands and the sandy lands, the glacial till and glaciofluvial deposits along with some of the better-drained lake bottom deposits. Spodosols are derived from acidic parent materials (such as the bedrock granite and gneiss of the region), in a moist climate, under forest. The New England climate supplies more rain than sun, more precipitation than potential evaporation—that is, water moves down through the soil more readily than up. Consequently, these soils are strongly leached: dissolved nutrients are carried down and away, rather than being held in the topsoil. Compared to the more arid grasslands of the Midwest that built deep, rich topsoils, there was relatively little accumulation of organic matter or nutrients in Concord's upland soils. These soils did not favor the farmer (whether native or newcomer) with stored fertility that could last for years, let alone decades, as the prairie soils did. They could be worked for only a short time before needing to be restored to good heart. Keeping them productive was always a problem, and a pressing problem at that.

The remaining 25 percent of Concord's soils are the moist lands; mucks and peats classified as *histosols*. These are soils formed in the presence of water near the surface, which retards the breakdown of organic matter. In other words, a good quarter of Concord is covered by wetlands. Wetlands were crucial to Concord's natural and cultural development, and this too was largely a consequence of glaciation. Glacial lakes filled the valley of the Sudbury and Concord rivers with a plain of lake bottom silt, hemmed to a narrow ribbon by flanking hills in some stretches, spreading far abroad in others, nearly level from end to end. Sluggish brooks empty across this low, irregular plain into an indolent, winding river. The river drops only two feet on its twenty-five-mile journey through these ancient lake beds, from Sudbury through Concord into Billerica—a mud bottom fact of life destined to absorb the vital energies of generations of inhabitants. Winter and spring floods inundate the low-lying meadows, and water at any season is never far underfoot. Under these conditions, waterlogged silts formed soils known to science as mucks—mineral soils laced with partially decayed organic matter, which cover about 10 percent of Concord's surface. In their wet native state, mucks supported the extensive meadows of native grasses that were the foundation of the first two centuries of English agriculture in Concord. Artificially drained, these soils could also provide prime cultivated land, being level, free of stones, easily worked, and rich in humus.

Mucks are not the only wet soils in Concord. In places where shallow water covers the land year round and supports marsh, swamp, or bog vegetation, peat forms. Peat is organic matter that accumulates from dead plants under constantly wet, anaerobic conditions, where decomposition is severely inhibited. In Concord, it has built beds as much as forty feet deep. Peat contains almost no mineral soil particles, aside from a small amount of fine sediment washed in from surrounding uplands. In its natural state, peat can support only specialized plants adapted to wet conditions. Drained, it provides potentially rich soil, although harshly acidic and deficient in some nutrients. It is also, paradoxically, very light and droughty when dry and has poor structure unless mixed with sand. Dug up and carted off by farmers as "swamp muck," it made an excellent "soil" for barn cellars and an amendment for sandy upland fields.

Peat formed not only in low spots in the river floodplain, but also in pockets strewn throughout the uplands as well. Like many postglacial landscapes, Concord is full of ponds, bogs, and swamps. When the meltwaters drained away, the exposed debris was dotted with kettleholes of every size. Many became ponds and bogs with no outlets, gradually filling with peat in the shallows where plants could take root, or beneath a mat of sphagnum. The uneven outwash plains laid against the skirts of glacial till uplands also impounded depressions, where more wetlands developed. Beavers curtailed drainage further yet. Many of the small brooks running into the rivers simply seep along from one swamp to another until they finally reach the level of the river floodplain. In their native state, some of these wetlands suspended throughout the landscape were covered with meadow grasses, some were sphagnum bogs, while others supported forested swamps of red maple, black ash, black spruce, or white cedar. In any case, they built peat underneath, which covers some 15 percent of Concord.

"The scenery of Walden is on a humble scale," wrote Thoreau, and indeed the landscape of Concord is composed of convoluted ups and downs, making an intimate rather than a grand, dramatic setting.[7] Like the weather, the Concord landscape is subject to abrupt change. It is often difficult to find one acre that is the same as the next. It may be helpful to think of Concord as a basin, with moist lands along the river at the center, bordered by sandy upland plains, overlooked by rocky glacial till highlands at the outer rim. On the ground, however, these elements are scrambled in a maddening way. Swamps are tucked between steep little rocky knolls on one side and sandy outwash on the other. Drumlins of glacial till rise like the backs of whales from the low-lying meadows. A five-acre field may contain five different kinds of soil, even though the farmscape was laid out long ago to enclose each distinct physiological unit as much as possible within its own fences, for consistency (if not ease) of management. This land is ruled by diversity, and inconsistently distributed diversity at that.

This was manifestly not a landscape predisposed for large-scale, uniform agricultural production. Many have been satisfied to call it marginal or hard scrabble, and let it go at that. Landscapes suitable for large-scale farming, however, such as the Great Plains or California's Central Valley, are often landscapes that require large-scale environmental manipulation to be brought into production and that are vulnerable to large-scale environmental disruption. The highly variable soils of Concord presented both Native and English farmers with limits on every hand, but also with welcome diversity in creating ecological systems designed to meet a wide range of cultural needs. The glacier left all future Concord inhabitants with both a curse and a blessing. We will follow their contrasting approaches to organizing these diverse soils, and the vegetation covering them, into a working whole.

Forest

Forest returned to Concord some 12,000 years ago, succeeding tundra as the ice withdrew from New England and the glacial lakes shrank and drained away. From that time to this, Concord's forest has changed repeatedly and in a sense continuously. Yet the patterns that formed and reformed were not random but operated within parameters that were defined by climate, soils, disturbances, and the ecological adaptations of particular creatures. Forest dynamics were driven by a number of interacting factors: climate warming and cooling, species coming and going, diseases, fire, and wind. The diverse soils left by the glacier provided a range of substrata upon which different combinations of species coexisted at different times.

The process by which the northern world was reclothed in vegetation after the retreat of the glacier, and then repeatedly changed its costume, has been discovered largely through palynology, the study of fossil pollen grains preserved in sediments at the bottom of lakes and bogs. From these and other plant and animal fossils, the science of paleoecology has developed a longer perspective on ecological dynamics than was available from the study of present-day communities alone. Where community ecology has shown how dynamic interactions among species maintain a homeostatic balance at the level of the ecosystem for years or decades, paleoecology has revealed the cyclical and directional changes that take place at the level of the landscape over centuries and millennia.

From the time the ice was melting away in Concord about 13,500 years ago until about 12,000 years before the present, the newly formed land bordering the glacial lakes was dominated by tundra—grasses, sedges, bog species, and Hudsonia, a northern creeping evergreen similar to heather in form. Then spruce and willow trees moved into the region, creating an open parkland. This led to a period of two millennia lasting until about 10,000 years ago, during which the landscape of much of southern New England came to be dominated by a boreal forest of spruce, jack pine, and alder. Toward the end of this period white pine arrived on the scene and began to replace spruce as the weather continued to improve.

The period from 10,000 to 5,000 years ago marked the postglacial optimum, during which the climate reached its warmest level to date (discounting the surge of the past century). For the first thousand years or so of this mild time the weather remained fairly wet but then turned drier than it is today. The boreal forest gave way first to a northern forest dominated by white pine along with hemlock, oak, and beech; and then to drier oak and pitch pine vegetation that became dominant by 8,000 BP. Hemlock remained prominent in the pollen record for this period and probably flourished on sheltered slopes and in ravines, while white cedar took up its specialized niche in swamps. Spruce and tamarack, along with other locally rare bog flowers and shrubs that persist in Concord to this day, hung on around the coldest kettlehole bogs. As time passed more hardwood species, including elm, ash, and maple, continued to migrate into the region after the long march from their glacial strongholds in the South, reshuffling the composition of ecosystems across the landscape.[8]

From 5,000 years ago until the past century the climate has cooled slightly, the weather growing colder and wetter and then ameliorating by turns. Meanwhile, two important nut tree species whose journey from the Mississippi Valley had been delayed by the difficult Appalachian Mountain crossing finally arrived. Hickory reached New England about 5,000 years ago (having long been established in the Midwest), possibly on the wings of passenger pigeons.[9] Chestnut, consigned to squirrels, made its way slowly up the Appalachian Chain to New England by this time but did not become widespread until about 2,000 years ago. Native people may also have played a significant role in planting these valuable nut species. Both trees became important codominants along with oak in the forest of southern New England during this period, and this oak-chestnut ecosystem probably became the most common forest type in Concord.

That is an outline of the major trends in climate and tree migration in Concord over the past 12,000 years. But what Concord's forests actually looked like at any particular moment was not a simple function of prevailing climate, distribution of soil types, and whatever species happened to be on hand. Ecological dynamics at the landscape scale are also driven by *disturbances*—events powerful enough to destroy or severely damage the dominant vegetation in some part of the landscape and to allow another ecosystem to grow up in its place. The pattern of disturbances is crucial to determining what kinds of ecosystems most commonly prevail, under a given regime of climate, soils, and available species. The most notable ecological forces to disturb Concord include pathogens, windstorms, floods, and fires. While such events are unpredictable and may occasionally trigger long-term changes in ecological direction, for the most part they occur in patterns to which species are well adapted. So while it now appears certain that the same ecosystems did not renew themselves again and again on the same spot, in the old climax sense, the paleoecological record does suggest that broadly similar suites of interacting species prevailed for long periods—what has often been called a shifting mosaic. At the landscape scale, these patterns appear to have changed in their fundamental makeup only gradually over thousands of years, as sketched above. Such change is rapid in geologic or evolutionary time but quite slow as time is measured by people.

This is important because of what it suggests about the ecological opportunities and limits encountered by human cultures. In the first place, while natural change has indeed been continuous throughout the 10,000 years that Concord has been inhabited by people, at the broadest scale that change has been gradual enough to provide reasonably stable ecological conditions over many generations of our kind. Certainly cultures have needed to be sufficiently flexible to deal with (and indeed become part of) ongoing natural disturbances, fluctuations, and trends; but at bottom those who lived here at any given time were dealing with an environment that was predictable enough so that adapting "sustainably" to it was a meaningful possibility. But it is also apparent that given a natural world of changing climate, chance disturbances, diverse soils, and migrating species, several different ecological realities are always possible. Just as nature prescribes no single ecosystem uniquely adapted to each part of the landscape but supplies material for a range of alternatives, so there may well be more than one human ecological adaptation that is workable.

In any case, we have no idea what a purely "natural" landscape in Concord would look like. Human beings have been modifying nature throughout the entire postglacial period. There have been people in Concord about as long as there has been forest. The most significant human impact on the landscape, at least until the past few centuries, has probably been the use of fire. But people have also been important foragers and predators of a wide range of plants and animals, exerting influences on many ecosystems through that selective pressure. For this reason, a detailed account of the natural systems that have prevailed in Concord must include people.[10]

The Native Landscape

If the United States had been settled from the Pacific Coast New England would not yet have been discovered.
—UNKNOWN

This Californian quip about the rigors of New England's soils and climate has actually been tested.[11] America *was* originally settled from the Pacific Coast. While it was a long time before icebound New England could even be explored by the first Americans, we know it was discovered and settled in the end. Even the Indians, though, found New England to be an out-of-the-way place, on the margin of the habitable American land mass. While civilizations rose and fell in Mexico and the mound-building Adena and Hopewell cultures flourished in the great central valley of North America, aboriginal New England society remained in a world by itself, remote and unmonumental, with only sporadic connections to the trade networks of the interior. On the other hand, the Indians of New England do appear to have put in place a succession of cultural systems that made it possible for them to thrive well enough in this landscape for long periods of time. The prehistory of New England suggests the limitations of this environment but also some of its most proven potential. And it was people as well as nature who produced the world that greeted the English invaders and who made that world more welcoming to farmers than it might otherwise have been. Only then, by an accident of geographic proximity to Europe, was this rubble-strewn corner of the continent built into the stoniest, most austere hearth of white American culture.

People arrived in Concord more than 10,000 years ago—soon after the sedges and before most of the trees. We have only the haziest notion of how these people lived; of how they shaped their environment and were shaped by it. Many prehistoric sites are known in Concord. Only a few have been professionally excavated, but artifacts from private collections (beginning with Henry Thoreau's) have been inventoried and mapped by the Concord archaeologist Shirley Blancke and fitted into the framework of prehistoric cultural evolution that has been elaborated for New England as a whole. By combining this evidence with the pollen record of vegetation throughout the region and projecting it upon the map of Concord's soils, we can at least speculate intelligently about the probable shape of Concord's inhabited landscape at various times in the past.[12]

Just as the New England environment has seen dramatic changes, so the Indian way of life here has changed several times. The ancient inhabitants' cultural systems developed not only in response to environmental changes, but also as a result of the adoption of innovations. Cultural developments changed the natural world in turn. It appears possible that at several periods Indians deliberately managed the landscape on a large scale to maintain favorable conditions, particularly in southern New England. Like societies everywhere, Native American cultures faced the long-term challenge of maintaining a working balance between their populations and their environment, and they had the power to both enhance and deplete their resources. By any reasonable judgment they succeeded very well, but at times they may have overreached themselves and become unable to sustain the systems they had created. In other words, we can point to at least three periods in which New England Indians *may* have outgrown their resources. The first of these occurred just after the Indians arrived, the last just prior to the arrival of the Europeans.

The first humans to live in New England, known to archaeologists as Paleoindians, probably arrived in strength just after 11,000 years ago as the open spruce parkland was thickening into a boreal forest. The Paleoindians are represented in Concord by a few spear points found at two undated sites.[13] Their way of life lasted about 1,000 years in our region. The earliest explorers may have walked the land while there were still glacial lakes and lingering blocks of ice buried in the New England earth—a cultural memory that may not be entirely lost. In *Walden*, Henry Thoreau recounted a Concord tradition, learned from the Indians, of a Walden mountain that once rose as high as the pond is deep. Perhaps it is only a coincidence that this is an accurate description of the origin of Walden Pond, and that such a mountain, constructed of ice (but possibly clothed with insulating soil and even vegetation), really once existed.[14]

The Paleoindians in New England are believed to have lived in bands moving over large territories, following the megafauna that grazed the open landscapes south of the retreating ice, especially along the broad coastal plain which emerged from the ocean as the land rebounded. While these foraging people ate a variety of small game and plant foods, they seem to have relied heavily on such big game as caribou, mammoths, and mastodons. There is no direct evidence of elephant hunting in New England but there is plenty elsewhere in North America, and these were apparently very similar people using the same tools.[15] The Paleoindians may have also used fire as part of their tool kit. Charcoal is found in some New England sediments from this period but not in concentrations as high as those from the period of the drier, oak/pine landscape to follow.[16] Whatever fires the Paleoindians set were not pervasive enough to stop the northward march of the forest that eventually drove this ecological and cultural system from New England.

By 10,000 years BP, the boreal forest had closed in and was being invaded by white pine. This coincided with the disappearance of many of the large animals that had inhabited New England. Some, such as caribou, moved further north into new territory being uncovered by the melting ice. Many others, such as mammoths, were famously going extinct across North America as the habitats available to them retreated and shrank and as hunting pressure continued. The Paleoindian way of life vanished from the region. Some bands may have gone north with the remaining large game, while others may have stayed in New England and adapted by hunting smaller game such as deer and collecting more plant foods, broadening their resource base. It is possible that the Paleoindians helped bring on their own cultural crisis by driving many of the large game animals (along with some smaller species) to extinction.

Whether human hunters were primarily responsible for these Pleistocene extinctions has been a controversy among scholars for decades. Times of rapid transition from glacial to interglacial conditions pose daunting challenges to many species of plants and animals. Only the nimble can rapidly migrate great distances, often across inhospitable barriers, to reach suitable environments in which to continue living—and so, we are learning, the species that make up our world are nimble. We might expect some extinctions during unsettled times whether humans were present or not, and so there have been—but nothing during the entire Pleistocene to match what happened 13–10,000 years ago. That is odd. The Pleistocene has been going on now for more than two million years, with interglacial interruptions and abrupt climate transitions as regular features. The species that have survived are those which have learned to move great distances and to persist, sometimes with expanded ranges and large populations, and sometimes with smaller, disjunct populations. It is telling that the previous great wave of extinction took place at the end of the Pliocene, just as the climate was deteriorating into these Pleistocene conditions—this presumably weeded out most of those species which could not cope with the new glacial rhythms. All the species that existed 10,000 years ago had already passed many rounds of freeze and thaw very similar to what they were experiencing again. The novel feature, this time around (especially in the Americas), was the advent of a very efficient new predator, *Homo sapiens sapiens*. The question is surely still open to research and debate, but the circumstantial evidence that our ancestors played a major, perhaps decisive role in many of the Pleistocene extinctions around the world remains hard to resist.[17]

In any case, once the big game was gone people living in Concord—like those across the continent—had to intensify their food gathering and rely on a wider range of locally available species. But the dense forest dominated by white pine that held sway over much of southern New England 10,000 years ago did not produce much of the kind of plant and animal food that people can eat. Faced with a dearth of known sites compared to later periods, many archaeologists have concluded that this period—termed the Early Archaic—was a time of light population in New England, perhaps even more sparse than the preceding era of large game hunters. It is uncertain whether new people migrated into the region to replace the Paleoindians or if a new culture was derived in situ among the former inhabitants. Only a few spear points from the Early Archaic have been found in Concord. This period of two thousand or so years may mark the town's low point in human occupation.[18]

By 8,000 years ago the oak forest that, in one version or another, was to dominate much of southern New England down to the present was established. This forest was influenced by fire and contained a large admixture of pitch pine, particularly on sandy soils. The mosaic also included fluctuating elements of northern hardwoods, hemlock, and white pine on more mesic sites such as glacial tills, along with wet meadows and several types of wooded swamps on the moist lands. This was a productive landscape for people, and it supported a cultural system that throve for some 5,000 years, until about 3,000 BP. We cannot be sure that a single culture was continuous in Concord throughout this long period, or that all its times were good. Still, on the face of it, these Archaic New Englanders can lay claim to maintaining a sustainable way of life for by far the longest period in the history of our region.[19]

New England Indians are thought to have organized themselves by river basins during this time, making use of a variety of food sources at different seasons along streams and in bordering uplands. In the Concord area, a cultural system encompassing the Concord, Sudbury, and Assabet drainages left evidence of a network of seasonal campsites. A large camp at Flagg Swamp near the Assabet headwaters was devoted to fall and winter hunting and nut collecting. A summer campsite was located at the Clam Shell Bank in Concord, where river mussels and turtles were the main meat—or at any rate, the meat that left the most enduring remains. The surrounding deciduous forest, dominated by oak, was key to the prosperity of the people of southern New England. This Mast Forest provided acorns, hazelnuts, and especially hickory nuts, which, with their high fat content, seem to have become great favorites once they arrived about 5,000 BP. The forest also contained a good supply of meat, including turkey and other small game, bear, and, above all, deer. Like the bison of the grasslands to the west, deer have a high reproductive rate. They were able to survive the Pleistocene extinctions, expand their population to fill vacant niches, and thrive in postglacial environments dominated by human hunters. This was not simply a world of natural bounty with people nibbling at the edges, but a complex natural and cultural landscape built on the intimate interaction of the forest, game animals, and human beings with their multifarious flints.[20]

Did the Indians deliberately use fire to maintain this landscape and enhance its productivity? Like the Pleistocene extinctions, the issue of Indian burning has long been a matter of debate and is far from resolved. Studies of charcoal and pollen in sediments offer some hard evidence of prehistoric fires reaching back to the Archaic period. Cores from Cape Cod reveal frequent fires throughout the past ten thousand years, with the heaviest periods of burning between 9,000 and 5,000 BP and again during the past 1,000 years. Of course, sandy pine barrens are particularly fire prone, but even in such dry conditions natural fires ignited by lightning are rare in the Northeast, where electrical storms are normally accompanied by heavy rain. Where fires were a consistent part of the forested landscape in New England, the odds are that people were chiefly responsible. White oak, hickory, and chestnut are well adapted to a regime of frequent light burning on dry soils and will also survive and reproduce better than northern hardwood species in a regime of occasional large fires on moister uplands. Fire would thus have been an effective tool to encourage acorn and nut production, to improve berry picking, and to increase pasture and browse for deer by creating burned patches and edges. Patterned burning at varying frequencies over a diverse topography would not have produced a uniform landscape.

We need not imagine a southern New England forest that was all open-grown, studded with extensive grasslands, and completely burned every few years from the Berkshires to the Cape. The extent of Indian burning in New England remains unproven, but the presence of significant fire is at least consistent with the available paleoecological and anthropological evidence.[21]

Archaic culture in southern New England apparently reached its zenith, in terms of both the number of sites and of population, about 4,500 years ago. A turn toward cooler and wetter conditions began about that time and has continued to (almost) the present.[22] This would hardly have been welcome to the Indians of the Mast Forest of southern New England. Cooler weather would have facilitated the expansion of northern hardwood species, along with white pine and hemlock (although hemlock dropped out of the picture for more than a thousand years, probably because of some kind of pathogen). The invasion of these species, if allowed, would have led to a decline in supplies of both nuts and game. We can imagine a gradual shift in the balance among landscape elements across the region: dense red oak and white pine forests replacing more open white oak and hickory on dry sites; beech, maple, and birch mixing into oak-dominated forests on mesic sites, hemlock spilling upslope out of ravines. The Indians had a strong incentive to beat back this transformation and to maintain their accustomed landscape, by using fire. However, they must have faced more years when fire wouldn't spread as readily in upland areas, broken by occasional dry spells during which more extensive burning drove fire-sensitive northern species back out of regions they had begun to occupy. Pollen and charcoal data show just such a seesaw battle under way on Cape Cod over the past 5,000 years. Archaeological evidence indicates that people were able to sustain their way of life until about 3,000 BP, after which the withdrawal of population from upland sites accelerated. Indian sites in Concord became sparser after that date, as people were gradually forced into new subsistence strategies or simply maintained smaller populations.[23]

Here was a second possible period of environmental stress in prehistoric New England, when population pushing an ecological system to the limit met deteriorating climate. This *may* have led to periodic regional depletions of key resources. We can imagine that deer became less plentiful, for example, and that not as many fat-rich hickory nuts could be gathered to bring communities through the lean months of late winter. We do not have data with sufficient resolution to judge the scale and severity of such episodes during this contraction of a formerly sustainable culture. The very persistence of the Mast Forest of oak, hickory, and chestnut over much of lowland southern New England for thousands of years down to our own time, in the face of a shift in climate that, in the absence of fire, would have favored more northern species, is a strong indication that people were able to do something to maintain a favorable environment. Nevertheless, people gradually developed new strategies, including first the increased exploitation of the coastal zone and finally the adoption of horticulture.

The Woodland period ran from about 3,000 years ago until the arrival of Europeans 400 years ago. The foraging ways of life adopted by the inhabitants of southern New

England during the first 2,000 or so years of this period, up to about 1,000 BP, were not radically different from what had gone before, although important changes were under way. There appears to have been a growing reliance on coastal shellfish, particularly common clams, whose increased abundance may have been tied to declining seawater temperatures. Ceramic pots began to appear, indicating that grain was being stored—foraged grains such as wild rice, amaranth, and chenopods. A native horticultural complex consisting of a dozen or more domesticated small grains, oil seeds, and cucurbits had become established in the Mississippi and Ohio valleys by this time, but to what extent such husbandry became an important feature of life in New England has not been established. This seems to have been a period in which old ways were adapted to fit somewhat reduced environmental circumstances.[24]

Sometime after 1,000 years ago, the picture changed dramatically. Perhaps encouraged by a warm spell lasting a few centuries (this was the same period when the Vikings colonized Greenland and reached the shores of Vinland), horticulture involving tropical cultigens (corn, beans, and squash) was adopted in southern New England, providing up to 65 percent of caloric intake by historic times. Although it did not replace the older foraging way of life, gardening fundamentally altered the relationship between the Indians and their environment. Stored grain made it possible to carry a larger population through the uncertain months of late winter. By AD 1600 population in southern New England ranged up to ten times the density found among foraging tribes of northern New England.[25] Horticulture was taken up within the region where 150 or more frost-free days were available to mature a crop of corn, a band stretching along the southern and eastern coasts as far as southern Maine and up the Connecticut River valley as far as the Vermont border. The densest populations were found where a 180-day growing season could be counted upon, a narrower band following the coast as far north as the Boston Basin.[26]

The Concord River drainage falls within the 150-day zone but not firmly within the 180-day zone. Growing corn was practical but not as consistently reliable as further south and east toward the ameliorating ocean. Horticultural Indians were living along the rivers when the English arrived in Concord in the early seventeenth century, but their artifacts do not suggest settlements as dense as those found along the coast south of Boston harbor.[27] They called the place where they lived Musketaquid, which means Grass-ground River, or Meadow River. Just who these people were—what regional group they belonged to, how many they were, how far their territory extended, and how their cultural system was organized—seems lost to us forever. Their world was in such catastrophic disarray by the time it was being recorded in our tongue that we have no clear understanding of how they once lived. They may have been associated with the Nipmucks to the west, the Massachusetts to the south and east, the Pawtucket group of the Pennacook Confederacy in the Merrimack drainage to the north, or all of these. Just as Concord lies at a transition zone between the oak-pine forest of southeast New England and the northern hardwoods to the northwest, it was apparently near a cultural frontier as well. Based on the political and marriage ties of the historic Musketaquid Indians who sold Concord to the Puritan settlers in 1637, it seems most likely that they were Pawtuckets. The Musketaquids' only known alliance to the east was with a village on the Mystick River, which was also part of Pawtucket territory.[28]

The Indians of AD 1600 probably still moved seasonally within the Concord River basin, but their settlement and migration patterns had changed since Archaic times because of their new reliance on horticulture. Historical sources suggest that the precontact Pawtuckets numbered about 1,500 or 2,000 in total. The Pawtuckets' main center was at Wamesit (present-day Lowell), some ten miles to the north of Concord, near the rich fishing grounds where the Concord River plummets into the Merrimack. Their territory apparently included the Shawsheen River and perhaps part of the upper Ipswich River to the east and the Concord drainage as far south as Sudbury. It seems most probable that the Pawtuckets wintered in several villages of a few hundred each, located at intervals along the rivers.[29] These winter villages generally consisted of a group of longhouses, pole frames covered with bark, with an extended family group of forty or fifty living in each. Winter houses were probably set up reasonably near clusters of planting grounds currently in cultivation but in a sheltered spot close to fuel supplies. Villages were moved about once a decade, either because of soil depletion in the surrounding fields, firewood depletion in the surrounding forest, or both.[30]

Let us imagine that during the years before the English came a band of as many as five hundred Indians, or about one hundred families, lived along the Musketaquid at the future site of Concord—a plausible if unprovable hypothesis. Forcing these people into a square box in the shape of an English town is an obvious distortion of a more fluid, largely unknowable reality of Native settlement up and down the valley. It does allow us to get a sense of what the native landscape might have looked like, by gathering several layers of information into a single frame (fig. 2.2). Given what is known from early town records and archaeological sites, and what is known ethnographically of similar groups in southern New England, how might these people have inhabited, utilized, and cared for the diverse landscape of Concord, with its rocky lands, its moist lands, and its sandy lands?

The sandy soils that cover about half of Concord's landscape lay at the heart of the ecological system of the Indians. These dry lands were maintained by fire, in two distinct but overlapping ways: some of the finer soils in a horticultural cycle, and the coarser soils in an open pine-oak forest that was rich in berries, nuts, and game. Indian women cultivated sandy soils because they were light, warm, and easy to work with clamshell hoes—heavy, stony till soils would have been next to impossible to dig. Glacial outwash that was *too* coarse, on the other hand, would have been forever plagued by drought and low in yield. Best were fine sandy loams on terraces flanking rivers and ponds. Families dispersed throughout such fields every summer, where the women tended their plots of corn, beans, and squash. Concord has these soils in abundance, and we know that some were cultivated in aboriginal times. How many acres were needed, where were they located, and how were they managed?

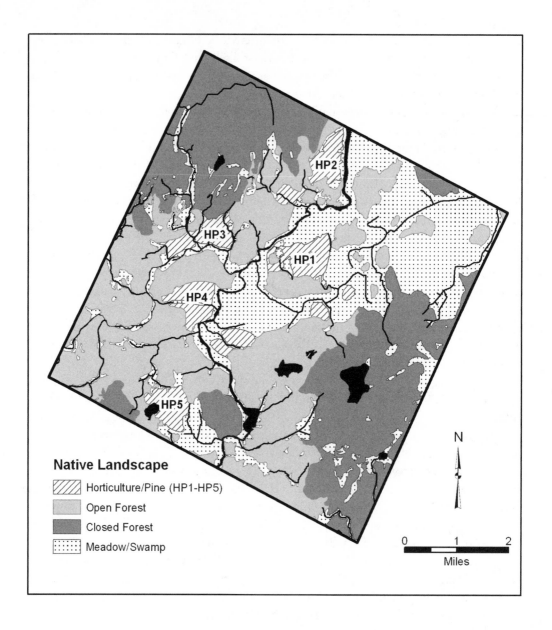

FIGURE 2.2.
Concord native landscape. Soils, climate, disturbances, and Native land use combined to order the vegetation of Musketaquid into a variety of ecosystems, a working landscape. This interpretive map shows five tracts devoted to cultivation, an open pine/oak forest across the sandy plains along the river, a denser forest on the rocky highlands, and wetlands grading from grassy meadows to wooded swamps.

Indian corn consumption in southern New England has been pegged at about eight bushels per capita, or forty bushels for a family of five. Corn yield has been estimated as high as forty bushels per acre. If so, an Indian woman could have supplied her family by cultivating only a single acre every summer. Forty bushels is a good yield, especially since the Indian fields were producing squash and beans from the same soil—forty-bushel corn was still considered respectable for Concord farmers on well-manured land in the mid-nineteenth century. Indian women are universally given high marks for weed control, and the milpa triad of corn, beans, and squash is crawling with polycultural virtue (including a nitrogen-fixing legume), but none of that alone could have maintained forty-bushel corn yields for very long on the dry, acid sands of Concord. Other sources put twenty bushels at the low end of the range of estimates of Native yields. Most accounts of Indian horticulture indicate that the women indeed tended one or two acres each. So a village of one hundred families at Musketaquid might have required somewhere between one hundred and two hundred acres planted in crops each year.[31]

But the fields being cultivated in any given year were only part of a longer cycle, a larger necessary pool of cultivable land. By most accounts, the Indians of southern New England used a slash-and-burn forest fallow system. Fields were cleared (by girdling or felling), burned, planted, cultivated by hoe, harvested, and stubble-burned for several years, then allowed to revert to woods. Nutrients that accumulated slowly in the biomass as the trees grew were made available at once by fire. According to many scholars, fields could be kept in cultivation for eight to ten years, a long time.[32] Fields may have been enriched by rotating them over abandoned sites of previous habitation, which concentrated nutrients from a wide area. Fields close to fish weirs would also have been relatively easy to fertilize with migratory alewives and shad.[33] How long the land was allowed to grow up in forest before being cleared and planted again is anybody's guess—estimates range from twenty-five to fifty years. If we assume eight to ten years in cultivation followed by thirty-two to forty years in forest, our hypothetical Musketaquid village of five hundred would have required a horticultural base of five hundred to one thousand acres in farm and forest rotation, roughly divided into five planting areas of one hundred to two hundred acres each. The village would have moved to a new area every decade or so, near freshly cleared fields. Needless to say, this is at best a grossly simplified model of what was undoubtedly a much more variable, messy reality.[34]

Whatever the exact yields and acreage, Indian women are said to have managed their gardens with consummate skill. Like experienced shifting cultivators the world over, they neither chose their fields from the forest at random, nor relinquished them without direction. They apparently exercised some control over succession when annual planting ceased, gaining a few years of strawberry and then blackberry picking from the fields as the forest returned. In southeastern Massachusetts, Indian fields often then grew up to pitch pine. Pitch pine is superbly adapted to dry sandy soils and frequent burning, which were the prevailing conditions across the horticultural land base. Pitch pine may have been preferred: according to some accounts, Indian men sometimes helped girdle trees and clear fields with stone axes. This would have been easier work in pine than in hardwood. In brief, the Indians created a set of rotating elements in the horticultural portion of their landscape, a successional complex cycling through cornfields, brambles, and pitch pine. Therefore, if anything like five hundred Musketaquid Indians were on hand, there should have been at least one thousand acres of planting fields and associated pine plains situated on light, well-drained soils along the rivers in the place the English purchased and called Concord.[35]

We can in fact identify five areas of several hundred acres each in Concord that fit this description. These tracts are shown on the map of the Native Landscape (see fig. 2.2). They have archaeological evidence of Indian occupation during the Woodland period, and several became the sites of the first English planting fields. Others were identified as pine plains in the early historical records of the town. Four of them are situated on stretches of fine sandy soils laid down during the low stage of Lake Concord and border the river and its meadows. These are (1) the Great Field lying just north of the Hill that backed Concord village and south of the Great Meadow; (2) an area north of the river opposite the lower Great Meadow; (3) an area by Dakins Brook and Spencer Brook north of the Assabet River where the North Quarter later had a general field; and (4) an area on the sandy plain between the rivers extending from the Clam Shell Bank north to the South Bridge, where the South Quarter had its New Field. The fifth area lies on outwash graded to the final stage of Lake Sudbury, on the plain east of White Pond. This area bordered not the river itself, but the pond and several brooks feeding into Fair Haven Meadow. All five sites were perfectly suited to the wedding of farm soils and fish weirs.[36]

Taken together, these five soil formations comprise slightly more than two thousand acres, although not every acre was fit for the hoe. Several were bordered by coarse soils and steep ice-contact slopes and had swampy sections embedded within them. Still, between them they supplied ample acreage, composed of the proper soils, to support a horticultural rotation for the postulated five hundred Indians. There are indications in the archaeological and historical records that these tracts were in fact used in this way. I am not suggesting that they actually made up a five-stage rotational land base for a discrete group of Musketaquid Indians corresponding exactly to the eventual political boundaries of the town of Concord. Real Indian horticulture was no doubt more diffuse and less regular throughout the valley, clearing and then abandoning planting fields in a manner that was apparently just beginning to take shape in the centuries prior to the arrival of the English.[37] In truth, many more acres that could have been cultivated by Native methods existed across the sandy plains of Concord, and some no doubt occasionally were cultivated. During the long part of the rotation when these horticultural lands were growing up to pine and were subject to light ground fires, they graded (probably across no sharp boundary) into the second, more widespread ecological component found on Concord's sandy soils: an open forest of oak and pine.

At any moment the great majority—something over 95 percent—of Concord's sandy soils would have been covered by a relatively open forest, as shown on the map. A large part of these sandy soils were simply too coarse or too inconvenient to be cultivated. Only the equivalent of one of the horticultural tracts (along with perhaps some land recently cut for fuel) would have been cleared at any given time; the rest back in forest.[38] This open woodland was probably composed of pitch pine, white oak, scrub oak, hickory, and chestnut, with an understory of blueberries, huckleberries, sweet fern, and bracken—similar to the Cape Cod forest of today. Both pollen data and documentary evidence suggest that pitch pine and white oak were the most prevalent trees in Concord in the early seventeenth century.[39] This forest may have graded into pitch pine barrens that were more frequently burned. Periodic ground fires (either deliberately set or escaping from land-clearing fires) were crucial to this ecotype, continually burning back and regenerating underbrush, discouraging fire-sensitive species, and favoring older, thick-barked, flame-tolerant tree specimens. If burned often enough to eliminate regeneration of almost all trees, parts of this forest might have come to resemble what were known as wood pastures or parks in England, with large trees scattered widely over an open, grassy plain. This is the way some coastal areas were described by several early English observers in southeastern Massachusetts. However, pollen cores have so far yielded little evidence of such extensive grasslands in New England, and—except for the wet meadows, of course, which were quite extensive—there is no mention of large grasslands in the historical record of Concord.[40]

An open forest rich in white oak, hickory, and chestnut had been the ancient core landscape of the Indians. It remained largely intact during the horticultural period, except that it may have burned more frequently, encouraging pitch pine. Pitch pine did nothing to supply food, but it may have supplied first-rate fuel—it burns hot and is easier to cut than hardwood. How much firewood did the Indians need? Early English observers reported that the walls of the Natives' longhouses and wigwams were thin and that the Indians burned wood prodigiously, year round, for heating and cooking. The English found these houses hot and stifling. Let us suppose (for purposes of calculation) that the Indians consumed as much wood per family as the English settlers who followed them, and that they cut it clear. Cutting one acre of young growth each year would have supplied fifteen to thirty cords, a generous allotment. At that rate, a village of one hundred families would have required on the order of one hundred acres for fuelwood each year. At the end of a decade as much as one thousand acres, or a strip about half a mile wide around the periphery of the planting grounds would have been cut, and it would have been time to move the village. The recovery time for such woods to grow back to optimum cutting size again would have been a few decades, very similar to the supposed horticultural rotation. How the Indians *really* went about gathering their fuel—what species they preferred, from how many sites the wood was gathered, whether it was simply deadwood or was deliberately cut, at what age it was cut, what lengths were preferred, what distance it was hauled—is apparently unknown. They may have procured wood by a combination of picking up sticks, coppicing hardwood, and felling pitch pine in the vicinities of their various seasonal dwelling places. But it is clear at a glance that even if the Musketaquid Indians required as much as five thousand acres—five one-thousand-acre areas—of forest to keep themselves sustainably warm, the necessary woods were readily to be found along the river, convenient to planting grounds and villages.[41]

Frequent light ground fires would have served to keep those parts of the forest that were dominated by oak, hickory, and chestnut relatively open and full of large trees. Spreading hardwood crowns that allowed light to reach the forest floor between them would have resulted in increased nut production for people and game, good blueberry picking, and plenty of browse for deer on the tender shoots that followed the fires. Nuts, berries, and game from this landscape element supposedly provided about 20 percent of Indian diet and probably more of their protein, not to mention their clothing.[42] Embedding cornfields within a landscape long devoted to bear, turkey, and omnivorous, all-conquering deer would have presented the horticultural Indians with an interesting new management dilemma. They presumably met it by having children guard the fields and by keeping hunting pressure high in the woods immediately surrounding their gardens. Indian women may have done the great bulk of the work tending corn and pumpkins *within* the fields, but had the deer not been kept from the crops, all their labor would have been in vain.[43]

Back from the river a mile or two, the sandy plains gave way to uplands dominated by till: the rocky land. Here, in all likelihood, stood a different kind of forest, more closed in structure and less frequently cut or burned (see fig. 2.2). This forest was dominated by red, black, and white oak, hickory, and chestnut. It probably also included white pine, white ash, red maple, yellow and black birch, and a few stands containing sugar maple, beech, and hemlock, especially on moist, protected northern slopes and in steep ravines. In other words, it looked in many ways like the transitional hardwood forest found throughout much of central Massachusetts today. This forest would have differed from the open oak and pine forest of the sandy lands in the greater diversity of tree species and in the darker canopy produced by a greater density of tall, straight, high-branching trunks. The crucial ecological factor allowing such forests to develop would have been the exclusion of frequent fire, a consequence of moister conditions and greater distance from the horticultural centers. This would have supported the growth of upright timber and even, in a few spots least prone to burning, the persistence of northern, shade-tolerant, fire-sensitive species. Outposts of these species are found scattered throughout Concord and surrounding towns today on just such sites. North and west of Concord the northern hardwoods become more common, as rockier uplands and cooler temperatures prevail.[44]

But fire was probably not altogether absent from the upland, closed forest of Concord. Indeed, stands containing northern hardwoods are so rare and discontinuous throughout southeastern Massachusetts that there is every reason to believe that fire played a role in holding back the spread of these trees over the past few thousand years, in spite of cooling climate. Mature deciduous forests on glacial till soils did not burn easily, but during long droughts with favorable winds, fires may occasionally have spread from the

more frequently burned sandy plains into parts of the moist uplands. Fire had a particularly good opportunity in the years following large disturbances such as hurricanes,
which passed through Concord on the order of once every century, felling a swath of
the mature upland forest. This would have provided dry fuel for a large burn. Tangled
or blackened tracts grew up quickly to blueberries, huckleberries, and brambles. During
the first years following a large fire, regenerating patches would have afforded excellent
berry picking and superb habitat for deer and bear. Thus, Indian women probably sought
out recently disturbed parts of the upland forest during the August berry season, while
men made their way to burned-over places for late fall deer drives.[45] The berries were
followed by thickets of alder and alternate-leafed dogwood and then by forests of either
white pine or hardwoods, depending on seed sources, timing and intensity of the fire,
and chance. Occasional fires within the rocky uplands may have contributed a few scattered white pine stands to the landscape, but in general they discouraged fire-sensitive
northern species and maintained a forest dominated by oak, hickory, and chestnut. But
this forest was different in structure from the more open forest of the sandy plains and
was shaped by a very different pattern of natural and cultural disturbances.[46]

These upland forests would have been found along both sides of the Concord River
basin, stretching off to the southeast and much farther to the northwest. They formed a
peripheral territory much more extensive than that shown on the map, which focuses on
the river valley. The uplands to the southeast of Concord in what are now Lincoln and
Weston probably served as a borderland between the Pawtuckets and the Massachusetts
people in the Charles River basin. The largest upland region available to the Musketaquid Pawtucket band probably lay to the west, up the tributary brooks of the Assabet
into the hill country of Acton, Littleton, Boxboro, and Stow, toward the divide with the
Nashua River drainage and Nipmuck territory. This was later the site of the praying village of Nashoba, to which many of the remaining Musketaquid people moved after the
sale of Concord to the English in 1637.

Emerging from the top of this landscape were dry knobs with thin soils and exposed
bedrock that burned more frequently and were visited every year for their crops of blueberries and huckleberries. The landscape of eastern Massachusetts is dotted with these
Hirtleberry Hills, Blueberry Hills, and Bear (bare?) Hills. It was on such a Hirtleberry
Hill, somewhere between Concord and Watertown, that three Christian Indian women
and their children received permission to go berry picking during King Philip's War in
1676 and were murdered by a passing troop of Puritan soldiers, including several from
Concord. Before the catastrophe that struck these people in the seventeenth century,
among the high points of the year may have been the summer berrying parties to the
breezy heights of the rocky lands that overlooked the forested landscape.[47]

Low, moist lands were also of great importance to the horticultural people living
along the Musketaquid, as were the rivers, brooks, and ponds themselves. The waterways
and wetlands offered another key set of resources to the Indians and were also modulated
by natural and human disturbances. The rivers were the Musketaquid Indians' princi-

pal transportation network to reach their seasonal foraging camps. They probably went down to the falls at Wamesit for the spring fish runs and perhaps upstream to the Saxonville falls as well. They also built numerous weirs along the rivers and brooks—indeed, the weir on the Mill Brook was part of the Concord land purchase in 1637, and here the English built their first gristmill. The Indians' main catch were anadromous fish such as alewives and shad that made their way upstream to spawn in the spring, along with catadromous eels, which came up as juveniles before heading downstream full grown a few summers later, to spawn in secret in the Sargasso Sea. There is no evidence of salmon in Concord waters. The Indians also caught and ate many resident fish and water creatures, including mussels, turtles, and snakes. They went in for winter ice fishing on ponds, too. In all, fish accounted for something like 10 percent of southern New England Indian diet and probably more of protein intake—they may have surpassed deer. Fish were thus a crucial protein source, though apparently not as central to native New England subsistence systems and culture as salmon were in the Pacific Northwest and to a lesser degree in Labrador and Newfoundland.[48]

Grassy meadows covered much of the former bottoms of glacial lakes Sudbury and Concord (see fig. 2.2). These meadows were most extensive where the river floodplain spread wide but could also be found upstream where brooks percolated out of wetlands at their headwaters. The meadows were rich food sources for the horticultural Indians, as they had been for their exclusively foraging ancestors. While the planted crops were ripening during the late summer, Indian women gathered from the meadows and marshes a wealth of plants including arrowhead tubers, cattails, and reeds for weaving mats and baskets. The juxtaposition of sandy upland planting grounds and low meadows thus created a convenient working landscape for Indian gardeners and foragers, just as it later would for English husbandmen and herdsmen, although in a different way. In the fall the meadows yielded the women wild rice and cranberries, while the men hunted the flocks of waterfowl that visited these prime feeding areas on their migrations. By some accounts, the hunters glided in their canoes among the sleeping birds at night, lit torches to confuse the creatures, clubbed them with their paddles, and had their dogs leap from the canoes to fetch them.[49]

Meadows were most common along the river and at the headwaters of brooks, but there were other kinds of wetlands in the native landscape of Concord. The ecology of wetland vegetation is complex and influenced strongly by the fluctuating seasonal height of the water table. Grassy meadow is only one of numerous possibilities—there are also marshes, swamps, and bogs. *Meadows* occur most often where there are regular winter floods, but the water drops below the surface during the growing season. However, if summer flooding does not occur now and then to discourage the trees, red and silver maple, elm, swamp oak, and black ash may be able to invade meadows, creating a wooded *swamp*. Where there is standing water year round, on the other hand, *marshes* containing sedges, cattails, reeds, and wild rice commonly occur—but such habitats can also give way to aquatic shrub swamps of buttonbush, willow, alder, and red osier dogwood. Finally, here and there in Concord, kettlehole ponds were covered with sphagnum, supporting *bog* vegetation such as leatherleaf, pitcher plant, cranberries, and rhodora.[50]

How these various ecosystems were distributed throughout the wetlands of Concord depended not only on flooding patterns driven by climate and topography, but also on disturbances introduced by at least two other creatures: beavers and human beings. Beavers were a powerful cyclical force in the landscape. Their dams created numerous ponds along the brooks, killing trees and providing nesting sites for birds such as great blue herons, osprey, and eagles in the process. The beaver lived in lodges within the protection of their ponds and dined on aquatic plants and surrounding vegetation, such as the inner bark of aspen trees. Eventually the beaver moved on, either because the pond had filled in, or the local food supply was exhausted, or perhaps the colony was wiped out by predators—including people, who hunted beaver for both their fur and their meat. The drained beaver pond then reverted to seasonal flooding, becoming a grassy meadow. In time, trees invaded the meadow and made a forested swamp, continuing the cycle until the beavers returned.[51]

But beaver were not the only ones with both the power and perhaps the will to influence wetland vegetation. The people of Musketaquid had a stake in what grew on their moist lands, so it was in their interest to drive out shrubs and trees if they could. Meadows and marshes provided reeds, grains, berries, tubers, and plentiful feed for waterfowl—more of what people wanted than did woody swamps, which provided little besides the makings of black ash baskets. As we have seen, the Indians had a ready tool for this task: fire. Did they use it? The abundance of open meadows that existed throughout the Concord landscape when the English arrived in 1635 is arresting and needs to be explained. The English kept the meadows open for centuries primarily by mowing and grazing, but by burning occasionally as well. Following agricultural abandonment during the past century many of these wetlands have not remained open but have instead grown up to buttonbush, black willow, and wooded swamps. This suggests that some force (beyond winter flooding alone) may have been keeping them open before the era of hay mowing. In all probability, the native moist lands of Concord were covered by a meadow and swamp complex that was predominantly open grass and marsh but that contained shifting elements of woody swamp that were periodically driven back by fire and flood.[52]

Another type of wetland found here and there in Concord was evergreen swamp, consisting of black spruce, tamarack, and Atlantic white cedar. This occurred mostly in isolated pockets that were seldom burned, primarily (although not exclusively) tucked away in upland areas near stream headwaters, or in bogs. A prime regional example was the Great Cedar Swamp at the headwaters of the Sudbury River in Westborough. Numerous smaller "spruce swamps" and "cedar swamps" were mentioned in the early records of Concord, including a local Great Cedar Swamp adjoining the lower end of the Great Meadow near the Billerica line. These dense coniferous swamps provided winter shelter for deer, and so may have been favorite winter hunting grounds of the Native inhabitants.

The subsistence ways of the Indians of southern New England, when projected over the landscape of a specific place like Musketaquid, thus reveal an intricate pattern of adaptation both to and of the landscape. The native people relied on five major ecosystem complexes: the waterways; the wetland meadows and swamps; the horticultural lands and pine plains; the dry, open forest of pitch pine and oak; and the closed upland forest. Each complex was related to the underlying glacial geology of the region. But each complex was itself made up of a group of component ecosystems that were successively related to one another and that shifted about. The distribution of these elements within each complex at any given time was determined primarily by the history of disturbances: flooding, windstorms, agricultural and fuelwood clearing, and, in all four terrestrial systems from the moist lands to the rocky, some degree of fire. As the keepers of fire, the Indians held the master key to much of this dynamic landscape, and they very likely employed it to improve those elements within each complex that they found most useful. The cultural knowledge, social arrangements, and spiritual relationship to the land by which this was accomplished can no longer be known.[53]

This—or something very like it—was how the landscape of Concord looked on the eve of English settlement. A people with no idea of the disaster that was about to befall them had been evolving a cultural relationship with the land for thousands of years but with the advent of farming had recently added a new wrinkle. Their world was an interesting combination of a horticultural core supplemented by a broad range of tended and harvested resources—a dual economy not uncommon in human history and certainly widespread in North America. We may well respect its flexibility and the detailed understanding of the surrounding landscape it required, especially when we compare it to the shortsighted environmental blunders of our own culture and our seeming inability to address landscapes in patient, holistic ways. It is very tempting to go from that respect (and that frustration) to an assumption that the Indians' relationship with their environment was perforce "sustainable" and to begin searching for the political, economic, and spiritual qualities of Native culture that made it ecologically superior to the European culture that replaced it. But how sustainable were the environmental practices of the horticultural Indians of southern New England?[54]

The short answer is that we don't know and probably cannot know. The key issue is population growth and the capacity of the land to both absorb an expanded shifting cultivation cycle and at the same time continue to provide a wide range of other resources. Researchers broadly agree that Indian population densities were much higher in southern New England than among the foraging peoples farther north, because ample underground corn granaries relieved the late-winter lean times that prevail in a purely foraging system. The question is whether horticultural Indian populations were pushing the limits of their environment, and perhaps beginning to degrade it, on the eve of the arrival of the Europeans. Some have suggested that southern New England Indians may have drawn all of their potentially arable soils into a tightening shifting cultivation cycle as their population expanded and were indeed in the process of degrading their land.

Periodic food stress may have been reflected in the growing importance of coastal shell-fish that could ensure survival in bad years and in the apparent political power of coastal groups controlling these resources.[55]

Expanded cultivation and increased population density change the ecological dynamics of any culture, often leading inexorably toward more intensive agriculture as other options narrow—and as political power swings toward the more populous cultivators. In southern New England, higher and perhaps growing populations may in time have put pressure on other key food resources such as game, forcing still greater reliance on horticulture. Demand for arable land may have shortened fallow periods, reducing yields. Excessive burning could have gradually driven forest composition away from mast-producing trees and toward less productive pitch pine barrens or to the nearly treeless plains reported by some early observers along the coast south of Boston harbor. Demand for fuelwood could have led to deforestation in some densely settled regions.

Perhaps, but it is hard to find evidence for any of this. Indeed, a look at the soils and forests of Concord has shown that there were ample croplands and fuel supplies on hand to support a village of five hundred people, even after making the most extravagant allowances. And that is granting the Concord River valley a population density as great as that suggested for coastal areas, more than were probably actually living at Musketaquid. It is possible that at this density some foraged resources were already being overexploited, but in fact early European observers were uniformly awestruck by the abundance of fish and game in the region. In short, the horticultural people of southern New England may have started down a path that would eventually have led to an environmental reckoning, but it does not seem likely that they had arrived there yet in 1600.[56]

The horticultural production system in southern New England was only a few centuries old and perhaps still expanding when the Europeans arrived. It may be that encountering limits and evolving the kind of relation with the landscape that had apparently allowed previous foraging systems to be sustained for thousands of years were challenges still somewhere in these horticulturalists' future in 1600. We don't know because that future was not to be granted. In a horror story that has been told too many times now to need to be repeated in detail here, epidemics of European diseases in 1616 and 1633 reduced Native New England populations perhaps by as much as 90 percent, "providentially" clearing the countryside for Puritan colonizers arriving from England.[57] Those Indians who survived (and whose population may even have begun to rebound) found it impossible to maintain their established production systems alongside their aggressively expanding English neighbors. Ultimately, all efforts at cultural accommodation failed, and following brutal warfare the remaining New England Indians became a submerged culture in Yankee society for the next three centuries. Native people by no means disappeared, but their impact upon the landscape became negligible, at least in Concord.

In 1637, several Indians took part in the sale of their fishing weir and planting grounds to English settlers who had been granted by the General Court of Massachusetts a six-mile-square plot of land at Musketaquid, henceforth to be called Concord. These Indians included a sagamore named Tahattawan who lived at Nawshawtuc Hill near the confluence of the rivers and a woman known as Squaw Sachem who was married to a sagamore at Mystick.[58] Many of the remaining Concord Indians eventually moved to Nashoba and other praying villages that were set up during the 1650s to accommodate Native people who had adopted Christianity. In spite of this, they were incarcerated at Concord during King Philip's War in the fall of 1675 and transported illegally to grim confinement at Deer Island in Boston Harbor several months later, where Indians from Natick were already scraping by on clams or starving. A few Indians returned to Nashoba after the war, but finally in 1686 sold this last piece of their homeland, which became part of the town of Littleton.[59]

For at least ten thousand years before the arrival of the English in 1635, Concord had been inhabited land. The people who lived on the sandy lands along the grassy river through that long time were not all of a single culture and did not always live by the same means. All of them altered the land in some way, although we will never fully understand exactly how much. There is some indication that at times these cultures experienced difficulty in maintaining their ways of life, sometimes partly because of their own actions. In every case these difficulties arose after centuries or indeed millennia of ecological success at sustaining a working relationship with the land. It should be obvious that, precisely because both nature and culture change, a term such as *sustainable* cannot be an absolute judgment but is a matter of direction and degree. We cannot simply ask whether a culture stayed inoffensively in balance with an unchanging environment forever—that would be to subject people to a false test by a false standard. We can try to identify what ecological changes occurred. We can attempt to specify how long a culture was able to maintain an ecological system without undercutting itself, and how well it was able to adapt to change, whether external or of its own making. What can we say of the various native cultures of Concord?

We can say that the first people, the Paleoindians, persisted in New England for a thousand years but then apparently—on a continental scale—had a hand in undermining their culture by overhunting their chief game. We can say that the next culture, the Archaic Indians, sustained their broad-based foraging way of life for more than five thousand years, although there may have been periods of environmental stress within that long span that are simply invisible to us. Toward the end of their time, the Archaic people may even have been able to defend their ecological system through several thousand additional years of cooling climate, using fire to defend the Mast Forest. Here, perhaps, was a case of culture striving for ecological continuity while nature persistently called for change. It appears possible that in time, the difficulty of keeping up their population or their prosperity in this changing environment played a role in stimulating these people to adopt new subsistence strategies, including horticulture. We can say that cultivation was very effectively integrated into the foraging way of life and the ecological landscape of places like Musketaquid, allowing a substantial rise in population over several centuries. This new combination of horticulture and foraging may have posed a long-term challenge sustaining a higher population, an adequate diet, and a productive environment, but we do not have convincing evidence that any such crisis was at hand by the early seventeenth century. Finally, we can say with certainty that the Native culture and its ecological system were overwhelmed by the invading English culture. The Indians were at a severe demographic disadvantage, both because of their susceptibility to European diseases and because they had lower population densities than a fully agrarian people to begin with. And so the English kept coming and waxed stronger once they came.

There were of course striking differences between the Native people and the English settlers, both in their ecological organization and in the social and economic forces that drove them. During the colonial period the English were to alter the landscape of Concord far more dramatically than the Indians ever had. There was also, at root, one important similarity: both had to survive and prosper primarily from what they could produce *locally*, within the diverse landscape of sandy soils, rocky soils, and moist soils. Each people had a deep tradition and a keen ambition to work the land in ways that would maintain their families comfortably for generations, *in the same place*. But the English had a very different agrarian agroecology and culture, and a radically different market economy was taking shape in their larger world. It remains to be seen what the settlers of Concord would make of the place the glacier, the forest, and the people of Musketaquid had prepared (as they saw it), by God's grace, for them.

CHAPTER 3

Mixed Husbandry: The English Ecological System

THE English settlers who ventured to Concord in 1635 came from a homeland that was at the beginning of an economic and ecological transformation. By the early seventeenth century, England was being revisited by the scourge of population growth, ecological imbalance, and scarcity that had undermined feudal society during the fourteenth century. Partly in response to these pressures, the English adopted a dynamic, modern economic system that eventually gave rise to the capitalist, industrial world power of the nineteenth century. Near the start of that transformation, they settled New England.

English Husbandry, Woods, and Water

In the broad sweep of world history, the establishment during the seventeenth century of a string of European colonies from the West Indies up along the eastern coast of North America was a crucial step in the global expansion of a market economy. Colonial New Englanders soon took a leading role in developing this Atlantic trade, as merchants of timber and fish, sugar and slaves. It would foreshorten New England's history to the point of distortion, however, to view the European invaders as simply free marketeers on the cutting edge of an expanding world system. By the time the economy of inland settlements like Concord became dominated by commercial production and consumption, another two centuries had passed. The people who settled Concord surely needed to find marketable products to supply themselves with critical imported goods, but they also brought with them deep traditions of locally adapted agrarian life. Those traditions, not the sale of commodities beyond the borders of the town, would be their mainstay until well into the nineteenth century. The story of Concord over that time revolves around the adaptation of that English heritage to a new environment. A new working landscape molded novel American elements into a familiar European agroecological pattern. To follow that story, we need to look at the world whence these Concord farmers came.

Characterizing that world is no simple matter. The Concord settlers came from several English regions, some of which leaned more heavily toward arable farming and others more toward livestock. The largest group came from a pastoral part of Kent in the southeast; another group of prominent families came from Bedfordshire on the fringe of the arable Midland clays; still others came from Derbyshire, on the pastoral northwestern edge of the Midlands. An American who begins to explore the dense copse of English agricultural history, or who simply wanders about the English countryside, soon discovers the remarkable regional diversity packed into that small island. This complexity comes from the making and unmaking of agricultural systems through thousand of years, played out over striking geographic variation in soils and topography. Ecological intricacy was nothing new to the English when they hit New England. In the early seventeenth century, this variety in the English countryside bespoke the long evolution of farming systems to meet the problems of local survival, as well as the beginnings of regional specialization to meet the emerging demands of the market. For the most part, early modern England as the settlers of Concord left it behind was still composed of a myriad of ways of wresting a living directly from the surrounding land.

With their multiple origins, Concordians had recourse to a range of pastoral and arable traditions. Like many New England towns, Concord first put in place a common field system that integrated livestock husbandry and grain cultivation in ways that were broadly familiar to most Englishmen. This was no simple matter of transferring a single regional model intact, but a complex ecological challenge. Each group may have arrived predisposed to replicate the local features of rural life with which they were most familiar, but they also needed to be flexible in adapting those aspects of their English experience — crops, livestock, farming methods, landholding patterns, means of water regulation — that seemed best suited to their new environment. They drew upon the many variations on a few basic themes afforded by their home regions.[1]

Mixed Husbandry

The success of agrarian life in England hung on the balance of three interlocking elements: husbandry, woods, and water. To begin with husbandry, the key to any English agricultural system was the fit between grain and livestock, between arable and grass. From ancient times until very recently European farming has been characterized by mixed husbandry, the marriage of herding and tillage. This union is not so universal as it may seem to those raised on English and American nursery rhymes. In many other cropping systems around the globe domestic animals play only a secondary role or (as with the New England Indians) no role at all. The world is also full of pastoralists, who do little or no cultivation — although neighboring farming and herding cultures often enjoy an uneasy symbiosis in practice.[2] In England, both the sympathy and the tension between grain and grass were deeply internalized. Parts of England may have been more pastoral while others leaned more heavily toward arable, but in all cases tilling and herding were tightly integrated within the same communities and were for the most part under the management of the same yeoman farmers.

Mixed husbandry had a number of advantages. It allowed farmers to augment the production of their tilled land by grazing rougher areas, thereby providing dairy products and a bit of meat. This added protein to the starchy diet of bread, porridge, and beer. Livestock furnished other useful materials such as leather and wool for shoes and clothes. By pulling the plow, livestock made it possible to cultivate heavy soils that were difficult to work by swinging the hoe and to incorporate weeds and grass sod. Livestock also afforded the means to cart great loads of stuff about the landscape. These were powerful synergies. But mixed husbandry had costs as well: it necessitated keeping the stock out of the crops, for one thing—"the sheep's in the meadow, the cow's in the corn" may sound idyllic, but in fact describes a big nuisance following a little boy's dereliction of duty. Keeping stock also obliged the husbandman to devote land and labor to feeding his beasts that might otherwise have been going more directly to feed his family. The conflict grew painfully acute when population rose, land became scarce, and the hungry arable began to gnaw into the acreage that was available for grazing. This undermined the entire system. Not only were the crucial resources provided by livestock put in short supply, but arable production itself was undercut even as it expanded: in mixed husbandry, livestock also were the principal means for maintaining the fertility of cropland, through their manure.

Maintaining the fertility of the soil is the central problem of any farming system. Soils are inherently variable in their physical and chemical makeup and thus in their ability to hold water and nutrients, to build organic matter, to generate nutrients through mineral weathering, and to bind or release those nutrients because of acidity. Some soils, in other words, are better than others to begin with. But all soils decline in fertility when they are simply cropped, because mineral nutrients—phosphate, potash, calcium, and others—are removed and consumed elsewhere. That is, after all, the purpose of growing the crop. Tillage also speeds the decomposition and oxidation of organic matter, allowing nitrogen (the limiting nutrient in most farm soils) to be removed by cropping and by leaching. If nothing is done to counteract these losses the level of fertility sinks, resulting inevitably in rock bottom yields. The soil arrives at a depressed steady state in which the outflow of nutrients removed by crops is roughly balanced by the natural replacement of nutrients through the weathering of tiny rock particles, the action of nitrogen-fixing microbes and other soil organisms, and the deposition of small amounts of nitrate in rain and snow— the poor man's fertilizer. Such yields make for a miserly return on the labor of growing the crop.

There are many ways to restore, maintain, or even increase soil fertility. Most systems combine several but revolve around one or another. A common approach, as we have already seen with the New England Indians, is to fallow cropland for several years or decades and allow the growth of natural vegetation to gradually rebuild the level of nutrients circulating in the system. This is fine where light population permits.[3] More settled forms of agriculture, usually associated with denser populations, need to find other ways to replace lost nutrients and go on cropping the same land year after year. Some of the densest populations in the world have relied on irrigation. Periodic flooding of fields provides not only water, which is of course its primary purpose, but fertilizer— from sediments and dissolved nutrients, from the growth of nitrogen-fixing blue-green algae (common in rice paddies), and often from the effluent of upstream communities, too. Farmers in many populous cultures have returned human excrement to the fields deliberately, as well. Such intensive methods have sometimes kept farmland productive for thousands of years, though often at great cost from debilitating diseases and periodic floods.[4]

Rain is plentiful in England, so English farmers used irrigation only in a limited way—although it played an important, somewhat hidden role in cycling nutrients, as we shall see. Some fraction of human waste made its way back to the fields, though it is hard to get a handle on how much.[5] Not enough, surely. So where else could farmers turn? Another approach to sustaining fertility is to enhance processes that rebuild soil nutrients within the fields themselves. I will discuss the use of legumes in a moment— though by no means absent, they were of limited importance in English husbandry before the seventeenth century. The last option is to deliver nutrients to the cropland from somewhere else, whether close by or far away.[6] Mixed husbandry in both England and New England relied primarily on bringing nutrients to the tilled fields from another part of the local landscape, the pastures and meadows. Essentially, livestock transferred soil nutrients from the grasslands to the arable land in their manure.

Let us stamp on this simple idea until it is thoroughly trodden into the ground. There is nothing magic about manure. Fertilizing nutrients are not created in the guts of animals. They come from somewhere. If either the stock or the feed moves from the point of uptake to that of deposition, one part of the world loses while another gains. If the manure returns to the very ground upon which the feed was grown, then it does not build soil fertility there but falls just short of keeping it flat. When we read of cows grazing a pasture and fertilizing it at the same time, we should shred that literature and use it for bedding. Grazing is a net loss, nutrient-wise. Livestock leave a large part of the grass they eat on the ground behind them, but (unless they are consuming supplemental feed) on balance they take soil fertility with them when they walk away from the pasture.[7] How many animals that pasture can continue to support is determined by both natural limitations and by the stockman's skill in managing livestock, pasture plants, and soil. A pasture receiving plenty of rain, on soils rich with limestone and flourishing with legumes maintained by good grazing practices can withstand heavy stocking and tolerate a slow, steady export of nutrients essentially forever. Another pasture may be inherently able to support fewer stock or may be degraded by poor management. In the case of land that is mowed for hay, larger amounts of nutrients are carted away to be fed to livestock elsewhere. How to resupply the soil that feeds the grass then becomes a major challenge to the husbandman. At the same time, the resulting concentration of nutrients from the grassland in animal manure provides an opportunity to fertilize other parts of the landscape. None of this manurology is a recent scientific discovery.[8] Early modern farmers in England and New England were keenly aware of it and organized their agricultural systems around it as a matter of course.

How complicated this mix between livestock and cropland became turned largely on how hard it was to find grass. In pastoral systems where population was light and the ratio of grassland to arable was high, keeping the tilled fields reasonably (if not often outstandingly) productive was not a great problem, at least in theory. Part of the arable land could be simply moved about within large areas usually devoted to grazing, a form of shifting cultivation known as outfield farming. Manure generated close to home, especially during the winter, could be concentrated on an intensely cultivated arable patch known as the infield, which never moved. While this infield-outfield pattern was most characteristic of the Highland Zone comprising much of Ireland, Scotland, and pastoral regions of England and Wales, it had also been an early feature in many British lowland regions as well and remained imbedded to some extent in almost all field systems.[9] But as English population grew during the Middle Ages and the arable expanded to supply more bread, a basic contradiction arose in more densely settled districts: grassland was put in short supply. This meant that fewer livestock could be kept, which in turn meant that less manure was available for the expanding arable. English husbandmen were in a bind. It became necessary to make very careful use of all available grazing and to closely regulate the entire system to keep grass and arable in working balance. Out of this ecological compression emerged the common field system.[10]

Keeping a balance between arable and grass was critical to mixed husbandry not only because livestock were necessary to the agrarian subsistence economy in their own right, but also because manure was necessary to the success of tilled crops. There were various means by which dung and urine could be delivered to the fields. The most straightforward, letting the stock deliver it themselves, was known as folding on the arable. Folding employed livestock (usually sheep) that grazed by day on common pastures and that were folded by night in movable pens upon fallow plowlands. The pens were typically made of hurdles woven of hazel and other smallwood coppiced in nearby woodlands. Part of the community's land—generally the poorest part—remained in permanent grazing. Depending on where you were, these lands lying outside the arable fields were know as the commons, or the waste, or the wolds, or the downs, or the heath.[11] The number of livestock that villagers were permitted to graze on this common land was stinted, or allocated in proportion to the amount of arable land that each held. In this way, soil nutrients were walked daily from the grasslands to the tillage. In some districts there were even sheep bred expressly for this purpose. Fields were not fallowed simply to "rest" them— that alone would have done little to restore fertility. They were fallowed to clean the plowlands within them of weeds by repeated stirring and to be replenished by systematic folding. An English agricultural historian aptly described the sheepfold in one region as "the sheet anchor of arable husbandry."[12]

To make matters even more complicated, there was also grass to be found within the arable fields themselves. A large English open field was not uniformly plowed from section line to section line like an Iowa cornfield. Instead, each field was composed of dozens (or even hundreds) of individual tilled strips, the intermixed arable holdings of the villagers. Between many of these tilled strips ran grassy baulks, while at the ends of some blocks of tilled strips lay grassy headlands upon which to turn the plow. This made for a substantial amount of grass: indeed, in many Midland villages by the crowded thirteenth and fourteenth centuries, most of the remaining grazing lay *inside* the fields, which had expanded to take in almost the entire landscape. In such cases, the fallow field provided a large part of the summer pasturage. In addition, fields were thrown open after grain crops were harvested, allowing livestock to forage through the fall and winter over the stubble and the grassy baulks and headlands. This was the key feature of the common field system: the large arable fields were regularly opened when they weren't growing crops to provide grazing for the common flocks and herds. In the process soil nutrients were redistributed from grasslands to tillage within the fields, as livestock wandered over the intricate, undulating landscape of ridges, furrows, and grassy borders.[13]

At seasons when forage growth slows or ceases because of cold or drought, the stock must be fed stored fodder. Grazing was generally possible for a greater part of the year in England than in New England, given the milder winters there. Animals could find adequate grazing from April or May through December or January, and even longer (verging on year-round) in some southern and western districts.[14] This still left a hungry gap in the late winter and early spring months. Livestock were then fed partly with fodder from the arable fields: straw, legume hay such as peas and vetches, and some oats for working horses and oxen.[15] But the key source of fodder was hay mown from meadows.

A meadow in rural England (and later, New England) was not just any old grassy place bigger than a lawn, as we use the word today. Meadows were a specialized kind of grass resource, a small but critical element in the landscape. They were grasslands that were mowed, and they were usually found along streams subject to seasonal flooding. As Oliver Rackham put it, "Hay had to be cut rapidly and labouriously by hand, and risked spoiling in bad weather. The best grassland, on which the grass grew thickest, was therefore reserved for hay."[16] Some hay was mown on upland if farmers were driven to it, but before the spread of cultivated grass in the seventeenth century, this did not amount to much. In the southeast Midlands, including Bedfordshire, "sufficient meadow to meet local hay requirements existed only in common meadows and closes along the flood-plains of the main streams."[17] In Kent, "within the Weald, meadows were prized for their good lush grass, often in contrast with surrounding pastures which tended to be thin and coarse on sandy soils or rank on the ill-drained clays."[18] Except for the "ill-drained clays," this could well describe seventeenth-century Concord, Massachusetts, settled largely by Wealden men.

Meadows would be absolutely crucial in New England, where the hungry gap in pasturage becomes a ravenous maw stretching to half the year, so it is important to understand their smaller but still critical role in English husbandry. According to one English historian "Nearly all the hay came from permanent, wet meadows set aside for this especial purpose. They could be mown down every year and still leave an aftermath for the animals to feed on. [W]et meadows gave about three times as much hay an acre as dry and upland ones, and gave it not just occasionally, but year in and year out. But they were seriously limited in their extent."[19] Meadows covered only about 4 percent of the English landscape by the thirteenth century, and they were the most valuable kind of land. The amount of meadow probably increased somewhat in the late Middle Ages and was much higher in certain low-lying districts.[20]

The meadows were an important source of nutrients to the fields, again via manure. But instead of being carried by the animals themselves, this manure mostly accumulated at home in the byre or the adjoining farmyard and had to be carted out and spread by the husbandman—a heavy labor aptly known as mucking. Farmers understood the importance of fertilizer and worked hard to conserve and augment it. "For manure, anything and everything was put to use," writes Joan Thirsk, ". . . the dung of cattle, sheep, pigs, horses and pigeons was employed with the utmost diligence and economy."[21] All manner of household and farmyard organic wastes made their way onto the dunghill as well, including such materials as bracken cut for livestock bedding, old roof thatch, and just about everything that went out with the garbage.[22] Field scatter of broken crockery that got spread with the dung reveals that arable holdings close to home were most often mucked, while more distant fields were folded—a commonsense vestige of the old infield-outfield approach.[23]

The meadows were a rich source of grass not only because they were well-watered, but also because, like all irrigated land, they were well fertilized. Lying at the bottom of the landscape, they collected nutrients leaking from land higher in the system, including runoff from the common grazing, erosion from the common fields, and effluent from the barnyards and streets of the village (or other villages upstream). From the meadows, the farmers brought the grass home to feed their livestock and from there transported the manure to the fields, closing the loop. The meadows and muck-heaps, sheepfolds and fallows, and the strips, baulks, and orderly rotations of the open fields were all part of a tightly integrated system. This complex form of husbandry had developed to provide adequate grass to the livestock and manure to the arable land, where those resources were in short supply. The same basic elements and circular flows could be found, in one variation or another, throughout traditional English agriculture.

Woods

Another important part of the agrarian world, preceding it and lingering within it, was the woods.[24] Like New England, old England was once covered with trees. The reduction of this ancient woodland began deep in antiquity—perhaps half of Britain was already cleared in 500 BC, at the end of the Bronze Age. By the end of the Roman era Britain was well cultivated, by 1086 England was only about 15 percent wooded, and by 1350 only 10 percent of the ancient woodland remained.[25] Woods had been reduced to a scarce element in the English landscape, found on poor land least suitable for agriculture—dry sands and gravels, irredeemably wet clays, steep slopes. Nevertheless, they were critical to the rural economy and were thus carefully managed. By 1270 woods averaged a higher annual return than arable land.[26]

Underwood, the principal product of woodland, was coppiced: sprouts were cut from self-regenerating stools on a rotation usually shorter than a decade. This smallwood was used primarily for fuel, but also for a host of other purposes such as thatching spars (large staples that hold roof thatch in place), woven hurdles used for temporary pens in the foldcourse, and wattle and daub paneling of timber-framed buildings. Trees grown for timber (generally oak) were scattered sparsely through the woodland to avoid shading the underwood too severely. Hewn timbers were used mainly in buildings and also in carts, bridges, and ships. A by-product of oak timber was bark, used for tanning leather. In short, local wood products were another key part of the village economy and were carefully managed.[27]

Enclosed woods were privately owned parcels of land, their products available only by purchase. More widely accessible were wooded commons. The primary use of commons was grazing, but commoners often had rights to use whatever wood and timber grew there, as well. The problem with this "wood pasture" combination was that the grazing animals inexorably destroyed the regeneration of the trees. To protect it from this fate, smallwood on commons was often pollarded rather than simply coppiced. Sprouts were periodically lopped from the top of a trunk ten to fifteen feet high, out of the reach of nibbling livestock. Clumps of thorn sometimes served to guard young timber seedlings on the ground. Still, after centuries of grazing the wooded portion of wood pasture tended to diminish, so that many commons simply became bare pastures or heaths, with few trees left standing.[28]

Pannage, the right to fatten swine on acorns in wood pasture, also steadily decreased in importance over time as woodlands shrank. One of the most remarkable declines in woodland during the Middle Ages took place in the Weald in upland Sussex and Kent, which had been 70 percent wooded in 1086 and famous as autumn hog pasture. By 1350 the Weald was largely reduced to grazing and scattered arable farming, although it retained more woods than much of the rest of the English landscape.[29] A few hedges, wood pastures, and woodlands remained, but throughout England agriculture expanded at the expense of wood resources. Both arable and pasture ate away at the trees, making them the most difficult element in the landscape to conserve. By the height of the Middle Ages some villages had no woodland of their own but only rights in woods elsewhere or else had to purchase fuel supplies, a heavy cost. Some of the least wooded districts, such as Norfolk, turned to burning peat.[30]

Woods and heaths (whether enclosed or common) also served as a small but steady source of nutrients to the arable fields. Bracken and heather were cut as bedding for stock, and in a pinch lopped branches of holly, ivy, and even mistletoe served as winter fodder—and so made their way from the byre to the dung heap to the plowland. The hearth yielded potash that had originated in the woods, and although some ashes were used to make lye and soap, the residue went on to the gardens and fields along with other household wastes. As with pastures, woodland soils were able to sustain a low level of nutrient export for centuries, if not millennia, without greatly undercutting their productivity. Woodlands and other uncultivated fringes of the landscape provided peasants even with a bit of surreptitious protein in the form of poached rabbit and deer.[31] All in all, woodlands were a vital but vulnerable part of the agrarian landscape, and their growing scarcity in the early modern period would drive monumental changes in the English economy.

Water

The flow of water through the English landscape was also carefully managed from top to bottom, serving a variety of overlapping purposes, including the recapture of nutrients. The first problem was to get the water out of the arable fields. Draining the heavy clay soils of the Midlands was accomplished by the ridge and furrow method of plowing. The long, narrow plowstrip (perhaps ten yards wide), built up over centuries, was high in the center and low at the sides, forming the ridge and furrow. The strips ran uphill and down, except on the steepest slopes. Thus, the furrows between the strips also served, where necessary, as shallow drains. Through the landscape of the great fields ran a network of drainage in the hollows of the land, collecting the water and conveying it to the brooks and streams.[32]

Where the land was perennially wet, especially along the margins of rivers and streams, lay the valuable meadows. Meadows posed a crucial challenge to water management. As we have already seen in Musketaquid, grassy meadows form in wetlands that are *seasonally* flooded. Meadows could therefore be improved, and even expanded, by two seemingly contradictory means: draining (especially in summer) and flooding (especially in winter). Drainage of floodplains, marshes, and fens created some arable land, but its usual purpose was to augment and make accessible the growth of natural grasses, along with special products like reeds for thatching. There appear to have been few unimproved swamps and marshes left in lowland England by the late Middle Ages. Virtually every square foot of wetland within carting distance of someone's cowyard was carefully ditched and mowed.[33]

Meadows and the upland immediately adjacent were also improved by irrigation. The fringes of the floodplain sometimes served as leat grounds, a leat being a canal that supplies water. An early-sixteenth-century account by a man named John Fitzherbert described this method of watering the grass, advising that "yf there be any rynning water or lande flode that may be sette or brought to ronne ouer the meadowes from the tyme that they be mowen vnto the begynning of May / and they will be moche the bettr and it shall kylle / drowne / & driue awaye the moldywarpes. . . . All manner of waters be good / so that they stand not styll vpon the grounde. But specially that water that cometh out of a towne from euery mannes mydding or donghyll is best / and will make the meadowes moost rankest."[34] The object of such watering was less to counteract drought (or even to kill, drown, and drive away moles) than to recover waste nutrients and to encourage the grass to grow earlier in the spring by warming the ground. The leat grounds often lay at the tail end of the village water system, by which part of a stream was diverted to supply millponds and farmpools and then returned to the head of the meadows, having picked up the runoff from middens, dunghills, and waste drains along the way. Meadows on level floodplains were sometimes drowned for part of the year simply by damming the stream during the winter, which brought on lush grass in the spring. Such practices formed the basis for the more elaborate floating of watermeadows, begun in the late sixteenth century and widespread by 1620, by which extensive meadows were overflowed by intricate systems of channels and drains.[35]

Water served other needs which had to be fitted into this landscape. One was to drive mills, the most important of which were gristmills, although there were other mills, too—for fulling cloth and for hammering iron, for example. Another was to provide fish and fowl. Weirs were erected in streams to trap fish, including eels that rubbed elbows with the eels of Musketaquid on their breeding grounds in the Horse Latitudes. Ponds were also constructed to catch and raise fish, ranging from eels to pike and trout—some have called the Middle Ages a golden age of fish. "Fish have long been recognized as an important element in the medieval diet," writes one archaeologist, but it is difficult to tell how important because their bones are poorly preserved. There were even ponds specially designed to decoy wild ducks. In short, just as in Musketaquid, waterways provided an important food source in their own right.[36]

The management of water was hardly a casual matter in the English agrarian economy. It was conscious and complex. The flow of water was carefully integrated into the ecological structure of the villages and fields. Seasonal flooding was encouraged in some places, whereas water had to be drained from others. Water was directed to drive mills, to carry away wastes from homesteads, and to return those nutrients to the grasslands. Rivers were used for transportation of bulk commodities during the Middle Ages, although how far upstream shipping was practical is debated—fish weirs and milldams presented serious obstacles to water transport until the coming of canals centuries later.[37] All this required intricate vernacular hydraulic engineering in each community, subject to constant regulation, given the wide range of uses and abuses to which water might be put. Routine conflicts and nuisances came before manorial courts, but by the thirteenth century formal "commissions of sewers" were being appointed to deal with the ongoing business of maintaining dikes and drains in the watery regions.[38] This legal custom was to be transferred to New England.

Growth of Population, Economy, and Agriculture

The traditional agriculture that dominated the English landscape in a wide variety of regional forms down to early modern times had evolved to maintain a working balance between arable land, grazing land, woodland, and water. These elements went together to provide the range of resources that people needed to live. Although in many ways they complemented one another within the landscape, they could also conflict at times, especially when population grew. To some extent, a region could specialize in certain products at the expense of others and then trade to supply the deficit and more—and there was a great temptation for those in positions of power to do so. For example, ma-norial lords often tried to confine sheep folding to their own demesne land, upon which they grew grain for the market. For common people, however, reasonable access to some combination of all these resources within the local landscape was crucial to survival, and they often resisted any turn toward market specialization from which the majority of them, lacking the financial means, would not benefit. Peasant economies were founded on broad husbanding of resources. The market economy that eventually replaced these local economies encouraged each place to focus its production on a few commodities, often at the expense of others. New England was settled by yeomen, artisans, and traders in the midst of the transition between these two worlds.

The common field system had developed in large measure to afford broad access to the complete means of subsistence, in a world in which individual holdings were small and resources scarce. It operated not so much to extract the highest productivity from the land through specialization as to maximize security through diversification. This "bet spread-ing" included scattering tillage strips within several fields and thus growing a variety of crops over a range of soils, opening fields for common grazing during fallow and after harvest, and allowing controlled access to grazing, wood, and other resources that could be found on commons. Gamblers who spread their bets are playing for survival, not a big strike. They are betting against disaster, but also to a certain necessary extent against themselves. Most peasants could not have survived on their arable holdings and livestock alone, nor in many cases could they profit from the market. In most regions commoners labored under the heavy hand of feudal lords who, through their military power, ran a protection racket that extracted from 25 to 50 percent of the peasants' gross income—not a social environment that particularly inspired entrepreneurial risk taking.[39]

Traditional English mixed husbandry was a conservative adaptation to the English environment. Together, the elements in the system allowed controlled use of a diverse landscape in mutually reinforcing ways. Mixed husbandry was sustained for many centuries and appears to have been theoretically sustainable for much longer because it did not fundamentally degrade, but rather renewed, the resources it relied upon. This was also a system that had scope to be sustainably intensified by the tighter integration of grain and grass and the incorporation of legumes, as we shall see. Unfortunately, in his-torical practice it encountered ecological difficulty on several occasions and met disaster once, in the fourteenth century. A system may be sustainable, but that does not guarantee that it will be sustained. What went wrong? Medieval agriculture was *not* done in by a supposed "tragedy of the commons," in the sense of an intrinsic collective inability to limit individual exploitation of commonly held resources that leads inevitably to their degradation. On the contrary, for the most part both common grazing land and the common arable fields were stinted and regulated quite effectively.[40] The devastating crisis of the fourteenth century was caused partly by obstacles to improvement within the commons system, perhaps, but much more by an apocalyptic convergence of four dark horsemen: the parasitic feudal structure of the economy, bad weather, population growth, and bubonic plague.

The medieval version of mixed husbandry might have been able to sustain itself indefinitely given a stable population, but it was not able to foster strong enough economic growth to support a steadily growing one. The demographics of this period of English history remain hazy, but in broad terms it appears that a population in the neighborhood of two million (and surely well under three million) in 1086 more than doubled to something like six million by 1300. That was a level it would not reach again until the late eighteenth century.[41] But it was already a level that could not be safely carried by the medieval economy. The ecological expression of population growth was the steady expansion of the arable fields to produce more grain, at the expense of other important elements in the interlocking agrarian economy. This imbalance resulted in scarcity and degradation and set the stage for disaster.

One place where imbalance appeared was in the decline of woodland. There were timber and fuel shortages across Europe in the fourteenth century, although historians are uncertain about the depth of the crisis in England. It may be that a considerable interregional trade in wood kept even the heavily cleared Midlands supplied. One wonders how many peasants could afford to buy wood, though. In some regions, especially in densely populated East Anglia, the deficit may have been made good by cutting peat, an excellent fuel. As for building material, increased use of stone in some regions may indicate a shortage of timber, but there is no widespread evidence of reduced timber construction. It appears that even at the peak of population in England fuel and timber supplies remained barely adequate to meet the demands of the medieval economy, although there was undoubtedly much cold and suffering among the commoners.[42]

Shortages of pasture, livestock, and manure posed a more severe challenge, but historians are not of one mind about how well that challenge was met. In many districts overexpansion of grain acreage appears to have tipped the balance between arable and pasture, leading to declining yields and leaving medieval society in an extremely vulnerable position by the opening of the fourteenth century. Arable expansion was forced onto marginal soils, making for poorer yields. Worse, the loss of grazing made it possible to support fewer animals, which meant less manure to spread over more tilled land. When peasants tried to maintain the number of animals needed for cultivation, the condition of both pasture and beasts deteriorated.[43] The wet meadows that supplied winter fodder were a finite resource that could not be easily expanded, either. Scarce grassland, lack of plow teams, and declining grain yields from inadequately manured fields have been documented throughout much of England during the late thirteenth and early fourteenth centuries. In some districts cultivators were able to intensify production and keep up with demand through tighter integration of arable, grass, and manure. But with such improvements barely staying ahead of hunger, England went into the fourteenth century in an impoverished and precarious condition.[44]

The first blow to the feudal world came between 1315 and 1322, when a run of wet weather destroyed harvests, resulting in the Great Famine. Even as the arable failed, epizootics struck both cattle and sheep, which were already in a weakened condition from the scarcity of grazing. Villagers died from starvation and sickness. Things continued to slide as the agrarian system came apart—in some areas population was already down 30 percent even before the Black Death. In 1348, the plague struck, killing at least one-third of the remaining people. Subsequent outbreaks of the disease brought the population to about three million or perhaps even lower by century's end—close to where it had stood in 1086, and half of its size at the beginning of the fourteenth century.[45]

The long period of depressed population that followed the collapse of the fourteenth century saw a pronounced shift from arable toward pastoral farming, as large parts of the English landscape were grassed down and devoted to raising sheep and wool. Aside from periodic recurrences of plague, life materially improved for the surviving lower orders because the slump in population raised the value of labor and depressed the value of land. Many peasants were able to enlarge and consolidate their holdings, to improve their living standards, to free themselves from dissolving feudal property relations, and to involve themselves more in the market—in essence, they became yeomen. This helped set the stage for a different agrarian response when population finally began to rise again in the sixteenth century.[46]

After two centuries of stagnation (or relief), population growth in England resumed about 1540 and continued for over a century, through the first half of the seventeenth century. By 1650, population was again approaching the level it had reached before the debacle of the fourteenth century, although it probably did not attain quite the same height, leveling off at about five million. Once again, the arable fields had to expand to fill more mouths with bread. As a result, the scourge of underemployment, poverty, and scarcity reappeared in early-seventeenth-century England. But the outcome of this crisis was quite different from that of the catastrophe of three centuries before. The economy of early modern England did not collapse under its own weight, as the feudal economy had done. The curve of population bent, but it did not break. Instead it held steady for one hundred years until the mid–eighteenth century, when rapid growth resumed. In the meantime agricultural production continued to increase, generating exportable grain surpluses, improving standards of living, and laying the base for the sustained economic growth of the industrial revolution.[47] How this happened is one of the key questions in modern history, and it is also crucial to this story, for it was during this period of transition that New England was founded.

The mounting pressure of people on the land was deflected from disaster and transformed into an economically creative force by three closely linked responses: demographic limitations on population growth, economic changes toward marketing and capitalism, and ecological changes in the way resources were utilized. This transformation was gradual; it did not overthrow the traditional relationships between agrarian communities and the land overnight, but in the end it did amount to what has been called an agricultural revolution. The birth of New England was part of this ferment, and the interplay of much the same forces would reshape life in Concord in turn, more than a century after that town was founded. The tension between demographic growth and ecological and economic response has generated a deep, recurring rhythm in human history, but the outcome has not been the same at each turn. The demographic, economic, and ecological changes in seventeenth-century England set in motion a great transformation in world history.

First, although population grew enough to create scarcity—enough to both stimulate the economy and to make life miserable for many people—it did not continue to grow as fast or as far as it had three centuries earlier. It thus stopped short of overwhelming the pace of economic growth and engendering catastrophic decline. English population began the sixteenth century at about 2.75 million people, still near the lowest point since its collapse in the fourteenth century. From about 1540 population grew rapidly until it crested near 5.3 million in the 1650s, slipped back to about 4.9 million in the 1680s, and then stayed essentially flat to the middle of the next century. Population control was achieved primarily by a drop in fertility, which declined from the mid–sixteenth until the late–seventeenth century and stayed relatively low through the first half of eighteenth century. The decline in births resulted from a rise in the age at which women married and a rise in the percentage of women never marrying. As economic conditions worsened in England, young people decided (or were forced by circumstances) not to marry as readily as they once had. This amounted to a negative, preventive check on population growth—much to be preferred to "positive" checks which had stalked the land under names like the Great Famine and the Black Death.[48]

This is not to say that life was long and merry in seventeenth-century England. In fact, life expectancy was short and getting shorter, sliding from somewhere near forty at the beginning of the century to the mid to low thirties by century's end. Towns and cities were very unhealthy places for young people arriving from the country, and historians blame rapid urbanization for keeping mortality high. Elizabethan and Stuart England was a society on the move, especially among poorer people trying desperately to find work or a bit of common land upon which to squat. Freedom of movement tended to relieve pressure in regions where scarcity was harshest and then to build pressure in new areas in turn. It also delivered a stream of victims to that great demographic leveler, London.

From the ports, increasing numbers took ship to America. Whether driven by the search for religious freedom or better opportunity, emigration removed more than half of the natural increase of the English population through most of the seventeenth century. In short, the combination of migration, declining fertility, and continuing high mortality kept population growth in check in seventeenth-century England, turning rapid growth in 1600 into shrinkage by 1650.[49]

This changing demographic pattern had a profound economic impact. From the mid–sixteenth until the mid–seventeenth century rising population drove strong demand for conventional agricultural products, particularly grain. After 1650, stagnant population growth meant one hundred years of agricultural depression, leading to intense competition to find the most efficient ways of growing grain and to a search for specialty crops.[50] In the seventeenth century the agrarian response to both rising and falling demand ran increasingly through the market and was not as constrained to simple expansion and contraction of local subsistence production as it had been during the feudal crisis of the fourteenth century. During the late medieval period husbandmen of all social strata, and not just large landowners, had become more involved in commercial production. The early modern English economy was thus able to respond more effectively to demographic pressure and to ease shortages.

Underlying this birth of agricultural capitalism were long-term changes in tenurial relations on English estates. Increasingly, landowners chose to lease their property to a few substantial tenants, rather than to many peasant smallholders, as had been the case under feudal society and as continued to be the rule in much of the rest of Europe. English landlords became more interested in incomes than in fiefdoms. The rents that landlords charged were increasingly determined by market forces and the productivity of the land, rather than by customary fines and fees. Through this long, slow process, a handful of yeomen were able to rise to become affluent middle-class farmers who rented large properties. The bulk of the rural population sank to become small cottagers. They scraped by on bare subsistence holdings and embattled rights in the diminished commons, on small-scale market production, and on the meager wages they earned on commercial farms or as pieceworkers in rural industries. A large class of landless rural laborers was slowly being born. These were the people who suffered most during the lean times, who roamed the countryside and gravitated to the towns in search of a living, sometimes going on from there as servants to the New World.

As this transformation took place the agricultural landscape was steadily enclosed. This was a long, uneven process that began in the late Middle Ages and was not completed until the nineteenth century. Enclosure refers to the conversion of commonly managed land—both grazing commons and open fields—to blocks of privately managed land. It also refers to the physical act of separating these parcels by hedges or walls, so that they could be farmed without being overrun by the common herd. Enclosure allowed private control of farmland to grow crops or raise stock for market. At many times and in many places it was fiercely resisted by smallholders who were losing cherished access to common resources, but at other times and places it was accommodated by small-holders who saw in it new opportunities for themselves. Participation in the market did not always require or automatically lead to enclosure. In many regions, common field parishes persisted for centuries alongside enclosed ones, and common field tenants marketed their produce. There was scope for new farming methods, new crops, increased production, and adaptation to market demands within the common field system. But in the long run enclosure, private management by fewer farmers, and production for the market went hand in hand.[51]

A larger and more fluid market in agricultural commodities allowed increased regional specialization. Each region could concentrate on those products for which it had a comparative advantage, responding to the consumer demands of other regions and to urban markets. By the seventeenth century there were already grain-exporting districts, market-gardening districts near London, large-scale sheep districts, smaller-scale pastoral districts specializing in dairy production (think Cheshire cheese), stock-rearing districts in the upland zone, and so forth. The region around Matlock in Derbyshire (which sent Thomas Flint, Henry Woodis, and Michael Wood to Concord) was a stock-rearing district on the fringe of the Midland plain, with large common grazings, small pasture closes, meadows, and limited tillage. The Weald of Kent (which dispatched Simon Willard, along with Joseph Meriam, James Hosmer, William Buss, and Thomas Stow) had a similar pasture and meadow economy devoted to rearing and fattening cattle and to large sheep flocks. The Weald was the heart of the English iron industry, as well. Cranfield and Odell in Bedfordshire (home of the Reverend Peter Bulkeley, William Hartwell, and the Wheeler clan) remained in the Midland mixed-farming, common field clays. There were production links between such regions as well, particularly the movement of stock from rearing to fattening. Regional specialization allowed for more efficient production across the country as a whole.[52]

The rise of the market economy undoubtedly helped increase overall production, but it was not the only, or even the most important, factor enabling England to weather the crisis of the seventeenth century. We should not exaggerate the degree of specialization in early modern agriculture, especially during the period of high demand for basic foodstuffs before 1650. The English survived the seventeenth century by expanding arable production at the expense of pastoral production (just as they had in the fourteenth century) and by curtailing births and shipping surplus people to America. There certainly was more marketing taking place, but this nascent development was more significant to shaping the future than to surviving the crisis of the present.[53] Specialized agricultural production in seventeenth-century England still revolved around mixed husbandry, especially in arable districts. There was still no way to grow grain without adequate manure. This meant that those producing grain (whether for themselves or for the market) still had to keep livestock and still had to find grass for their beasts to consume.[54] Increased specialization could not have made for more than marginal economic growth until it incorporated new methods for substantially improving the productivity of farmland. The most important of these improvements came largely after 1650—after the critical period of rising population and looming disaster and after the departure of the Puritans who settled New England.

Agricultural improvements were spreading before 1650. Bare fallows were replaced with legumes such as peas and vetch, a practice gradually adopted during the late Middle Ages in both enclosed and common fields. By the early seventeenth century the pressure for grain production led to pastures being plowed, sown, cropped for a few years, and then returned to grazing—the beginnings of convertible husbandry, the regular rotation of grain and grass. Farmers began to make better use of marl and lime—those growing grain for market were in a better position to purchase such inputs than smallholders growing the same grain for subsistence. The development of the floated watermeadow boosted the productivity of low-lying grasslands by capturing suspended nutrients and managing the growth of grass more efficiently. By providing more food, these improvements undoubtedly saved lives.[55]

The most important innovation in English farming was the introduction of clover into cropping rotations, and that took place largely *after* 1650. According to Thirsk, "The need for grassland improvement attracted urgent attention as a result of several years of bad weather from 1646 onwards. Hay rotted, fodder was short, and meat prices rose alarmingly."[56] The use of clover and other legumes like alfalfa and sainfoin spread rapidly in the following decades. Clover rotations scored their greatest success on light, chalky upland soils, where convertible husbandry often replaced permanent pastures. Farmers discovered that by plowing up pastures for a few years of cropping and then laying them back down to grass and clover, they could produce grain more cheaply than their lowland neighbors. Faced with this competition, many lowland arable parishes grassed down their clay and got into dairy and fattening, and also into protoindustries such as textiles and metalworking, in order to support themselves. As a result England continued to increase its agricultural production into the eighteenth century in the face of slackening demand, becoming a net exporter of grain. A dual economy combining new commercial patterns in agriculture and widespread rural industry got under way. Here was a profound economic and ecological shift in the English landscape, a world turned upside down.[57]

We should take a moment now to plow this convertible husbandry concept carefully under because it will come up again later. There may be nothing magic about manure, but there is deep magic in clover. Rotating arable crops with grass and legume crops is in most cases the most productive, ecologically stable form of mixed husbandry known. Legumes such as clover fix atmospheric nitrogen into a form that other plants can use, and many also have deep foraging roots that bring mineral nutrients from the subsoil. Legumes typically are most vigorous during their early years, after which the legume content of a sward tends to gradually run out. Convertible husbandry replants clover every few years into soil whose nitrogen content has been depleted by cropping—an environment in which legumes thrive. At the same time, grain crops are moved into soil whose organic matter, nutrient level, and good tilth have been replenished by a legume which has passed its prime and been plowed under. The soil-improving power of clover is optimized by timely incorporation. That is why grain and grass rotation has lain at the heart of the best husbandry for the past four hundred years, and why organic farmers still worship at its leguminous altar.

Not all land is suitable for convertible husbandry because not all land is suitable for tillage. Some soils are too poor, too rocky, too steep, too dry, or too wet to be safely tilled. Permanent grassland presents a separate, although parallel challenge to maintaining fertility. By good grazing management it is possible to encourage vigorous perennial legumes such as white clover, which help keep pastures in good heart. Periodically applying lime, phosphate, and potash to the grassland also helps.[58] Nor does convertible husbandry make good management of manure any less necessary. If anything, it makes conserving the nutrients in manure and returning them to the fields all the more important because in many cases more of the grass crop is harvested as fodder and fed to the stock at home and less is consumed in the field as forage. Legumes are beneficial for both stock and soil, but the augmented nutrients they generate must be conserved and returned in abundance. Clover does not replace manure, it increases manure—and so increases the importance of good manure handling.[59]

Lime is crucial to most legumes, whose productivity declines markedly in acid soil. The naturally high calcium content of many English soils was a great blessing. But beyond that, the increased use of marl (limestone mixed with clay in natural deposits) and lime during the seventeenth century was critical to the successful development of convertible husbandry. New England farming, by contrast, was severely handicapped by nature. Soils were mostly acidic, marl and limestone deposits scarce. Clover, although adopted during the eighteenth century as a nutritious fodder crop, proved disappointing as an improver of soil during the colonial period and indeed through most of the nineteenth century. Mixed husbandry in New England thus continued to revolve around manure more than legumes, with grassland and tillage remaining largely separate.

The expansion of agriculture in seventeenth-century England had a profound effect on woodland, and this drove revolutionary change. The late sixteenth and early seventeenth century was an age of wood scarcity, an early energy crisis. Farming cut into woodlands just as demand for firewood was rising, both for domestic heating and for expanding industries—ranging from burning lime and brewing beer to forging iron and firing bricks. The sixteenth and seventeenth centuries saw sharp rises in the price of fuel wood and building timber.[60] Villagers were faced with a stark choice between devoting land to producing food to eat or growing wood to keep warm. As usual when bellies grumble, trees lost out. People will generally shiver before they starve. England had been only 10 percent wooded at the height of the Middle Ages, and woodland made only a limited recovery after the collapse of the fourteenth century because wool growing remained profitable, keeping land open. With renewed pressure on woodland in the sixteenth century, England soon faced a serious ecological imbalance between agriculture and woods that might have frozen a subsistence economy as the colder weather of the Little Ice

Age set in. The dearth of fuel would certainly have cut off the tender young shoots of an industrial economy, had an alternative not come to hand.

The way out of this difficulty was down. Here was a key departure in modern environmental history, a veritable revolution and no question. It gave sprouting capitalism infernal roots far beneath the soil, roots that have sustained its growth for four hundred years. The English began to burn coal: first to heat houses; next to burn lime and brew beer; then, with some technical advances, to make bricks and glass; then in the eighteenth century to mine more coal (through the invention of the steam engine to pump water out of mines); and finally to smelt iron, power textile mills, and drive railway trains. All this was the fateful consequence of firewood scarcity in a nascent capitalist economy that chanced upon a plentiful supply of fossil fuel beneath its feet.[61]

Timber also grew scarce in the seventeenth century. This was the age of the Great Rebuilding in England, in which many medieval buildings were either handsomely remodeled or replaced as they fell into the hands of those who were making good in the new economy. The timber deficit was met partly by building methods that utilized smaller timbers and partly by the increased use of coal-fired brick from the mid–seventeenth century onward.[62] The English also continued to import timber from the Baltic and, increasingly, from America. Domestic scarcity that might have crippled a self-contained economy instead stimulated further commercial growth in an expanding one. One may debate the social and ecological consequences of capitalism, but no one can deny its dynamism.

Beginning in the seventeenth century England was able to rationalize the production of its own countryside partly by genuine improvements, partly by replacing scarce firewood with boundless coal, and partly by drawing on its new colonies and on trade with the wider world, whether cattle from Ireland and Scotland, sugar from the West Indies, tobacco from the Chesapeake, or cod from the Grand Banks exchanged for Madeira wine. All this marked the birth of twin forces that have since transformed the relationship between people and their environment all around the world and that have remained closely related. The first was the increased substitution, via the market, of imported resources for local resources. This has allowed specialization in the way land is utilized wherever it has appeared, with ecological implications—some beneficial, some baneful—that always need to be explored. The second was the unlocking of fossil energy, which has provided the power to both fetch and manipulate all other resources on the modern industrial, global scale and which also has had enormous environmental consequences.

New England was settled by English colonists during the Great Migration of the 1630s as part of an early wave of this capitalist expansion. The Puritans came famously for religious reasons, but also in search of resources that they could turn into commodities and export back to England or sell in other markets. The New World was full of experiments in market agriculture which were extensions of the search for specialties going on back in England.[63] Some American colonies did develop staple economies that revolved largely around the export of a single commodity, such as tobacco or sugar. New England participated in Atlantic trade by exporting fur for a short time, livestock and timber for a longer time, and fish from the great banks offshore until the past century. New Englanders also developed a shipping industry that helped tie the rest of the Atlantic economy together. The search for commodities played a small role in determining early settlement patterns in New England: Concord got its start partly as a fur-trading post. The growth of a capitalist economy throughout the expanding European world was one important influence on the development of New England during the colonial period, and it would become even more important after the Revolution.

But this was not the most important economic motivation that the settlers of New England brought to their new home. Their productive focus was still more local than international. Emigration to New England occurred during the first half of the seventeenth century, before the final push into convertible husbandry and commercial farming took place in England. Many of the original proprietors of towns like Concord were deeply imbued with the traditions of village agriculture. They were hardly ignorant of the market, but for the most part what they knew was mixed husbandry, designed to make integrated use of diverse local resources largely to satisfy local needs and to secure a comfortable subsistence, with perhaps a marketable specialty or surplus that varied according to circumstances. Many things about New England served to reinforce this inward-looking tendency. New England's climate and soil did not lend themselves to any major staple export crop. The religious convictions of the Puritans led them very deliberately to create self-contained communities and to embrace communal life, a far cry from the sprawling settlement patterns and ambitions found in many other American colonies to the south. Partly by choice and partly by necessity, New England settlers were thrown back upon past traditions of local self-reliance. They set about recreating a familiar landscape of English agrarian villages in a new world that was in many ways comfortably familiar, but that presented its own special trials and afflictions, too.

CHAPTER 4

The First Division and the Common Field System

The povertie and meannesse of the place.
—CONCORD INHABITANTS' PETITION

ONCORD, MASSACHUSETTS, has the great distinction of being the home of three American creation myths. The most famous was the birth of the American nation: the "shot heard round the world" was fired at the North Bridge on April 19, 1775, marking the start of successful armed resistance to British imperial rule and the beginning of the American republic.[1] Another celebrated conception took place 70 years later in the woods at Walden Pond, when Henry Thoreau borrowed an ax and cut down six pine trees for his cabin, thus launching the American environmental movement.[2] These creations are widely regarded as new, exciting things in the world, born of the encounter between European culture and the American environment. But 140 years before the exchange at the bridge, and more than two centuries before Thoreau sojourned at Walden, there first had to be the birth of Concord itself. That event took place in the fall of 1635. It is too little remembered to rank as a national founding myth, yet less than 20 years after it occurred it had already been recast as a tale of mythic proportions, prefiguring the great American frontier saga of transforming and civilizing the West. This might tempt us to ask, Is there anything to celebrate, ecologically, in this early attempt to set down an English agrarian village in the New England wilderness?

In his *Wonder-Working Providence,* published in 1654, Edward Johnson chose Concord as the setting for an "Epitome" of the Puritans' errand unto New England. Concord was the first inland settlement in the colony, and it seemed to Johnson (himself among the founders of neighboring Woburn, established a few years later) that the trials overcome in planting and peopling this remote place exemplified the hand of God at work in the new world. Here is a small taste of what Johnson wrote:

Yet further at this time entered the Field two more valiant Leaders of Christs Souldiers, holy men of God, Mr. Buckly and M. Jones, penetrating further into this Wildernesse then any formerly had done, with divers other servants of Christ: they built an Inland Towne which they called Concord. . . . Now because it is one of the admirable acts of Christ providence in leading his people forth into these Westerne Fields, in his providing of Huts for them, to defend them from the bitter stormes this place is subject unto, therefore here is a short Epitome of the manner how they placed downe their dwellings in this Desart Wildernesse, the Lord being pleased to hide from the Eyes of his people the difficulties they are to encounter withall in a new Plantation, that they might not thereby be hindered from taking the worke in hand. . . . With much difficulties through unknown woods, and through watery swampes, they discover the fitnesse of the place, sometimes passing through the Thickets, where their hands are forced to make way for their bodies passage, and their feete clambering over the crossed Trees, which when they missed they sunke into an uncertain bottome in water, and wade up to their knees, tumbling sometimes higher and sometimes lower, wearied with this toile, they at the end of this meete with a scorching plaine, yet not so plaine, but that the ragged Bushes scratch their legs fouly, even to wearing their stockings to the bare skin in two or three houres; if they be not otherwise well defended with Bootes, or Buskings, their flesh will be torne. . . . Their further hardship is to travell, sometimes they know not whither, bewildered indeed without sight of Sun, their compasse miscarrying in crouding through the Bushes, they sadly search up and down for a known way, the Indians paths being not above one foot broad, so that a man may travell many days and never find one. . . . This intricate worke no whit daunted these resolved servants of Christ to goe on with the worke in hand.[3]

Here already were the now-familiar elements of the American frontier myth, the story of people on a mission from God to redeem the wilderness. It would be easy to accuse Johnson of exaggeration (or at least of artistic license) in telling his tale of the founding of Concord. The nineteenth-century Concord historian Charles Walcott discounted Johnson as "a writer whose imagination sometimes lends too strong a coloring to his facts," and he has often been similarly derided by literary critics.[4] But however mythically imaginative Johnson's prose may be, there is no sign that he made anything up. The physical discomforts and sense of isolation in the early years of the new community would be hard to exaggerate. Concord lay a good ten miles from the nearest tidewater village on the Charles River at Watertown. The exact route of the first trail is unknown, but whichever way it went it surely did have to make many intricate detours to avoid wetlands—as those who try to walk any distance across this landscape even today will soon discover. By Johnson's account, some who took the trail to Concord lost their way and wound up floundering through the swamps. As for bitter storms, devastating hurricanes in 1634 and 1635 may well have blocked the path, which would explain the tangled tree trunks the settlers had to clamber over. These hardy souls were mercifully unaware that they were in fact heading into some of the most extreme winters that New England has ever known.[5] The newly arrived servants of Christ faced a severe test of faith in making this land yield them the barest subsistence, let alone the comforts of home. They went to their wilderness work without the specialized pioneering skills that would be perfected over the generations to follow. In short, Concord is indeed an apt setting for the drama of a pioneering English community in the American wilderness, even if it has no exclusive claim to that mantle.

Just as many pioneers in the American West two centuries later may have believed they were fulfilling a national manifest destiny, so many of the Puritan settlers of Concord may have shared Edward Johnson's faith that the everyday labor of making a homeland was also a wonder work of preparation for the second coming of Christ.[6] But it is worth remembering, first, that all of these pioneers had practical problems to solve in learning to live in particular places, and, second, that Concord is *not* the West. It is a place much more like England than the land beyond the Mississippi, and it was settled before the newcomers had ceased being English and become American, at a time when the modern market economy was just being born—so the terms of the ecological and cultural encounter were very different from those that prevailed in nineteenth-century America. Furthermore, no matter what they thought, the English settlers of Concord were not really on their way to transform a wilderness. Their errand was rather to replace one culturally modified landscape with another, as scholars sitting in their studies today put it. Concord was not entirely covered with mature forests (although it had plenty of those), but in many places entangled by the surge of adolescent growth that had followed recent abandonment. Large stretches of the sandy lands that had been kept open by the Native people were no doubt overgrown by impassable thickets of brush and brambles, just as Johnson described. Native care had all but collapsed following the epidemics of 1616 and 1633, leaving much of the land in a feral state. Coming from an agrarian countryside densely settled by Christians, the English may have initially responded to the native landscape as a howling, devilish wilderness, but we need not continue to think of it that way. The founders of Concord encountered a land that had been arranged into an ecological order by one combination of natural and cultural forces, and they set about remaking it into another order that fit their own ways of fending for themselves. To do this, they would have to work out a new synthesis of both native and imported elements. The question is, Did they have the cultural capacity to make a durable marriage of English husbandry and American land?

Although the English may not have confronted a raw wilderness, they still faced no easy task. They had to "discover the fitnesse of the place," tumbling higher and lower. They also had to discover their own fitness *for* the place. They had to learn which of their familiar crops, domestic beasts, and arts of husbandry had a useful role in the new world and which were useless baggage. They had to discover how far they should modify their farm practices to suit the new land, and how far they could modify the land to conform to their desired agricultural order. They had to take unfamiliar elements from the new land and make them their own—or change them. In Concord, it took some four generations, nearly a century, for this task to be completed, although the main lines of the work were well in hand by the end of the second generation, fifty years after the town was founded.

It took time, trial and error, and (as Johnson put it) the labors of Hercules to drain and improve meadows, establish productive pastures, bring the most suitable soils into cultivation, and master the New England woodlands. These elements had to be integrated into a working ecological whole, one that achieved a balance among the kinds of land available and the seasons of labor throughout the year. The English settlers eventually did create a New England version of the English countryside and way of life in Concord, but it was a long, arduous struggle. For a while at the start, they truly wondered if it pleased God that they would succeed.

The Settling of Concord

> *Upon some inquiry of the Indians, who lived to the North-west of the Bay, one Captaine Simon Willard being acquainted with them, by reason of his Trade, became a chief instrument in erecting this Town, the land they purchase of the Indians.*
> —EDWARD JOHNSON, *Wonder-Working Providence*

In one way Concord's settlement did foreshadow what would become the American formula of frontier development: even when the West was less than a day's walk from the saltwater tide, the trappers were the first to appear. Capt. (later Maj.) Simon Willard, a soldier involved in the fur trade with the Indians, is credited with discovering the site of Concord in the early 1630s.[7] By then, Puritans arriving in the colony from England were finding that land was already in short supply in established communities around the bay like Watertown, and they went looking for new places to settle towns. On September 2, 1635, the General Court granted Willard, the Reverend Peter Bulkeley, and a number of other proprietors a petition for a plantation at Musketaquid. A congregation for the new plantation was gathered at Cambridge on July 5, 1636.[8] A year later, in August 1637, the land was purchased from several surviving Indians for an assortment of hatchets, knives, cloth, and clothing—at least that was the testimony of the few remaining English and Native witnesses who could be summoned to swear out depositions about the sale half a century later in 1687, when Concordians were trying to shore up the title to their land against the depradations of Governor Edmund Andros.[9] We know next to nothing about how many Indians were living at Musketaquid in 1637 or how they coexisted with their new neighbors in the decades that followed. It is thought that most of those who had survived the epidemics eventually moved northwest to the praying village of Nashoba in 1654.[10]

Whether any English settlers actually moved out to the new plantation in the fall of 1635 is unclear, but they were certainly there by 1636 because an order of the General Court dated September 8 of that year refers to "the inhabitants of Concord."[11] Simon Willard, soldier and fur trader, was granted land at Nawshawtuc Hill near the confluence of the Assabet and Sudbury rivers that forms the Concord River (or, as they were called by the early English inhabitants, the North, South, and Great rivers).[12] It is not clear when beaver disappeared from Concord, but we can be sure they soon became scarce.[13]

Although the fur trade may have originally led Willard to this part of the country, and the location at the forks of the river may have suited his trading purposes for a time, the attraction of Concord for most of its English settlers had little to do with fur. It had everything to do with planting ground and, above all, with meadows.

The new town centered on a stretch of fine silt loam laid down at the bottom of glacial Lake Concord. These soils supported large expanses of grassy meadows in their lower parts, all along the meandering rivers and brooks. The village itself was laid out within easy reach of these meadows, which were immediately necessary to maintain livestock. As Howard Russell has pointed out, all the earliest settlements in Massachusetts Bay colony were located near plentiful natural hay sources—first salt marshes along the coast and then large river meadows inland: Concord, Sudbury, Dedham, and Andover.[14] The village houselots were clustered near the meadows and also in the midst of a collection of already cleared Indian planting grounds, no doubt in various stages of regrowth. The Native corn grounds were located on sandy glacial outwash plains that flanked the meadows. These provided tillable fields around the perimeter of the village, a pattern familiar to many English settlers. The houses backed up directly to these fields. "Their buildings are conveniently placed chiefly in one straite streete under a sunny banke in a low level," Johnson reported.[15] Most of the narrow houselots were laid out along the glacial kame slope called the Hill, which bounded the Great Field. Houselots across the Mill Brook butted on the South Field, while a few over the North Bridge formed their own general field. The bulk of the landscape outside the fields and meadows, largely forested, was left as commons upon which to pasture livestock and cut wood. Very much the same configuration was employed when Sudbury was settled a few years later several miles up the same river to the south.

With these major elements—village houselots, clustered planting fields and meadows, and large, undivided commons surrounding the rest—the settlers of Concord put in place a commons system of husbandry with a tightly nucleated central village. This organization was familiar to many of them, it offered the convenience and security of living and working closely together in an uncertain and sometimes dangerous environment, and it reflected the intense communal ideals of these fervent Puritans.[16] Many of the settlers of Concord came from Bedfordshire, a mostly open field district in the English Midlands. Others came from Kent in the southeast and Derbyshire in the northwest, where pastoral systems that combined commons and closes were the rule. They brought with them a range of experience concerning how arable fields, livestock, and grazing could be integrated, but the ecological principles were broadly similar. How were these traditions adapted to a new set of environmental circumstances in Concord?

The Commons System in Concord

No matter what part of England the Concord settlers hailed from, Derbyshire, Bedfordshire, Kent, or even London (though not many who came from London were born in that town), they would have shared some basic expectations about how a system of husbandry should work. At their houselot, they would have expected to have a house, barn, and cowyard, along with a garden and perhaps an orchard. On their tillage lands, they would have expected to grow their bread grains and perhaps their beverage grain (barley) as well. They would have expected to dress these crops primarily with livestock manure, by mucking or folding. They would have expected the commons to supply grazing for the summer, and they would have expected the meadows to supply hay for the winter. They would have expected to secure fall and winter grazing on the aftermath of the meadows and on the stubble and grassy balks within the tillage fields. They would have expected to get fuel and timber, clay, sand, and gravel from the commons. Finally, they would have expected the streams to run their mills, to supply them with fish, and to stay well behaved and out of their meadows during the haying season. How well did Concord meet these expectations? How fit was this place?

It may be that of any part of America, New England is most like old England and hence well named. Certainly compared to the Far West, the West Indies, or the Canadian North, New England did not confront northwestern Europeans with novel environmental conditions. The topographical scale of New England is on the same order as Britain. The seasonal swings in climate are similar in rhythm, though somewhat wider in range. The flora and fauna are closely parallel. Along with these comforting similarities between England and New England, however, the husbandmen encountered several unsettling environmental differences: colder winters, hotter summers, thinner soils, denser forests, and unruly rivers. These posed great difficulties to the colonists in their attempt to establish an English system of husbandry in Concord.

A bird's-eye view of landholdings in Concord about 1650 is a bit startling (plate 1).[17] Something very much resembling a classical English common field village, in all its intricate complexity, had been set down in the midst of the American forest. Some fifty long, narrow houselots formed a tight nucleus within a mile of the meetinghouse, obeying the spirit if not the letter of the 1635 order of the General Court in Boston.[18] Beyond the houselots lay the upland tillage lots of the town, mostly collected into one great and several lesser general fields, with only a few detached lots about the outskirts. Nearly two hundred mowing lots flanked the watercourses throughout the town, clustering along Elm Brook, Mill Brook, Spencer Brook, and of course, the rivers—culminating in the Great Meadow. Surrounding everything else, covering three-quarters of the land in the infant community, were the commons, composed of upland hardwood forests, pitch pine plains, and spruce and cedar swamps.

We have no direct record of how the First Division of land in Concord was made, how the agricultural business of the town was conducted, or how the system evolved over its few decades of existence. We have only the pattern it made on the earth by the 1650s and 1660s, when First Division grants were recorded in the course of making the much larger Second Division. We also have a handful of orders passed by town meeting later in the century—some parts of the commons system continued to operate even after most of the land had been privatized. From these clues we can deduce something of the way in which the commons system functioned.

Houselots

The first settlers to reach Concord had to shift with simple digs. As Johnson told it, "They burrow themselves in the Earth for their first shelter under some Hill-side, casting the Earth aloft upon Timber." In time, "which ordinarily was not wont to be with many till the Earth, by the Lords blessing, brought forth Bread to feed them," they moved above ground, and the village grew.[19] By the 1650s fifty or so houselots were clustered along several roads leading away from the meetinghouse. The Bay Road ran east toward Cambridge, Sudbury Way led south along the east side of the river with a fork leading to Watertown, the Lancaster Road headed west across the Sudbury River at the South Bridge and then over the Assabet River, and another way crossed the North Bridge into the north part of town. Houselots flanked these roads and the lanes connecting them. At the nucleus of the village were the meetinghouse, two burying places on either side of the brook, a training field, and a millpond with a gristmill belonging to Peter Bulkeley, the minister.

Concord houselots were ordinarily anywhere from 3 acres to 10 acres in size—6 or 8 acres was typical. The largest single parcel surrounding a house was Thomas Flint's 119½ acres on the northern fringe of the village, across the Great River. A wealthy merchant from London who made substantial investments in the fledgling town, Flint may have bought out several neighbors who left Concord during the hard times of the 1640s. Peter Bulkeley accumulated property in this way, as did several others.[20] Concord founder William Spencer received a First Division grant of 108 acres in the northwest part of town and left his name on a brook there; but it does not appear that Mr. Spencer ever lived on his grant.[21] At the other end of the spectrum, poorer families in Concord owned little more than a small houselot or simply rented a dwelling, and some were not even recorded in the land divisions.

Houselots typically contained a house and barn, outbuildings, a cowyard, and a garden.[22] It appears that most houselots in the North and South Quarters and some of those in the East Quarter contained tillage land.[23] Depending upon soils and terrain, houselots might also have included pasture, meadow, and sometimes even woods, if part of the lot happened to be steep or swampy and so unimprovable. By the 1650s and 1660s many houselots already boasted an orchard. These lots were larger and more agriculturally robust than the cramped houselots packed into many English villages, which had often contained little land beyond a garden, cowyard, and midden. Nevertheless, none of these Concord houselots, however big or small, was expected to comprise a complete, working landholding for a yeoman. Each was but the focal point of a much larger constellation of lands and privileges, including planting grounds in the general fields, mowing lots in the meadows, and grazing and other rights in the commons.

Tillage

The husbandmen of Concord granted themselves tillage land in three different ways, although there were no hard and fast lines between these categories. First of all, most had plowland within their houselots. These lots often ran together with those of several neighbors within a common fence, forming a general field—this was definitely the case with the lots on the north side of the river.[24] In addition to fields that were extensions of adjoining houselots, there were several freestanding general fields in Concord in which men owned planting lots, including the Cranefield, the Brickiln Field, the Chestnut Field, and the New Field across the South River bridge. Finally, many men owned detached pieces of upland that could be broken up and sown. Sometimes called "great lots," these upland areas bordered one of the general fields or at times were simply carved out of the commons.[25] How was this arable land distributed among the early settlers?

William Hartwell came to Concord from Bedfordshire, a region of common field villages and private estates in the English Midlands. His steep nine-acre houselot against the Hill toward the east end of the village may have included some tillage land below the highway and perhaps even some on top of the hill behind his house, but not much (plate 2, table 4.1). Hartwell also owned a five-acre lot of upland, a ten-acre lot of upland, a nine-acre lot of upland, and a swamp all located half a mile north of his houselot within the Cranefield, "beyond ye Hills." His sons Samuel and John were granted similar lots nearby. On top of that, William Hartwell owned another four-acre lot within the Cranefield on the Far Plain, this one a mile northeast of his house.

But that was not the end of Hartwell's planting ground. He also owned a four-and-a-half-acre lot of upland and meadow within the Brickiln Field, one mile to the east of his house along the way to the bay. Beyond that he had a three-acre lot in the Chestnut Field two miles by road to the southeast, in the hills above Elm Brook on the shores of Flint's Pond. In addition, Hartwell was granted a seven-acre parcel of upland near the Chestnut Field. Such a scattering of upland inside and outside of fields was typical of Hartwell's East Quarter neighbors who lived under the Hill.

Deacon Luke Potter's tillage land was even more dispersed. Potter was a tailor who migrated to Concord from Newtown Pagnell in Buckinghamshire, just over the border from Bedfordshire.[26] His six-acre houselot by the millpond may have contained some plowland (plate 3, table 4.2). He also owned five acres of upland in the field south of Town Meadow, a mere quarter mile from home—in all likelihood this was the former houselot of a family who had left Concord. But Deacon Potter's most substantial piece of tillable land lay a good two miles from his door, at the far northeast end of the Cranefield. He owned another upland lot beyond the Cranefield, on Rocky Island near Elm Brook Meadow. And he owned three-and-a-half acres of upland on the east side of Brickiln Island, adjacent to his meadow there—another two-mile oxcart drive from home.

Like many others, Humphrey Barrett came to Concord from Kent. Barrett's land-

TABLE 4.1   William Hartwell

First and Second Division		
First Division	*No. of Acres*	*Location*
Houselot	9	
Meadow	5	Great Meadow
Meadow	13	Elm Brook Meadow
Meadow	4	Elm Brook Meadow
Meadow	2	Elm Brook Meadow—at Brickiln Island
Meadow	5	Rocky Meadow
Meadow	4	(Rocky Meadow)
Meadow	1.5	by Brick Kiln Field
Upland	4	Cranefield
Upland	6	part of 9 acres upland and swamp
Upland	10	(Cranefield)
Upland	1	both sides Elm Brook Meadow at Pine Hill
Upland	3	Brickiln Field
Upland	0.5	by Brickiln Island
Upland	5	(Cranefield)
Upland	3	Chestnut Field
Upland	7	near Chestnut Field
Swamp	3	part of 9 acres upland and swamp
Total	86	
Second Division		
Upland	6	near meetinghouse frame
Upland	5.5	ox pasture
Upland	1	by Rocky Meadow
Upland	33	by Rocky meadow
Upland	72	by Watertown bounds
Upland	2	on Pine Hill
Upland	72	on hill near Hog Pen Brook
Upland	8	Elm Brook Hill
Total	199.5	

First and Second Division **Total Lands:** 285.5 acres

Note: Information in parentheses was not found in land division records themselves, but deduced from other sources or mapping context.

holdings were more consolidated than Hartwell's or Potter's, although by no means contiguous. Barrett had a substantial twelve-acre houselot just beyond the meetinghouse on the way to the North Bridge (plate 4, table 4.3). He owned two lots in the field across the road within easy reach of his house and another lot that included both meadow and upland at the head of the Great Meadow not far beyond. Just a bit farther east was a twenty-acre lot of upland and swamp meadow which later records indicate Barrett used

TABLE 4.2 Luke Potter

First Division	No. of Acres	Location
First and Second Division		
Houselot	6	
Meadow	4	near Pine Hill (in Elm Brook Meadow)
Meadow	8	(in Elm Brook Meadow)
Meadow	4	in Elm Brook Meadow
Meadow	4	east side Brickiln Island (Elm Brook Meadow)
Meadow	2	in Elm Brook Meadow—at Brickiln Island
Meadow	14	near Fair Haven
Upland	5	near Town Meadow
Upland	10	Cranefield
Upland	8.5	(by Elm Brook Meadow)
Upland	3.5	east side Brickiln Island
Total	69	

Second Division

	No. of Acres	Location
Meadow	4	near Mt. Tabor (in Long meadow)
(Meadow	6	near Watertown bounds)
(Meadow	7	Dunge Hole)
Upland	8	South Field
Upland	56	pine land
Upland	4	west side Pine Hill
Upland	4	bounded west by own meadow (at Pine Hill)
Upland	29	near Flint's Pond
Upland	8	joining meadow at Fair Haven
Woodland	16	by Walden
Woodland	4	in swamp near Pond Meadow
Woodland	21	by Flint's Pond
Woodland	11.5	by Goble and Dane farm
Swamp	6	(Virginia Swamp)
Total	184.5	

First and Second Division Total Lands: 253.5 acres

Note: Information in parentheses was not found in land division records themselves, but deduced from other sources or mapping context.

primarily for pasture. All of Barrett's tillage land lay little more than half a mile from home.[27]

Such holdings were typical of Concord's settlers. Most of these men owned planting lots in several general fields, some more widely dispersed than others. What distinguished these "fields" from other uplands was not that all the land within them was tilled, but that it all lay within a single common fence. The largest of these fields was the Crane-

TABLE 4.3   Humphrey Barrett

First and Second Division		
First Division	*No. of Acres*	*Location*
Houselot	12	
Meadow	5	at head of Great Meadow
Meadow	5	in Great Meadow
Meadow	17	in Great Meadow
Meadow and Upland	20	at the head of the houselots
Upland	14	in field across from houselot
Upland	2	in same field
Upland	7	by head of Great Meadow
Upland	14	
Total	96	

Second Division		
Upland	100	at Annussknet (Annursnack)
Upland	100	by Robert Blood
Upland	20	in Twenty Score
Total	220	

First and Second Division Total Lands: 316 acres

field, lying beyond the Hill to the northeast of the village houselots and running down to the Great Meadow. This field apparently drew its name from the village of Cranfield in Bedfordshire, very near the place of origin of the numerous Wheeler clan and the Reverend Peter Bulkeley.[28] In later years the Cranefield and several smaller fields and uplands bordering it were enclosed in a larger fence and called the Great Field.[29] Within the Great Field lay at least one pond and several swamps.

Farther to the east along the Bay Road lay a smaller general field called the Brickiln Field, in which perhaps a dozen men had First Division tillage lots. Somewhere within the field or nearby (no one is sure exactly where) were clay pits and a brick kiln.[30] Also in the east part of Concord was the enigmatic Chestnut Field, located several miles from the village near Flint's Pond. The South Field lay at the back end of the South Quarter houselots, southwest of the village. Later (presumably) the New Field was laid out farther west on the sandy plain between the rivers, near a handful of houselots just over the South Bridge. General planting fields in the North Quarter were found across from the houselots on the way to the North Bridge and surrounding the more distant cluster of houselots beyond the bridge.

Meadows

The meadow lots of Concord's husbandmen were even more widely dispersed than their tillage lots. This was true in 1652, and it remained true for the next two centuries. The first generation of farmers in Concord granted themselves some thirty to fifty acres of meadow each, usually in half a dozen separate pieces, thus creating this enduring pattern. Like tillage, some meadow was close to home. Hartwell, Potter, and Barrett all had houselots bounding on the Mill Brook, while others had houselots running down toward one of the rivers. Most of these houselots included some mowing land at the bottom end.

The home meadow was no more than a small piece of most settlers' First Division meadow holdings. Four to six meadow lots were typical, and some men owned even more. The lots were widely scattered among Concord's many meadows, from those that nearly encircled the village out to the far corners of town. William Hartwell owned one lot toward the lower end of the Great Meadow three miles from home, and three lots strung along Elm Brook Meadow (plate 2). He also had a small piece of meadow adjoining his upland in the Brickiln Field. In addition, he owned two lots in Rocky Meadow, two-and-a-half miles away beyond the Suburbs. In all, Hartwell had some thirty-four acres of meadow to his name.

Once again, Luke Potter stands out as a husbandman who was willing to travel a good deal in Concord. Most of his thirty-six acres of meadow lay in five separate pieces along a mile-long stretch of Elm Brook, the farthest some three miles from home (plate 3). But his largest mowing lot, fourteen acres, lay more than five miles from home against the Sudbury line in the opposite direction, following the ways and wents over the upland to Fair Haven Meadow. And Deacon Potter was not alone: at least ten South Quarter men owned meadows at Fair Haven. Others traveled three miles to reach their meadow lots in Nut Meadow or in Pond Meadow on the east side of the river.

Humphrey Barrett's meadow lots, by contrast, were almost as consolidated as his tillage lots. Barrett owned about thirty-five acres of meadow (plate 4). His largest mowing lot, seventeen acres, was two miles away at the lower end of the main stretch of the Great Meadow. His other three meadows lay closer to home—one in the midst of the Great Meadow, another at the head of the Great Meadow, and a parcel of swamp meadow against the Cranefield. Other denizens of the North Quarter owned meadows at the foot of their houselots by the North River, along Spencer Brook a couple miles to the northwest, or farther upstream at Fifty Acre Meadow or at Bounds Meadow, a good four miles or more from the houselots. In addition to lots in these larger meadows, Concord men granted themselves smaller patches of mowing land in low places scattered here and there throughout the commons, wherever it might be found: Joseph's Meadow, Wigwam Meadow, Birch Meadow, Muddy Meadow, Stick Fast Meadow, Bear Garden Meadow, Dung Hole Meadow (of which there were two), World's End Meadow, and Mentoo.[31] Virtually no Concord First Division grant was without its complement of meadow.

Houselots, uplands, and meadows comprised the First Division grants of private land to Concord's proprietors. Most received a dozen or fewer of these parcels, making well under one hundred acres each. The rest of Concord's land, some three-quarters of the town's surface, remained as undivided commons until the Second Division.

A few small pieces of these commons were enclosed as "pastures." Some of these enclosures may not have been part of the general commons but instead belonged to smaller groups of common owners—the impression they left in town records is too faint to be sure. They were the Ox Pasture and Horse Pasture in the East Quarter, the Ox Pasture and Sheep Pasture in the South Quarter, and the Calves Pasture and Hog Pen Walk in the North Quarter. There may have been others—a Hog Pen Brook beyond Flint's Pond in the East Quarter, for example, hints at another Hog Pen in that direction. We know of their existence only because their acreage was parceled out during the Second Division or because mention of them crops up subsequently in deeds or other records. We have no description of their function except their names.

Until the 1650s, the great bulk of Concord's land lay as an unenclosed commons. In addition to their private lots, men such as Hartwell, Potter, and Barrett were undoubtedly granted commonage—rights to enter this open commons to graze livestock, to cut wood and timber, and to dig sand, clay, and gravel, all in regulated fashion. In Concord we have no full account of these regulations, but only a few clues derived from orders that were passed years later when the commons was much diminished, after the Second Division.

Ecological Challenges Facing the Commons System in Concord

It is obvious from the layout of Concord's First Division and from the telltale scattering of lots that belonged to individuals such as William Hartwell that a working commons system had been put in place. This is not surprising given the world from which these husbandmen came, although their widely dispersed holdings may look surprising—even profoundly irrational—to modern American eyes. These men were used to thinking of husbandry as a collective as well as an individual pursuit. For them, agriculture took place across the entire landscape of the town, within which they owned various private parcels and rights. Few of them had any experience of a farm—a term reserved in seventeenth-century New England for a fairly large, consolidated tract of land meant to return an income. Instead, their deeds often recorded that they owned a "messuage or tenement"—a homestead with a set of associated landholdings, rights, and (although not recorded in the deed) responsibilities within an orderly agrarian community. Of course, within this community structure these husbandmen could be fiercely competitive and protective of their own interests as well as cooperative. As this system moved through time, each proprietor acted as much to secure his *family's* position as his own individual station. That is, these men thought collectively in a generational and a geographic dimension. Their chief ambition was to ensure the security and standing of their families, now and in the future. That is the way landholding and husbandry had taken place across much of England, time out of mind.[32]

The practice of mixed husbandry within a commons structure had evolved slowly in England over those many centuries. It had become intimately adapted to the social and ecological circumstances of particular places in that country. It could not be summarily imposed upon a six-mile square of New England simply by laying out village houselots, fields, meadows, and commons. The settlers had to find the right combination of crops and livestock, integrating them into a working system of husbandry. That system had to encompass the seasonal growth of crops, the annual round of labor, the feeding, movement, and breeding of livestock, the transfer of nutrients, the flow of water, the mastery of the New England forest. It had to take shape on Concord's soils, under Concord's sky.

The challenge facing the founders of Concord was twofold. First, could any system of mixed husbandry combining old England and New England elements be made to work at all in Concord? Second, did a *commons* system still make sense under these changed circumstances—was it workable and was it desirable? These two questions were initially closely linked, but they were separable—and as time passed they became increasingly separated. The answers that emerged in colonial Concord were yes to mixed husbandry but only partly yes to commons. Now that we have seen the typical distribution of landholdings in Concord and bearing in mind the five differences from England— colder winters, hotter summers, thinner soils, denser forests, and unruly rivers—let's take a closer look at these ecological challenges and how Concord's husbandmen sought to answer them.

Among the first problems Concord's settlers faced was finding soils that could be tilled and grains that would grow in those soils to provide their daily bread. As Edward Johnson wrote,

> The Earth, by the Lords blessing, brought forth Bread to feed them, their wives and little ones, which with sore labours they attaine every one that can lift a hawe to strike it into the Earth, standing stoutly to their labours, and teare up the Rootes and Bushes, which the first yeare beares them a very thin crop, till the soard [sward] of the Earth be rotten, and therefore they have been forced to cut their bread very thin for a long season. But the Lord is pleased to provide for them great store of Fish in the spring time, and especially Alewives about the bignesse of a Herring; many thousands of these, they used to put under their Indian Corne, which they plant in Hills five foote asunder, and assuredly when the Lord created this Corne, he had a speciall eye to supply these his peoples wants with it, for ordinarily five or six grains doth produce six hundred.[33]

What was needed was planting ground that could be immediately broken up with the broad hoe or the plow, planted, and fenced against damage by livestock and deer. The settlers granted themselves lots within large Native planting fields which were already cleared or in the early stages of forest regeneration and went stoutly to work. It was more convenient to build a single fence around a large block of land than to build many fences around small parcels, even if all the land within the large field thus enclosed wasn't yet under the plow. The primary purpose of the fence was to keep cattle and hogs, which were foraging on the surrounding commons, from getting in and destroying the crops. The secondary purpose was to enclose creatures within the fence at those times when the entire field was opened for grazing. As time passed, more lots were granted in new fields and beyond the fields, and more ground was broken up and brought into cultivation.

In time, some First Division land that had been more heavily wooded was cleared and cultivated. For example, in 1654, Peter Bulkeley sold William Hunt fifty acres on the north side of the Great River, reserving to himself the right to the wood on the property. Hunt, however, was granted the "liberty to cut that wood which stands on the upland when he breaks up the land by plowing—to prevent it from being an annoyance to the grain planted or sown upon it."[34] By the end of the First Division, most of Concord's husbandmen had accumulated much more tillable upland—typically thirty or forty acres—than they could possibly have cultivated in any given year. A family having more than ten acres planted and sown in grain was unusual in colonial Concord. Presumably not all of this potentially arable land was broken up yet in 1650, and men were looking toward providing for their numerous offspring.

Because for convenience they turned first to the existing Native planting grounds, the English in Concord found themselves tilling predominantly dry, sandy soils. These ranged from (at best) the fine silt loams of the old lake bottom underlying the South Field to the coarse sandy loams and (at worst) loamy sands of the Cranefield and the pine plains beyond. These were excessively drained, droughty soils, very low in organic matter. Having adopted Native soils, the English took very quickly to planting Indian corn as their principal bread grain. They are to be credited for not spurning this productive crop—in fact, as Johnson tells us, they concluded it was God's gift to them. They put up with it for almost two centuries before reverting to wheat. A plant with tropical origins, maize was well suited to New England summers, which are typically hotter and drier than those of England, and did well on sandy soils. Planted in hills and cultivated by hoe, maize was also well adapted to land recently cleared of brush and forest, which "til the soard be rotten" was difficult to plow and prepare for broadcast English grains. Indian corn was indeed a godsend to the English in Concord, and without it they might have starved or fled.

Most English grains did not perform well in early New England. The early varieties of winter grains whose seeds were brought over were often not hardy enough for New England, and wheat was soon plagued with several pests; but the most serious underlying problem was the coarse tillage soils of the Indian fields. As elsewhere in New England, Concord favored drought-tolerant rye as the English grain that could best make a crop in poor, sandy soil.[35] Some oats and peas were also grown, but Indian corn and rye were far and away the grains most frequently mentioned in seventeenth- and eighteenth-century Concord inventories and widows portions.[36] In 1668, the town agreed to pay Capt. Thomas Wheeler two shillings per head to keep the dry cattle herd, payable one-third in wheat, one-third in "rie or pease," and one-third in Indian corn.[37] Day in and day out, Concord's husbandmen settled on corn and rye. Aside from the Connecticut River valley, few places in Massachusetts succeeded with wheat, and johnnycake, or "rye 'n' Injun," soon became the daily bread of the province.

The wide scattering of tillage lots in Concord was certainly reminiscent of the English open fields. But this ancient practice lacked a strong ecological foundation in Concord. There had been two principal reasons for scattering strips among large common fields in England: spreading crop risk over a variety of soils and making it possible to open the fields for winter grazing and summer folding of livestock. The first had some utility in Concord, the second hardly any. Concord's tillage fields had little to offer in the way of supplementary forage. Given New England's cold winters, there was limited scope for fall and winter grazing within the fields; and summer folding of stock was also impractical, as we shall see. The maps indicate that no sustained effort was made to set up a regular system of field rotation in Concord. The tillage lots of William Hartwell and his neighbors were much too unevenly distributed to suggest that entire fields were being devoted to winter crop, spring crop, and fallow by turns. By the 1650s, we seem to be seeing simply the outcome of a process by which a succession of fields were created over a number of years as more tillage land was required and additional lots were granted. Once established, some general fields did persist for several generations, but they had been created out of a combination of experience and expedience that became less compelling as a New England system of husbandry developed. Their strongest attraction was that they saved on fencing, but beyond that they could play no central ecological role, as they had in England, in integrating crops and livestock. Concord's general fields were loosely bound, not tight-knit collections of tillage lots.

The location of a few of Concord's general planting fields suggests that at least some husbandmen did seek diversity in their plow lands. The Brickiln Field and the Chestnut Field were located on soils very different from that of the sandy Cranefield. Both of these fields contained between half a dozen and a dozen small lots belonging to as many owners, mostly from the East Quarter. Why were these sites chosen for fields? They may have been open and available when the English arrived, but this seems unlikely—heavier soils were not preferred by the Native gardeners, and neither site is a noted place of Native occupation in Concord. The Brickiln Field was a fine silt loam, underlain by a tight boulder clay. This may have been desirable in drought years or in an effort to grow wheat, barley, and oats. Part of this field may have been cleared in the course of digging clay for bricks and cutting wood to fire them—in any case, it was prime tillage soil and remains in cultivation today.

The Chestnut Field is another story. Nearly three miles from the meeting house, it occupied rough land hemmed in by stony slopes of glacial till. Deeds indicate that by the early eighteenth century much of it was growing trees.[38] Why locate a tillage field in this absurd spot? The Chestnut Field may well have been exactly what its name suggests—a grove of chestnut trees divided into private lots and fenced from pigs—and *never* cleared and tilled. It was not rotated with other arable fields in the ancient English way, but was instead an entirely different sort of subsistence resource altogether. It may, in fact, have been another legacy of Native management for precisely this purpose, an ancient American crop that was not scorned by the newcomers, who had to cut their bread thin through those first cold winters. Their descendants were still collecting chestnuts in these woods into Thoreau's time and beyond. In short, the evidence suggests that Concord's husbandmen made a real effort to diversify their basic sustenance holdings beyond grain into nuts, but that wide scattering of plowland among systematically rotated fields did not become a permanent feature of the town's agriculture.

Neither the New England forest nor the Native farmers bequeathed the English plowland that was particularly fertile. Compared to the famous prairie soils (and some deciduous woodland soils) farther west, Concord soils in their native state were thin and sour. Clearing and burning the forest could sweeten the soil for a few years, but cropping, by removing nutrients, drove it back toward an impoverished, acid state. That was exactly how the Native cycle of shifting cultivation worked—it was not designed to build rich, permanent arable soils, and it did not. The condition of the soil was an obstacle to perfecting an English farming system in New England. On the other hand, it did remind these husbandmen to look to their dung.

The chief drawback to Concord's planting ground as the English encountered it was that it lacked organic matter. As far as English husbandry was concerned, the land needed manure. The normal English fertilizer was livestock manure, but this was scarce in Concord's earliest years, so according to Johnson the settlers turned to the alewives arriving in their rivers and brooks "in great store" just at planting time. Perhaps the surviving Indians showed them how to use fish for fertilizer, perhaps they didn't—in any case, it was a simple trick that would have occurred immediately to any corn mother or husbandman, native or alien. That rotting flesh fattens crops has never been a mystery to farmers. Fish were particularly well suited to maize cultivation because maize is grown in individual hills, and fish fertilizer comes neatly packaged in individual fish. The biggest problem with alewives was probably keeping dogs and other scavengers away from the hills. How long fish continued to be planted with corn we cannot tell. The spring fish runs were jealously guarded against downstream encroachments through most of the eighteenth century, but there is no further reference to their use as fertilizer in Concord.

Whether or not fish continued to be used, we can rest assured that as soon as they could, the English began relying mainly on dung. Manuring had been the basis of English tillage for centuries (if not millennia), and nothing the English encountered in Concord would have given them any reason to abandon this worthy practice. Quite the contrary. In Concord, the tillage land was innately poor, and English husbandmen doubtless did what they had always done, which was to dung it. They carried on in the way of their fathers. Sometimes their fathers made it legally binding: when Edward Wright passed his estate to two of his sons in 1684, the deed recorded their pledge to "till six acres for the comfort of our ffather & mother where he shall appoint it in our land yearly so long as God shall continue their lives together to plow, sow, *dress,* gather in and thresh it for them."[39] Dressing the corn with manure may not have been as carefully done in 1684 as it would be in 1846, but it was certainly never neglected. Concord simply could not have been farmed in the English manner without it.

How was the manure delivered to the tilled fields? In England this had been accomplished partly by folding livestock on fallow cropland overnight and partly by carting dung from the barnyard. There is evidence that some folding was practiced by New England farmers.[40] Since livestock in seventeenth-century Concord were herded, it would have theoretically been possible to pen them (or simply turn them out) on fallow fields in the evening, and perhaps this was done to some extent. Yet there were limits to the effectiveness of this practice in Concord. In England, folding was integrated into a system of field rotations that included a stirred summer fallow, whose primary purpose was to clean the field of weeds before the next small grain crop. At the same time, this periodic summer cultivation worked in the dung of the folded stock. But the first purpose of a fallow, cleaning the field, was largely accomplished in New England by growing Indian corn, which is a hoed crop. Furthermore, in England the foldcourse normally preceded the planting of the most demanding and valuable crop, the winter wheat. But in New England it was the maize that most wanted the manure, and wanted it most in the spring, deposited directly in the hill or furrow. Moreover, Concord had relatively few sheep in the seventeenth century. Sheep were everywhere the preferred stock for folding in England and apparently wherever folding was practiced in New England as well.[41] Concord lacked sheep because there was little suitable grazing for them early on and because dangerous predators remained—wolves were still being killed in Concord as late as 1735.[42] Sheep were of far less importance to the economy of the community than cattle and were only required in small numbers. In short, the amount of folding that was done in Concord remains an open question, but it was surely never as important a manure delivery route as it had been in many English regions. Folding is not mentioned in the records of the Great Field in Concord.

Nutrients deposited directly by folded livestock could not have been as important as in England, for this additional reason: the grazing season is much shorter in New England. Winter forage was virtually nonexistent in New England from mid-November until late April or early May. Around the year proportionally more of a beast's feed was eaten in the form of stored fodder, which meant in turn that more manure accumulated at home in "ye yards and mitten."[43] Significantly more: whereas in England livestock might graze three-quarters of the year or longer, in New England they could graze only about half the year at best. This meant an effective doubling or even tripling of the amount of hay and other fodder that needed to be provided, and hence a doubling or tripling of the amount of dung that collected in the barnyard, was trodden in with soil and other wastes, and needed to be carted. This barnyard muck was slow, heavy stuff to shift. Given the increased necessity of carting dung, it seems quite reasonable to suppose that husbandmen had plenty of time, while treading alongside their teams down the long ways to their cornfields, to reconsider their old common field practices—particularly the wide scattering of tillage lots. They doubtless began to think seriously about bringing their plowlands closer to their barnyards. At any rate, the next few generations in Concord were spent working through precisely that transition.

Concord's meadow lots were even more widely dispersed than its tillage. But this seems to have posed less inconvenience and conferred more ecological advantage than scattering the plowlands. Thus for two centuries and more the hay land not only retained many of the distinctive features of common farming—including the frequent clustering of numerous small, privately mowed lots within larger, cooperatively managed meadows and common grazing during the off-season—but also something of the feel of commons.

The pivotal function of meadowland in the Concord system of husbandry has already been suggested: the meadows supplied the bulk of the fodder that wintered the stock, and through them supplied nutrients to the tilled land in the form of manure. This required much more meadow in Concord than in England. Only 4 percent of the overall English landscape was in meadow at the height of the Middle Ages.[44] But in England, fodder needed to be supplied for only a month or two, while in New England two to three times as much hay was needed to last the long winter. So we would expect at least two or three times as much meadow in New England, and this seems to be roughly borne out.[45] Meadow occupied over 15 percent of the land in Concord until the mid–nineteenth century.

In their First Division, Concord's husbandmen granted themselves thirty to fifty acres of meadow each, in about half a dozen pieces. Not all of this meadow was initially mowed, however. The first generation of Concordians simply did not have enough labor to cut that much hay. But in any case it is unlikely there was that much hay worth cutting over the entire run of meadows they owned. Large parts of the meadows were too wet underfoot to support the mowers, let alone their oxen and carts; and those along the river were liable to ruinous flooding in wet summers. A wide range of vegetation grew in these meadows, some of it more palatable to stock than the rest. Over time, a handful of species emerged as the principal constituents of native meadow hay. By the nineteenth century the meadows were dominated by cordgrass (*Spartina pectinata*), reed canary grass (*Phalaris arundinacea*), fowl meadow grass (*Poa palustris* and *Glyceria striata*), blue joint (*Calamagrostis canadensis*), red top (*Agrostis alba*), and a number of other grasses and sedges. Taken together these grasses made serviceable, if somewhat coarse hay.[46]

Exactly what plants dominated the meadows as the English found them, before the advent of annual mowing and systematic drainage, is anybody's guess. A large part of the natural growth of wetlands in New England is not useful stock feed. What grows in meadow muck is determined partly by the depth of the water and how long it stands throughout the year and partly by other factors such as the frequency of burning or mowing. Stands of native grass worth mowing no doubt lay in irregular patches across the unevenly drained lowlands that flanked Concord's rivers and brooks, intermixed with stretches of less valuable wetland vegetation. The edible grass was hemmed in everywhere by marshes thick with sedges, bulrushes, and cattails; by shrub swamps full of button-bush, sweet gale, and red osier dogwood; and by wooded swamps of black willow, black ash, and red maple. Something of this complex mosaic of meadow and swamp seeps into the early land division records. Humphrey Barrett's meadow at the head of the houselots bounded against Nathaniel Ball's swamp, but both lots were later described as "swamp meadow."[47] The little brook at the foot of South Field arose in Muddy Meadow, flowed west through Muddy Meadow Swamp, and passed once again through meadow lots in Mentoo before it entered the river, suggesting that this wetland graded from grass into woody vegetation and back again. The northern bulge of the Great Meadow that lay within an oxbow of the river was a woodland known as the Holt, which suggests a willow swamp.[48] It appears that thanks to the providential efforts of Native people and beavers more of Concord's wetlands were growing meadow grass in 1635 than might otherwise have been the case, but that this growth was highly patchy and variable.

Aside from the uneven quality of the hay, the chief difficulty with the meadows was that they were often too wet to mow at all. Reliable meadows needed to be well ditched and sometimes even diked, not to mention free from summer flooding in all but extraordinarily wet years. This meant that not only the brooks traversing the meadows, but the river itself needed to be improved. The English came from a land where this work had been largely accomplished centuries before and only needed to be kept up, so the unruly behavior of Concord's streams was quite an affront. Here they found that while summer flooding along the river may not have been the rule, it was not that unusual, either. The Great River ran—walked, crawled, oozed—through the meadows with almost no current, until it reached the Fordway Bar several miles north of Concord, in what would become Billerica. Here a bedrock ledge blocked the channel, controlling the flow of the river. Numerous sandbars upstream also slowed the current. Winter and spring flooding was an annual occurrence. Even in summer, a spell of heavy rain would fill the river much faster than it could empty over the bar below, with the vexing result that wide stretches of floodplain for twenty miles upstream through Concord and Sudbury were sometimes submerged for weeks on end.

When the river got up into the meadows during the haying season, it was a great calamity. High water made it impossible to mow standing hay, or spoiled that which had already been cut but not yet safely carted to hard ground. No sooner had the Concordians arrived than it was firmly impressed upon them that they needed to improve their river. A stream of petitions that would run unabated for two centuries began to flow from the inhabitants of the Concord River valley toward the Great and General Court in Boston, seeking relief.

The first petition went down promptly in the summer of 1636. We can only imagine the new proprietors of Concord setting out to mow their meadow hay for the first time and finding much of it unexpectedly covered by water. This affliction was to bedevil their progeny even unto the seventh generation. The General Court heard their prayer and ordered that "whereas the inhabitants of Concord are purposed to abate the Falls in the river upon which their towne standeth, whereby such townes as shall here-after be planted above them upon the said River shall receive benefit . . . : It is therefore ordered that such towns and farms as shall be planted above them shall contribute to the inhabitants of Concord proportional both to their charge and advantage."[49] It is in-structive that Concordians were prudent enough to bind hypothetical future upstream neighbors to this river improvement scheme of theirs, before they would even embark upon it themselves. Sure enough, a few years later in 1644 another petition was granted appointing Thomas Flint and Simon Willard of Concord, Peter Noyes of Sudbury, and Herbert Pelham of Cambridge as commissioners of sewers "to set some order which may conduce to the better surveying, improving, and draining of the meadows, and saving and preserving of the hay there gotten, either by draining of the same, or otherwise, and to proportion the charges layed out about it equally and justly."[50] Sudbury had indeed been planted above Concord in 1639, with Noyes as one of its leading founders. Squire Pelham had been granted a large, private farm along the river nearby, known today as Pelham Island. These were all men of high standing in the colony. A Commission of Sewers, as noted in the previous chapter, was a long-established English legal body for carrying out community drainage projects—an institutional means to compel all those meadowowners who might benefit (rather than ratepayers in general) to contribute to improving their soggy property. Sewer commissions charged with removing blockages in the river would continue to be appointed in the Concord valley well into the nineteenth century.[51] We will visit them again later.

These public drainage efforts in the river channel, along with the many smaller, mostly unrecorded private and common ditches dug within the meadows themselves, gradually rendered the hay crop more productive and reliable, although it probably still suffered from summer flooding from time to time. But it must have been tough to get enough hay in the early years, especially during wet summers. In time, as more water was gotten off, the meadows served their proprietors less badly.

From the very earliest days of the town, Concord's husbandmen carted their hay homeward from nearly every wet spot in the landscape. Scattering the meadow lots in this way was ecologically sensible, and so it persisted as the generations passed. Hay is light, compared to manure, and so could reasonably be carted a greater distance. Owning a wide range of mowing lots ensured having enough hay to cut no matter how the season turned out. River meadows yielded the lushest grass but might be rendered useless if the river came up in a wet summer. Higher-lying brook meadows supplied lighter, but more reliable growth. In addition to supplying hay, meadows provided some of the best grazing in Concord, especially in the late summer and fall. So it is no surprise that men availed themselves of a broad selection of mowing lots and plenty of meadow acreage. The mead-ows were a critical resource but, in English eyes, a woefully underdeveloped one. The settlers mowed grass wherever they could find it and meanwhile set about systematically improving their meadows. This would take them generations.

With common grazing, as with tillage and meadows, the English settlers faced a formidable challenge in establishing their accustomed system of husbandry. The commons, which still occupied three-fourths of Concord by the end of the First Division, were expected to support the stock of the town during the growing season. Unfortunately for the newly arrived English stockmen, Concord's uplands in their native state furnished precious little in the way of useful grazing. After the Native planting grounds and most of the meadows had been distributed, what was left for commons was mostly woodland, which made very poor forage. Perhaps some commons back in England had also been of questionable grazing quality, but only hogs found much that was filling in New England forests—and even swine had to scratch for a living on the barren pine plains that dominated much of Concord. Finding adequate grazing would become a great recurring problem in New England agriculture. It was a problem that became pressing the moment the English arrived with their cattle, not to speak of their more finicky sheep and horses.

For millennia, New England's uplands had supported not grazers such as cattle and sheep, but browsers such as deer. The signature American grazer, the bison, was confined mainly to the prairies a thousand miles to the west. Consequently, there were few native grasses in Concord's uplands that responded well to constant grazing. A few tufts of native grasses were probably found among the thickets and brambles on the open, dry, sandy plains: hairgrass (*Agrostis scabra*), beardgrass (*Andropogon scoparius*—later known on the western prairies as little bluestem), poverty grass (*Sporobolus vaginiflorus*, *Aristida dichotoma*, and *Danthonia spicata*), and panic grass (*Panicum* ssp). Analogs of the reliable, productive, sod-forming cool season grasses of the domesticated grazing regimes of Western Europe were conspicuously lacking. Instead, the understory of the pine plains was probably dominated by blueberries (*Vaccinium* ssp), huckleberries (*Gaylusacia baccata*), sweet fern (*Comptonia peregrina*), and bracken (*Pteridium aquilinum*)—miserable forage for English beasts.[52] The denser forests on the glacial till uplands, meanwhile, provided the stock with little more than an abundance of shade.

The paucity of grazing in Concord was made worse by the summer weather, which the immigrants soon discovered was hotter and drier than at home. It wasn't so much that it rained less in Concord, but that it baked more. Average rainfall is similar in New England and old England, above forty inches—in fact, parts of southeast England actually receive less rain than New England. Precipitation is evenly distributed around the year in both Englands, as well. The crucial difference is that summers in England tend to be consistently cooler and cloudier than in New England, where the summer rain often comes in brief cloudbursts, followed by clear, hot weather for days or weeks on end. The resulting higher rate of evaporation can dry the soil by early July, after which the growth of the principal cool season pasture plants that the English brought over along with their stock, such as bluegrass (*Poa pratensis*) and white clover (*Trifolium repens*), shuts down until fall.[53] Even after these plants began to establish themselves, their productivity was greatly limited on Concord's pine plains. On such sandy soils even a well-managed pasture frequently becomes parched and bare by midsummer.

The first generation of Concord farmers were faced at the outset with a dearth of upland grazing, confronting a choice between dense hardwood forests or barren pine plains. As one petition to the General Court put it bluntly, "our land much of it being pine land . . . affords very little feeding for cattle."[54] The one part of the landscape that did afford grazing was the meadows. Large stretches of meadow existed in early Concord, both on commons and in private lots. Many of these lowlands were not yet sufficiently improved to mow but nevertheless contained palatable grass mixed in with coarser vegetation. Livestock were herded to choice spots like Fair Haven Meadow everyday. In 1666, when the town laid out a "drift way for [the] herd to fair haven," it also confirmed all the existing ways to Fair Haven as public ways, showing a long-established practice.[55] The cattle were not being driven to this remote corner of town simply to get them out of the way—the attraction was the plentiful grass.

If the custom of other English and New England common field towns is any guide, different kinds of stock were herded separately in Concord—the cow herd, the dry herd, the sheep flock, and so forth. If the *experience* of other towns is any guide, stock also often wandered loose, wreaking havoc. Hence the urgency of keeping the general tillage fields fenced.[56] Concord did set up a number of enclosures for stock during the First Division, mostly on low, meadowy ground: the Ox Pasture, the Sheep Pasture, and so forth. These pastures were far too small to have accommodated the entire stock of the town, except perhaps in the earliest years. They were probably used for limited short-term purposes such as penning work animals to graze near the fields to avoid bringing them home at night. The Sheep Pasture and Calf Pasture were tucked in close to the confluence of the rivers, perhaps in hopes of keeping them secure from wolves. The Hog Pen Walk was a different case—an enclosure of several hundred acres beyond Annursnack Hill in the northwest corner of Concord. This appears to have been a joint undertaking of several men to fatten swine on oak mast. Many hogs ran loose, however. When John Hoar sold his homestead to Edward Wright in 1672, he stipulated that he was "to have his hoggs and swine yt he hath here, to be lookt after and drove into ye woods, where they used to go, when they do at any time come home, by ye sd Edward Wright, untill ye next winter."[57]

Another place where Concord's husbandmen may have looked for grazing was within their general tillage fields, after the crops were harvested. But they wouldn't have found much there. The primary purpose of having open fields in England was the ability to open the fields regularly for fall and winter grazing, especially in regions where common grazing outside the fields had grown scarce. Grazing was certainly scarce in Concord, but unfortunately it was just as scarce within the fields as on commons. In late seventeenth-century Concord, the proprietors of the Great Field were "prohibited the turning in of any cattle on the sayd land after the fences are up in the spring until the 29 of September"[58] (October 9, by our calendar). But this does not mean that the cattle were left in the field all winter—there would have been precious little for them to eat after November. Later practice was to open the Great Field for a mere two weeks in mid-October, after the Indian corn harvest was in.[59] In seventeenth-century Rowley, Massachusetts, the cattle were to be *out* of the common fields by November 20 (the thirtieth, by our calendar). They may have been fall grazing the young winter rye.[60] Concord's husbandmen did get some amount of late grazing from their tillage fields. But it wouldn't have amounted to anything like the grazing available in an English open field, through an open English winter.

No matter which way they turned, the English in Concord were hard-pressed to find sufficient grazing. There was plenty of acreage for the stock to wander over but not much grass yet growing beneath their hooves. The possible solutions to this problem were, first, to acquire still more poor grazing land and, second, to improve the quality of the grazing by clearing forest and establishing European pasture grasses. The first solution could and did include expanding the commons. The second could have taken place on commons and probably did to some extent, but for the most part it was to take place in private, enclosed pastures, as we shall see.

If Concord's upland commons proved inferior for grazing largely because they were forested, one would suppose that their second main function, providing wood and timber, was reasonably well served. For the most part, it probably was. This was good, because New England winters were cold. That was something of a surprise to the English. After all, New England lies about ten degrees latitude south of old England, so the newcomers logically expected a mild climate somewhat like southern France or northern Spain. For a time they believed that clearing the forest and cultivating the land would warm the climate and correct this anomaly, but eventually they just had to learn to live with the cold.[61] This meant that a larger fuel supply was essential to survival and comfort in New England than at home. By the same token, it meant that cultivation was curtailed for a greater part of the year because the ground was frozen or snow-covered, and nothing much could be done by way of husbandry. As a result, New England farmers found themselves spending a good part of the winter in the woods cutting trees, establishing a seasonal rhythm of farming and woodlotting that persists, in attenuated form, to this day.

Concord's native forest was by no means uniform, mature, and dense: it contained horticultural openings, grassy meadows, and large areas dominated by open growth of pitch pines. Nevertheless, the encompassing New England forest was like nothing the English had seen for many generations. As they were quick to learn, the forest at least solved the problem of cold winters, as William Wood observed: "It may be objected that it is too cold a Countrey for our *English* men, who have been accustomed to a warmer Climate, to which it can be answered, . . . there is Wood good store, and better cheape to build warm houses, and make good fires, which makes the Winter lesse tedious."[62] It did take the English a while to become adept at dealing with the forest they found in New England, which was full of large trees, whereas their experience was mostly with smaller coppice wood. Once they developed the tools and skills to make use of the forest, they discovered they had a resource of great domestic and commercial value.

How was this largely common resource to be managed? The settlers needed wood for fuel; they needed timber and boards to build their houses and barns, their meetinghouse and mill, their bridges; and they needed posts and rails for the fences around their planting fields—not to mention stuff for carts, barrels, turned ware, furniture, and the like. Much of this they may have cut in the course of clearing their private lots, but they no doubt also had rights to harvest wood for various domestic purposes from the commons. No regulations governing wood and timber cutting have survived from Concord to shed any light on this. They do exist for other towns. Orders restricting the cutting of trees on commons in neighboring Sudbury by the 1640s have suggested to some that firewood was already growing scarce.[63] It seems unlikely that there could have been much difficulty in gathering firewood so early, though. A careful study of orders regulating wood and timber cutting in Ipswich indicates that the real issue was not whether trees could be cut, but where trees could be cut and which trees could be cut; in other words, the real issue was white oak.

White oak (*Quercus alba*) is very similar to English oak (*Quercus robur*) and provided the settlers with the same familiar and desirable qualities for timber framing, shipbuilding, joinery, cooperage, and fencing. White oak is flexible but tough, watertight yet breathes, and splits well. Conflicts arose in Ipswich because oak was being stolen from the commons for private profit, supplying a growing trade for pipestaves in the Wine Islands, for ship timber, and even for firewood shipped to treeless Boston. The Ipswich selectmen moved to protect the supply of white oak for the "necessary use" of the community. At the same time, selected areas were being cleared for extensive sheepwalks. All in all, Ipswich provides a picture of English settlers attempting to regulate their common forest and grazing resources in a very deliberate manner but already struggling with strong private demands being placed on white oak as a commodity.[64]

Many of these same concerns must have applied in Concord, as the settlers of that inland town went into the woods each winter both to harvest wood and timber and to attempt to shape their developing agricultural landscape. It is likely that much the same kind of collective ordering of how the forest was to be cut went on in Concord as in Ipswich. Shortage of wood was probably not one of the ecological hurdles facing the commons system in Concord, as were poor tillage fields, scarce manure supplies, waterlogged meadows, and scanty pastures. The problem of how best to control the use of white oak and other particularly valuable species, though, may have played a significant role in the decline of the commons system. We will see that when the Second Division got under way in 1653, the men of Concord turned first to privatizing their woodland.

The Decline of Commons in Concord

The commons system got the plantation at Musketaquid through its first two decades, but only just. The 1640s were a hard time in New England, and nowhere more so than in Concord. Puritan immigrants stopped arriving soon after the English Civil War broke out in 1642, which crippled the barely developed New England economy. The steady stream of new arrivals looking to buy stock and provisions with their savings no longer constituted a ready market for the older settlers. New Englanders lacked sufficient means of exchange for English goods. Gradually, merchants began to develop new markets for fish, timber, and provisions in the Wine Islands and in the sugar plantations of the West Indies.[65] But by and large settlers in towns like Concord were cut off from English re-supply and had to make shift with what they could find close at hand. To make matters worse, the decade saw some of the most bitter weather ever known in New England—in 1641/42, Massachusetts Bay froze to the horizon.[66] The colony was in a hard way.

In 1642, it appeared that the hard-pressed community at Concord might be abandoned, as Governor John Winthrop recorded: "Some of the elders went to Concord, being sent for by the church there, to advise with them about the maintenance of their elders, etc. They found them wavering about removal, not finding their plantation answerable to their expectation, and the maintenance of two elders too heavy a burden upon them. The elders advice was, that they should continue to wait upon God."[67] In 1643, a group of Concord settlers who had arrived four years after the town was founded complained that they found "the lands about the town very barren, and the meadows very wet and unuseful, especially those we now have interest in," and they requested land to the northwest of Concord for a new plantation of their own.[68] This came to nothing until a decade later, too late to help the petitioners. In 1644, the Reverend Jones and fifteen other families comprising "a 7th or 8th part of the Towne" grew tired of waiting upon God in Concord and went instead to Fairfield, Connecticut. "Onely the reverend grave and godly Mr. Buckly remaines," reported Edward Johnson, and Bulkeley was not entirely happy about it. "I lose much in this retired wilderness in which I live," he lamented to a friend in 1640.[69] Johnson encouraged him thus:

> Riches and honours Buckly layes aside
> To please his Christ, for whom he now doth war,
> Why Buckly thou hast Riches that will bide,
> And honours that exceeds Earths honour far.[70]

In 1645, the remaining inhabitants, citing these setbacks as evidence of Concord's natural shortcomings, petitioned the Court for tax relief: "We have lived . . . at Concord since our coming over into these parts, and are not conscious unto ourselves that we have been grosly negligent to imploy that talent God hath put into our hands to our best understanding; Neither have wee found any special hand of God gone out against us, only the povertie and meannesse of the place we live in not answering the labor bestowed upon it, together with the badness and weetness of the meadowes, hath consumed most of the estates of those who have hitherto borne the burden of charges amongst us, and therewith the bodily abilities of maney."[71] This petition was granted. Bulkeley and others chose to live out their lives and raise their families in the struggling young settlement, and in spite of the hardships of the climate, the flooding meadows, the barren pastures, the "povertie and meannesse of the place," Concord survived and eventually flourished. But before that could happen, the agricultural system initially put in place had to be further modified and adapted to fit the new environment.

Like the founders of many other early New England towns, the proprietors of Concord began with a common field system for a combination of good reasons. It was a form of husbandry with which many of them were familiar. It suited their communitarian impulses and their need to feel secure in the "wilderness" to which they had come. Most of the land granted by the Massachusetts Bay colony went not directly to individuals, but to groups of proprietors with the expectation that they would form a church and a tightly knit community, a town. In practical terms, the commons system seemed a natural fit in a place where almost all of the immediately available planting ground tended to be clustered in a few areas, meadow was similarly concentrated, and the rest of the landscape was covered with forest that could serve as wooded commons. Under these circumstances, it made sense to begin with a commons system, and many towns did.

But the commons system was not ecologically necessary to successful mixed husbandry. In England, the common field system had appeared during the Middle Ages, when sufficient population pressure made the tight regulation of field rotations necessary, to maximize grazing within the expanding arable fields and to integrate the delivery of manure by the sheepfold. In other words, the system was the child of resource constraints in a certain kind of environment. Of course, it had acquired a social context within feudalism and agrarian village life as well, and like any system once in place it gathered a great deal of economic and social inertia, enough to carry it across the ocean to New England. But as the planters of Concord looked to expand and improve their transplanted agricultural system, to put more tillage land into cultivation, and to clear more pasture from the forests surrounding their fields, they found they were dealing with circumstances that differed from those of their ancestors in England. In particular, they no longer had any good reason to go on scattering their tillage lots across a far-flung collection of common fields. The compensating advantages of winter grazing, summer fallow, and livestock folding within those fields did not amount to much in Concord. They did not amount to a hill of corn.

As a new generation began to look for land of their own to cultivate, the people of Concord had to consider how best to expand or reorganize their developing system of husbandry. Unlike some towns, Concord left us little record of these possibly wrenching and discordant deliberations.[72] The inhabitants could have continued to clear common pasture and to run common herds, to manage their woods as commons, to assign more mowing lots in undeveloped swamps, and to create new general tillage fields—they had already laid out new fields in at least two places. They could perhaps have allocated new houselots in small hamlets near the new fields, reducing the distance between barnyard and tillage lots in that way. They could have followed the familiar English pattern of hiving off new villages, leading to a countryside with nucleated settlements two or three miles apart. But that is not what they chose to do. As the second generation came of age, the proprietors of Concord (like those of many other New England towns at about the same time) took another approach: in 1653 they decided to distribute nearly all of their common land to themselves and their heirs in private lots.

This would appear to be a decisive turn away from the commons system, and so it proved in the long run. Indeed, the English world of which Concord was a part was moving inexorably in that direction, as we have seen. By dividing their commons among themselves the proprietors assured that this valuable resource would not be diluted as newcomers arrived in town and placed in their own hands the means to arrange inheritances for their progeny. But not all aspects of commons were immediately abandoned in Concord. The Second Division led to both change and continuity in Concord's agroecological system as the new divisions were taken up and the town was settled out to its borders over the following century.

CHAPTER 5

The Second Division

Every man shall have some quantity of upland adjoining his meadow where it is in common.
—CONCORD TOWN RECORDS

ON January 2, 1653, the householders of Concord assembled in town meeting and agreed to a sweeping division of the commons. Each man was to receive three acres of Second Division for every acre of First Division he owned. By this action almost all of the common land in Concord passed into private hands. At the very same meeting the town approved a regulation allowing landowners to run a large number of stock on . . . the commons! In this ambiguous and seemingly contradictory way, the proprietors of Concord set the pattern by which the next four generations of their descendants would settle and improve the land.[1]

Like many other early New England towns, Concord moved away from communal control of land and resources and toward more consolidated, nucleated private farms. The turn from common to enclosed farming, however, was not as abrupt and the contrast between the two systems not as dramatic as might be supposed. Old habits die hard— especially when they are still useful to the living. In practice, several generations passed before common herding disappeared, even though much of the *land* over which the stock grazed had become private. The Great Meadow and the Great Field retained their identity and some degree of common management for generations as well. Even where common management was abandoned, the dispersed pattern of landholding that had been the hallmark of English mixed husbandry was only gradually consolidated and never did disappear. In making their Second Division, Concord's husbandmen had good ecological reasons for adopting more consolidated, enclosed private farming in some aspects of their husbandry while retaining older practices in others.

To make the Second Division, the inhabitants of Concord divided their village into three quarters: the North, East, and South. Each of these three "companies" of village householders was apportioned a corresponding third of Concord's territory to divide among themselves, stretching out to the town's boundaries. The North Quarter included the houselots north of the meetinghouse and all the land beyond the Great (or Concord) River and the North (or Assabet) River (fig. 5.1). The East Quarter consisted of the land east of the Great River and of a line that ran by Town Meadow and the gutter to Flint's Pond. The South Quarter (which sometimes called itself the West Quarter) comprised the land from that line to the South (or Sudbury) River and all the land between the North and South rivers. It is evident from the way house locations were described that the great majority of Concord's residents were still living in the village at this juncture and had not yet dispersed. In fact, one of the purposes of the Second Division was undoubtedly to enable them to move out to new holdings if they wished.[2]

Each quarter was empowered to lay out its own division by majority agreement, or an "indifferent man" could be appointed to do it. The ratio of three acres of Second Division to every one acre of First Division was established. Each proprietor was to receive upland adjoining his meadow, but the new divisions were not to hinder highways to men's properties. That is, numerous unofficial ways had already been hacked through the forest, winding across pine plains and around spruce swamps so that proprietors could reach their widespread First Division mowing lots. These cartways were not to be cut off by the Second Division transfer of the underlying commons into private hands—many were subsequently "laid out" as town ways, as they had been anciently trod. Although it was not recorded, at about this time each quarter must have been made responsible for maintaining the roads and bridges in its part of town because bridges soon became an issue.

Not everything went smoothly with the division. Two years later, on March, 7, 1655, the inhabitants were back in town meeting to talk it over again. After "much agitation" about expenses for bridges and highways and about the "hog pen walke," a nine-member committee (three men from each quarter) was delegated to end the dispute. Each quarter was assigned one major bridge to maintain, with the odd result that the East Quarter became responsible for the North River bridge at the west end of town in the South Quarter.[3] The controversy about the hog pen walk (a large, enclosed foraging area) in the North Quarter concerned a block of land near Annursnack Hill that had been left out of the Second Division but that the North Quarter men now wished to divide among themselves. Unfortunately the quarter did not own it—it was not common land. An agreement was struck whereby the hog walk was resigned to the North Quarter by its owners in exchange for some enlargement in their land elsewhere. Twenty acres of the hog pen were then reserved as ministerial land (land whose produce or income went to help support the pastor), along with another twenty acres nearby. At the same time, twenty acres of plowland from the South Quarter and twenty acres of woodland in the East Quarter were set aside for the same purpose.[4]

The agreement further stated that several men who were granted enlargements were

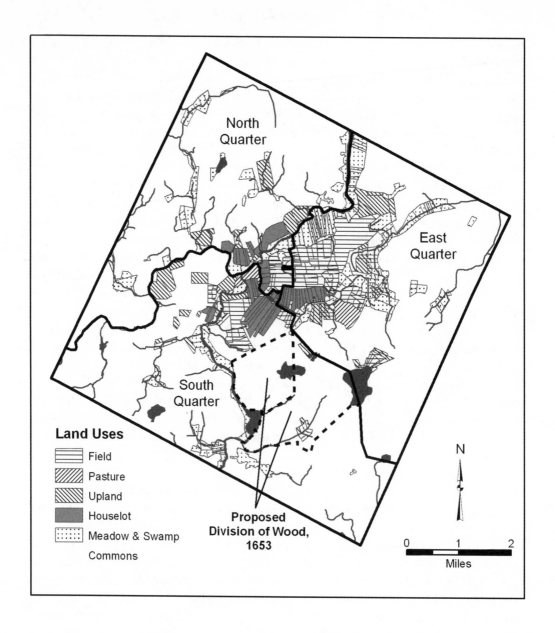

North
Quarter

East
Quarter

South
Quarter

Land Uses

☰	Field
▧	Pasture
▨	Upland
▰	Houselot
⬚	Meadow & Swamp
	Commons

Proposed
Division of Wood,
1653

N

0   1   2

Miles

FIGURE 5.1.
Second Division. Concord was divided into three quarters for the purposes of making the
Second Division in 1653. Just before that, the South Quarter decreed but apparently
did not complete a "Division of Wood," also shown.

to pay "12d per acre as others have done, and 6d if the Town consent." This suggests that men purchased their Second Division grants at a set rate.[5] One shilling per acre was a bargain price, since unimproved land in Concord fetched two or three shillings on the open market through the next few decades.[6] Still, it brought something into town coffers. The town needed money, and selling land was one way to raise it. Once the land was in private hands, it could also be taxed. The rate was set at two pence on the pound.[7] Later, in 1667, it was decided that waste lands should pay "only" two shillings six pence per one hundred acres to public charges every year, easing the burden on those holding unimproved land.[8]

Land was not distributed equally in early New England towns—people with more wealth and standing generally received more land. These towns were communitarian, but not egalitarian. The wealthy could not just buy as much land as they pleased directly from the town at will, however. The community as a whole controlled how much land a man was allowed to acquire by pegging it to his First Division holdings. The Second Division proceeded by lengthy negotiations among the members of each quarter, rather than by an auction or sale at which ready cash and prompt action talked. Concordians went to great lengths to ensure that their land divisions were fair and equitable, if not equal. Once the land was divided, of course, it could be bought and sold freely on the open market.

In practice the division of land in Concord was very deliberate, not to say ponderous. Although the Second Division was enacted in 1653, it took several decades for divisions to be laid out and "taken up" by the proprietors. The partially preserved records of the South Quarter give us some insight into the lengthy process of negotiation. During the 1650s and 1660s, the South Quarter met periodically during the winter months to work away at their land divisions. Each man made several separate requests for additional land over the years. On January 1, 1656, the quarter recorded a long series of every man's "propositions" for land. Several of these were disallowed and crossed out. Two days later the quarter clerk, John Scotchford, recorded that the company "granted all those propositions that are not defaced this 3rd of the 11:mo:1655."[9] The quarter met again the next winter on February 24, 1657, and proposed for more land. All recorded that day were granted.[10]

The records preserve something of the flavor of these negotiations. George Hayward and Michael Wood desired "that parcel at the north side of the brook beside the houses with this condition: if John Miles come thither to dwell he shall have a 3rd part of it paying them for their cost they have been at by proposing—Geo Wheeler, James Hosmer Jr. & St. Buss to lay it out."[11] When men made their propositions they were not talking about vague, ill-defined areas of land. They had already put some expense into staking the parcels, so that the rest of the company knew exactly what they were claiming. What the clerk took the trouble to record was no doubt the result of hard bargaining both at meetings and in the field, and even that was not necessarily the end of it. In this case the quarter took the precaution of specifying three men to lay out the possible subdivision of the parcel in question, in order to forestall a dispute should Miles exercise his option (he never did). In other instances the quarter changed its mind and renegotiated how a particular area was to be divided, even after the grants had been recorded.

In 1664, the records of the Town of Concord were transcribed into a new book, with the following preamble: "This Towne of Concord in these latter grants of land to particular persons being only written in paper bookes (as granted) and not recorded in a register book, the Inhabitants of the Towne; desired the selectmen to consider of a way to record them, and all other lands that men now do hold; and record them in a new booke: the thing tending to pece and preventing strife in the Towne."[12] It was also ordered that those who had not yet surveyed the land due them should "gitt it laid out by an artist between this and the last of March" of the next year. Accordingly, John Flint of the North Quarter was busy surveying lots for South Quarter men during February 1665. In all, Flint laid out thirty-one lots that month, measuring over one thousand acres.[13] But the town's self-imposed deadline came and went, and land divisions continued into the 1670s and beyond. Some land was left as commons and stayed that way for another half century. The long process of dividing land into private ownership was not really completed until the 1730s, when an inventory was made and the last remaining commons—mostly dry pinelands and dismal swamps—were finally sold off.

It is striking that the men of the South Quarter chose to make their land divisions in such a piecemeal fashion. As Concordians went about their Second Division, they could look to several models around them. Most land in Massachusetts Bay colony was granted to groups of proprietors to form towns like Concord, but at the same time a good deal of land was granted directly to individuals. Establishing the colony involved great costs, and many men who contributed greatly were well rewarded with land. Several of these large grants appeared on the borders of Concord, including Governor Winthrop's farm north of Concord on the east side of the Great River, which eventually became part of Billerica. Several unincorporated grants across the river from Winthrop were later acquired by the Blood family, becoming a thorn in Concord's side for generations. Major Willard received land northwest of Concord, while the Reverend Bulkeley received three hundred acres "towards Cambridge," before that town was expanded to reach the Concord line.[14] The neighboring town of Sudbury was similarly hedged about with private grants.[15] These large, contiguous blocks of land (often intended to be leased or sold by the grantee for income) were often termed farms, as distinct from the more dispersed agricultural holdings of the common yeoman.

Many towns, including Concord, granted leading men such farms within their boundaries as well. As a result, the landscape of seventeenth-century Massachusetts became a hodgepodge consisting of common lands, small scattered private holdings, and larger consolidated private estates, all lying cheek by jowl both inside and outside of towns. But this is not all that surprising. A similarly complex and fluid situation prevailed in many regions of seventeenth-century England. The Wheelers and Rev. Bulkeley came from Bedfordshire, which was open field country, but just down the road from the village of Cranfield there was a large, enclosed private estate called Virginia.[16] If anything,

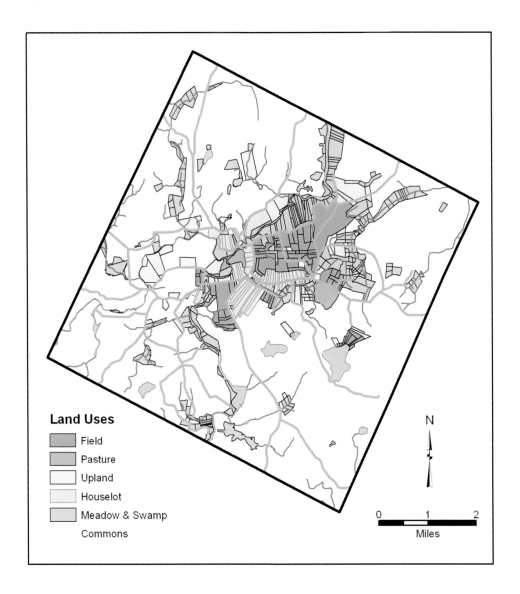

Land Uses

- Field
- Pasture
- Upland
- Houselot
- Meadow & Swamp
- Commons

N

0   1   2
Miles

PLATE I.

The First Division. By 1652, about fifty houselots clustered around the meetinghouse and millpond. Most of the plowlands were grouped within several general fields (some of which encompassed the backs of many houselots as well). A few parcels of detached upland were granted. Meadow lots flanked the rivers and brooks. A few small common pastures were enclosed, but about three-quarters of the town was left open as common grazing and forest.

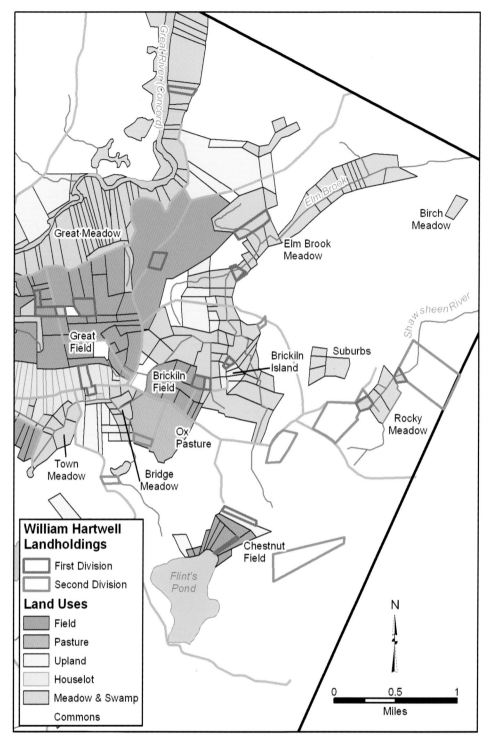

PLATE 2.

William Hartwell: First and Second Division. Hartwell's First Division included his houselot;
lots in the Great Field, the Brickiln Field, and the Chestnut Field; and mowing in the Great
Meadow, Elm Brook Meadow, and Rocky Meadow. His larger Second Division grants
were mostly on the uplands near Rocky Meadow two miles east of his house,
and were settled by his grandsons nearly half a century later.

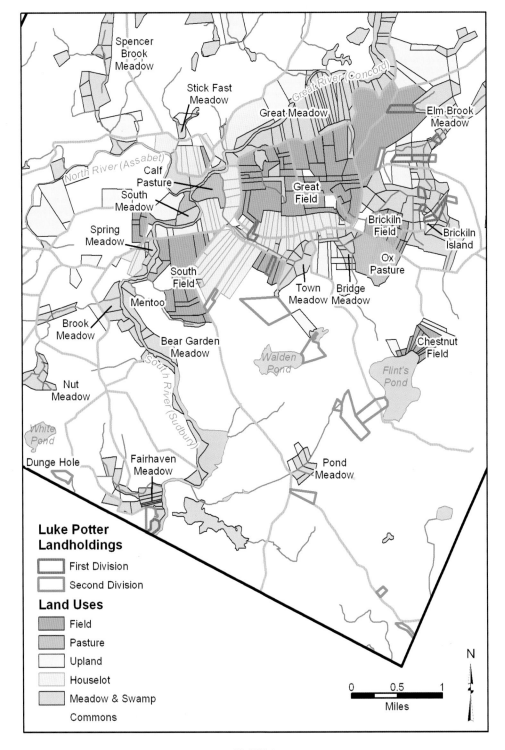

Spencer
Brook
Meadow

Stick Fast
Meadow

Great River (Concord)

Great Meadow

Elm Brook
Meadow

North River (Assabet)

Calf
Pasture

South
Meadow

Great
Field

Brickiln
Field

Brickiln
Island

Spring
Meadow

Ox
Pasture

South
Field

Mentoo

Town
Meadow

Bridge
Meadow

Brook
Meadow

Bear Garden
Meadow

*Walden
Pond*

*Flint's
Pond*

Chestnut
Field

South River (Sudbury)

Nut
Meadow

*White
Pond*

Dunge Hole

Fairhaven
Meadow

Pond
Meadow

**Luke Potter
Landholdings**

First Division

Second Division

Land Uses

Field

Pasture

Upland

Houselot

Meadow & Swamp

Commons

N

0   0.5   1

Miles

PLATE 3.

Luke Potter: First and Second Division. Potter's First Division included his houselot by the
millpond; tillage lots in a small field across Town Meadow and in the Great Field; and mowing
in Elm Brook Meadow and distant Fairhaven Meadow. His Second Division upland,
woodland, and swamp grants were widely scattered, mostly across the South Quarter.
Luke's grandson Samuel relocated the family homestead to the upland between
the South Field and Walden during the early eighteenth century.

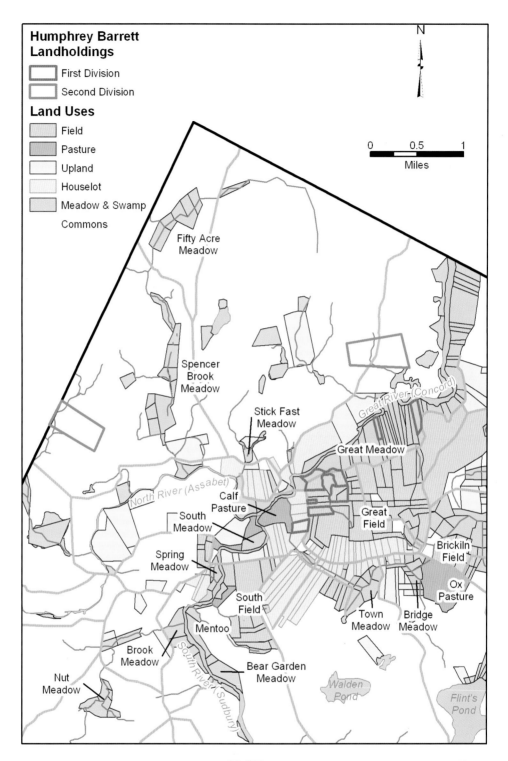

Humphrey Barrett Landholdings

First Division
Second Division

Land Uses

Field
Pasture
Upland
Houselot
Meadow & Swamp
Commons

N

0   0.5   1
Miles

Fifty Acre Meadow

Spencer Brook Meadow

Stick Fast Meadow

Great River (Concord)

Great Meadow

North River (Assabet)

Calf Pasture

South Meadow

Great Field

Brickiln Field

Spring Meadow

Ox Pasture

South Field

Mentoo

Town Meadow

Bridge Meadow

Brook Meadow

South River (Sudbury)

Bear Garden Meadow

Walden Pond

Nut Meadow

Flint's Pond

PLATE 4.

Humphrey Barrett: First and Second Division. Barrett's First Division included his houselot, and a set of tightly clustered plowlands and mowing lots in the small field across from his house, the Great Field, and the Great Meadow. His Second Division consisted of two solid 100 acre blocks, settled by his grandson and great-grandson in the eighteenth century.

PLATE 5.

Meriam and Brickiln Land Ownership. John Meriam's homelands were split into intermixed parcels by his three sons John, Joseph, and Ebenezer. By 1749 more homeland had been acquired, and was split among Ebenezer (who owned in common with his son) and his three nephews. 1771 shows no further subdivision, or consolidation. Aside from the Minot farm, the Brickiln Field neighborhood was not aggregated into homesteads until the third generation, and did not change much through the rest of the colonial era.

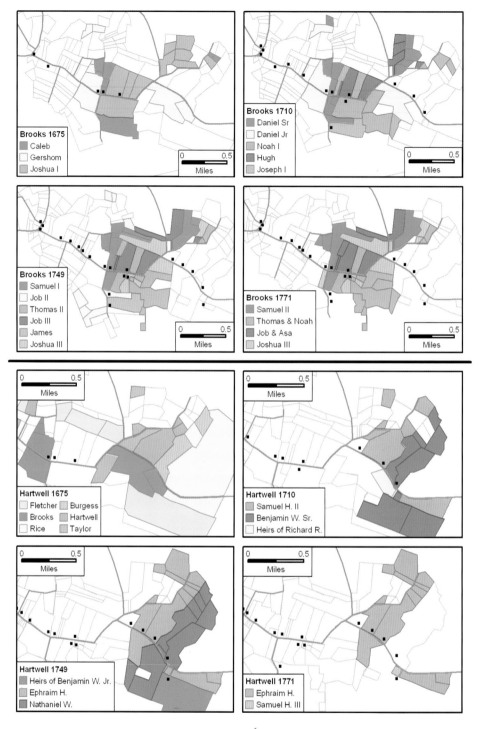

PLATE 6.

Brooks and Hartwell Land Ownership. Thomas Brooks's large grants were homesteaded by his three sons Caleb, Gershom, and Joshua. Over the next two generations this land grew increasingly subdivided and crowded, but by the end of the era in 1771 fragmentation had ceased. Farms on the uplands to the east, settled by Samuel Hartwell and Benjamin Whittemore of the third generation, were each subdivided once but remained in solid blocks. By 1771, Samuel Hartwell actually occupied about half of his father Ephraim's farm (but the land was not confirmed to him until 1794), while the two Whittemore farms had been sold from the family.

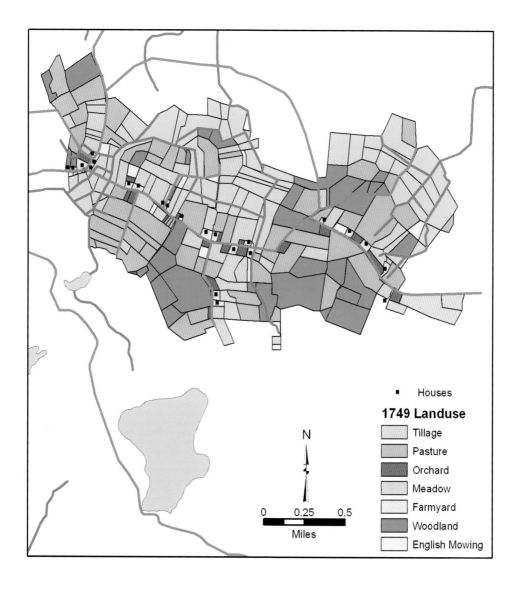

PLATE 7.

1749 East Quarter Land Use. Clusters of homesteads (often tied by kinship) were located near patches of tillable land that were subdivided among them—except on the eastern uplands, where settlement was more dispersed. Orchards were strung along the road, and broad meadows, divided into many small lots, flanked carefully ditched brooks. Very little upland "English" hay had been planted at this date. Pastures were spread throughout the landscape on a wide variety of soils, and a good deal of woodland remained, often on rocky hills and in remote swamps.

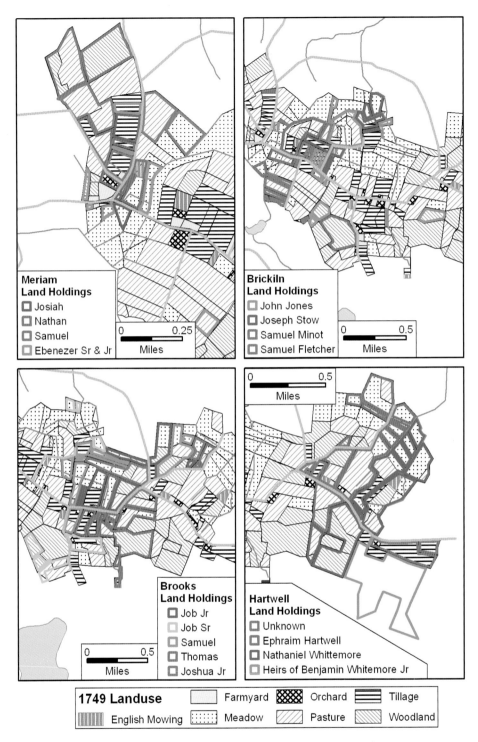

Meriam Land Holdings
- Josiah
- Nathan
- Samuel
- Ebenezer Sr & Jr

0   0.25
Miles

Brickiln Land Holdings
- John Jones
- Joseph Stow
- Samuel Minot
- Samuel Fletcher

0   0.5
Miles

0   0.5
Miles

Brooks Land Holdings
- Job Jr
- Job Sr
- Samuel
- Thomas
- Joshua Jr

0   0.5
Miles

Hartwell Land Holdings
- Unknown
- Ephraim Hartwell
- Nathaniel Whittemore
- Heirs of Benjamin Whitemore Jr

| **1749 Landuse** | Farmyard | Orchard | Tillage |
| English Mowing | Meadow | Pasture | Woodland |

PLATE 8.

1749 Land Use: Meriam, Brickiln, Brooks, and Hartwell. These maps show details of landholdings spread over different kinds of land. Note the intricately subdivided tillage and meadow lots in the first three neighborhoods, and the more consolidated pattern found on the uplands surrounding the Hartwell Tavern.

compared to the English system land tenure in New England was simplified and stream-lined because the cumbersome feudal superstructure had been cast off, and most land was owned in fee simple, whether by an individual or the community as a whole.[17] In time, free-floating farms outside municipal boundaries were absorbed into the maturing Massachusetts town structure, but only after a great deal of acrimony concerning property lines and payment of taxes.[18]

It took revenue to establish and maintain towns like Concord. There were taxes and fines to pay to the General Court in support of the colony. There were also the costs of surveying boundaries, of clearing title with the Native people, of building the meeting-house and keeping the minister, and of maintaining roads and bridges. The townspeople could satisfy some of these demands with their own sweat, but others required another kind of liquidity. In Concord, many of these cash expenses were initially met by a few wealthy individuals. Like the colony itself, the town repaid its underwriters with the principal asset it held, which was land.

The leading investors in Concord were Thomas Flint, a London merchant originally from Derbyshire, and the Reverend Peter Bulkeley. In return for the fortunes they laid out in establishing the town, these men received large, private farms in the southern part of Concord (fig. 5.2). Flint was granted a 750-acre farm on the Cambridge line, in addition to his holdings in the North Quarter. He returned to England, where he died in 1653, leaving his North Quarter land to his son John, his large farm in what would become Lincoln to his son Ephraim, and his perfect Yankee surname on a pond for Henry Thoreau to lampoon in later years.[19] Elder Bulkeley received a Second Division farm of 750 acres south of the Flint farm, along with substantial holdings elsewhere in Concord.[20] Several other Concord men (as well as some investors who never lived in Concord) assembled similar farms during the course of the Second Division, as shown on the map.[21]

Surrounded by examples like these, Concord's yeomen would not have surprised historians had they divided the bulk of their commons among themselves in similarly solid blocks, or farms. But that is not what most of them chose to do. They did not lay out rectangular grids of "ranges" and "squadrons" and simply draw lots, as happened on the outskirts of surrounding towns such as Sudbury, Watertown, and Cambridge. Instead, the proprietors of Concord went through intricate negotiations over the span of several decades, which ended with most men holding several hundred acres in not one or two but dozens of parcels scattered throughout their quarter. Driven by a desire for equity in the way land was distributed, by an already well-developed sense of the diversity of Concord's land and of the need for a combination of several kinds of natural resources to make a viable holding, and by a desire to create many such opportunities for their off-spring into the future, they cut the landscape into a baffling jigsaw puzzle of irregular pieces. Let us return to William Hartwell, Luke Potter, and Humphrey Barrett and look at how these men took up their Second Division.

William Hartwell received almost 200 acres of Second Division to go with his 86 acres of First Division in the East Quarter (plate 2 and table 4.1).[22] His Second Division

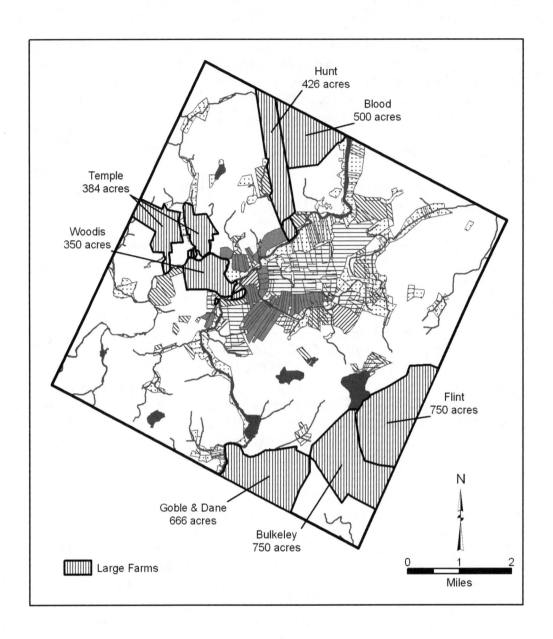

FIGURE 5.2.

Large farms. Several of Concord's wealthier proprietors received large, consolidated landholdings sometimes called farms. Many of these were subsequently sold (or leased) to actual settlers. Henry Woodis acquired his farm from Simon Willard, and Richard Temple purchased the farm of investor William Spencer.

lay in eight pieces—five of them less than 10 acres, three of them more substantial. Still, the largest two pieces were only 72 acres each. Like many of his neighbors, Hartwell was granted a small piece of the ox pasture and a parcel of woodland "near the meeting house frame." His largest Second Division grants of upland adjoined his First Division mowing lots in Rocky Meadow near the Cambridge line. Although a few neighbors like Thomas Brooks were granted holdings that were more consolidated, Hartwell's wide scattering of Second Division was quite typical of the East Quarter.[23]

South Quarter grants were, if anything, more scattered. Again, Deacon Potter was indicative. For his 69 acres of First Division, Potter received 187 acres of Second Division in fourteen separate pieces (plate 3 and table 4.2). Potter picked up a few small pieces of upland and swamp near his East Quarter holdings. The remainder of his Second Division lands fell across the South Quarter like a handful of rye broadcast into a stiff northwest autumn wind. In Potter's case the heaviest fell nearest to hand—the largest single piece was 56 acres of pineland on the hill leading to Walden Woods, just south of the village. As if his First Division mowing land had not been dispersed enough, in the Second Division Deacon Potter collected still more distant scraps of meadow, lying all the way from Dunge Hole near White Pond to remote Watertown Corner. If Potter and his sons mowed in all of his meadows, they must have driven their oxcarts five miles to the southwest, three miles to the northeast, and five miles to the southeast to fetch home their hay.[24]

Luke Potter's dispersed landholdings were not at all unusual, especially for the South Quarter. The same pattern ran through small and large landholders alike. By the time all the divisions were recorded, John Scotchford, for example, counted 120 3/4 acres in fourteen parcels. Michael Wood received 225 acres in fourteen lots, William Buss 319½ acres in nineteen pieces, and John Miles 458½ acres in twenty-six widely scattered parcels. George Wheeler's portfolio was the most diversified of all, comprising some 477 acres spread over thirty-five lots—and he was owed still more. Many of these men thus became landowners as substantial as some of those granted large, consolidated farms, but they chose instead to receive their property in dozens of pieces, all mixed up with those of their fellow townsmen.

Things looked a bit different in the North Quarter. For his 96 acres of First Division, Humphrey Barrett received 220 acres of Second Division (plate 4 and table 4.3). But rather than accepting a wide scattering of parcels, Barrett was granted two 100-acre blocks, along with a 20-acre undivided interest in the remaining North Quarter commons known as the Twenty Score. Most of his neighbors living north of the meetinghouse were like Barrett. Although many North Quarter men retained small patches of First Division meadow that were widely dispersed, they mostly chose to take their Second Division in a few large pieces. Indeed, as a result, several of them wound up owning odd bits of meadow that lay entirely within someone else's Second Division upland. The North Quarter did make a few smaller divisions of upland, however—for example, just past the mouth of Spencer Brook they created a new tillage field that enclosed at least nine small lots. So even in the North Quarter there was no uniform drive to consolidate the land into large, private holdings and be done with it.[25]

In summary, a few leading men in Concord acquired large farms, and many residents of the North Quarter, like Humphrey Barrett and William Hunt, received the bulk of their Second Division in one or two large blocks or long strips. But most townsmen, like William Hartwell and Luke Potter, took up their Second Division grants in pieces spread far and wide across the landscape. Many of these yeomen were not far behind the largest farmers in cumulative acreage. Clearly most of them were not inclined to set themselves up with large, neatly enclosed, private estates. They were concerned with providing their progeny with access to workable agricultural holdings, and they were content with the idea that in much of their husbandry they would be rubbing elbows with their neighbors for years to come.

Land Division Strategies

Concord's proprietors no doubt had several reasons for dividing their land the way they did. A driving force behind divisions in neighboring towns was controversy over who had access to common land and who was to be granted more land, especially as the second generation came of age.[26] We find hints of this in Concord. When First Division land was sold in Concord, sometimes the rights to commonage (and hence the subsequent right to Second Division) went with it; but other times the seller retained the commonage and sold only the bare parcel. This no doubt led to confusion and rancor over who had rights in the commons and who didn't, as newcomers bought land and settled in town. As they saw the commons being whittled away, the original proprietors may well have grown eager to get a firmer grip on their share by privatizing it. At the same time, even as the Second Division was being made Concordians deliberately retained some commons and set up mechanisms whereby nonproprietors could gain access to it. In fact, as we shall see, the town acquired a great deal *more* commons at this time and kept it more or less intact for several generations. So these men were hardly washing their hands entirely of the commons system.

Some men undoubtedly wanted to acquire Second Divisions so that they could move to more propitious, often more consolidated landholdings outside the village. We can surmise as much because that is precisely what several of them immediately did. Others stayed within the village and continued to work their scattered holdings. Almost all of them, however, had to provide land for several children, and it would have been awkward to shoehorn many more households into the existing common field structure. Therefore many older sons were either established on their father's Second Division land or were granted Second Division of their own as one generation gave way to the next. Providing land for children was a powerful incentive in this intensely patriarchal society, and men took up their Second Division with this end in view. We can see it in the way land was chosen, in the way land was subsequently settled, and in the language used when land was passed on. I will discuss this process of inheritance and settlement in Concord more closely in the next chapter.

Another motivation for the way the land was divided was the desire to gain secure access to a wide range of critical natural resources. Understanding these very particular needs helps clarify the otherwise baffling pattern of irregular, often minute subdivisions. In part, this pattern was simply an extension of the selection of specific kinds of land for plowing and mowing we have already seen in the First Division. To this were added in the Second Division a concern to secure certain kinds of woodland and a refined sense of how land resources could be configured to the best advantage—in particular, arranging for plowland that was closer to the barn. Concord's husbandmen were not dividing a flat, featureless checkerboard but a diverse and by now familiar home landscape; and they aimed to fulfill distinct needs. There was careful, precise parceling of land across the nearly forty-five square miles of Concord. These men had already come to know the intricate topography, soils, and vegetation of their town well—this land was no hostile, howling wilderness to them. Which resources were they most avid to secure?

Woodland

Concord's Second Division had a prologue—a division of woodland in the South Quarter. This dress rehearsal took place in 1652 or early 1653 and may even have preceded the order of January 1653 dividing the town into quarters.[27] It was entitled the "first division of wood" for the west part of town on the south side of the Mill Brook. The limits of the woodland to be divided reached from Town Meadow to Pond Meadow and west to the river (see fig. 5.1). A line cutting this area "in half" (actually closer to one-third and two-thirds) was to be run south over Walden Pond and Rocky Hill to another kettlehole bog called Dongye Hole and then west along the wooded valley of a small stream and a parcel of wood called the "short swamp" to Fairhaven Bay. Twelve men were to take woodland east and south of this line, and seven west of it. These nineteen men were largely the same as would appear in subsequent South Quarter division records.[28]

The individual grants made under this division were not recorded and may never have been taken up. This early attempt at a large division was either aborted or subsumed by the much more sweeping Second Division that soon followed. Nevertheless, it does tell us some things. First, gaining control of forest resources was on the minds of the proprietors of Concord. Second, these husbandmen already recognized that the droughty uplands south of the village were generally inferior for tillage and were ecologically better suited to be set aside as woodland. And that is in effect what happened. This tract (or at least the northern portion of it) became known as Walden Woods and for centuries provided woodlots for villagers and farmers in the South and East quarters. The woodland role that was destined to make this dry upland famous was forced upon it by the exceedingly coarse sand and gravel that had been dropped there thirteen thousand years earlier during the final stage of Lake Sudbury and by the ruggedness of bedrock escarpments like Fair Haven Hill and Rocky Hill (later known as Emerson's Cliffs) that poked up through the glacial debris. This part of Concord must have recalled the Weald from which many of Concord's Kentish settlers hailed, and the woods and pond were doubtless named Walden accordingly. Thoreau's *Walden, or Life in the Woods* was thus aptly titled.[29]

The men of the South Quarter became much more discriminating in their use of the term *woodland* when they made their actual Second Division. Although the land distributed in the Second Division was overwhelmingly forested, most parcels were simply termed upland, if anything at all. Those few that were designated woodland were doled out sparingly, often in very small pieces. Pineland and pine plain were identified somewhat more often and were usually granted in much larger chunks. The terms *woodland* and *pineland* were rarely, if ever, used interchangeably in the Second Division records. It appears that to these English settlers, pineland simply did not rate as woodland. For example, Samuel Stratton received a lone acre of woodland in what was described as a "round hole bordered with the pine hills."[30] In a landscape dominated by scrubby pitch pine plains, there were evidently other kinds of woods that had something special to offer.

Luke Potter's grants are good examples of this careful parceling of woodlands. Potter received four lots designated woodland, the largest twenty-one acres by Flint's Pond. The smallest was a mere four acres in a swamp running into Pond Meadow. Four other men received similar small woodlots in this one little valley. In all, forty-two parcels were designated woodland during the division of the South Quarter (fig. 5.3).[31] These woodland parcels tended to cluster in distinct locations. Significantly, most of them were found on the rocky lands, the mesic glacial tills. Most of the rest were along narrow valleys and in swamps—all places protected from fire and suited to growing good timber. Many of the parcels that made up these clusters were very small indeed: nineteen were three acres or less, thirteen were two acres or less. When people in the process of subdividing tens of thousands of acres of forested land cherry-pick two-acre tidbits of choice woodlands, we should take notice. It is unlikely that these were meant to be general all-purpose woodlots—firewood wasn't going to be in short supply for generations. It is more likely that they were stands of mature white oak—or perhaps, not to put too fine a point on it, of particularly valuable timber of any sort.

The East Quarter called only two parcels woodland in its land division records.[32] A subsequent deposition reveals very clearly, however, that the East Quarter did make at least one larger, deliberate division of woodland. In 1681, Joseph Wheeler and Nathaniel Ball recalled that "more than twenty years since there was a range of woodlotts, granted by the East quarter of the Towne of Concord to severall persons which sd lots were to run to Mr Fflints farme or Pond."[33] A dozen of these lots were granted (see fig. 5.3), ranging from two acres to eighteen acres in size—William Hartwell received six acres, in the lower-middle range as usual. Francis Fletcher's nearby ten-acre lot was described in 1694 as lying "in a place known by ye name of ye Meeting House Frame upon which the old sawmill stands," apparently where the timbers for the second meetinghouse had

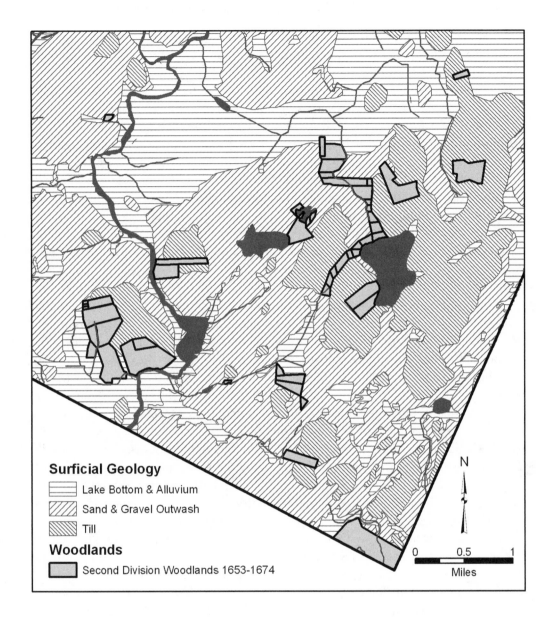

Surficial Geology

Lake Bottom & Alluvium

Sand & Gravel Outwash

Till

Woodlands

Second Division Woodlands 1653-1674

N

0   0.5   1

Miles

FIGURE 5.3.

Second Division woodlands. Only a handful of the more than eight hundred parcels distributed in Concord's land divisions were designated woodland. The map shows the location of most of these parcels in the South and East quarters, clustered on rocky till, in swamps, and in narrow valleys.

been cut about 1668. Because white oak was the framing timber of choice, it follows that what the East Quarter most likely divided here was a valuable stand of oak winding along the upper Mill Brook.

No woodland parcels were specified in the land division records of the North Quarter.[34] Like the other quarters, the North Quarter did mention several *swamps* in its record of divisions—for example, the Ash Swamp in what is now Estabrook Woods. These wet places were far from worthless. Certain swamps were subdivided into minuscule lots, although they lay several miles from the village. James Hosmer, Michael Wood, John Miles, George Wheeler, and Obadiah Wheeler all received splinters of another Ash Swamp in the South Quarter, near Fair Haven Meadow—Hosmer and Miles got three acres, the others one acre apiece. The ash in question may have been black ash (*Fraxinus nigra*), and the English may have already learned from the Indians of its superior qualities for making baskets.[35] The South Quarter Ash Swamp was still alive and well in deeds until the middle of the eighteenth century, when it was finally cleared and drained for farmland.

Black ash was not the only desirable species found in wetlands. The Great Cedar Swamp lay on Concord's border with Billerica, just east of the river. Cedar swamps were also minutely subdivided, and the tiny lots passed down for generations. Like many of his neighbors, William Hartwell acquired a vanishingly slim right in the Cedar Swamp and then split it: he left one acre to his son Samuel and another one and three-quarter acres to his son John.[36] In 1695, Joshua Brooks left his sons Joseph, Job, Daniel, and Noah an acre apiece in Cedar Swamp.[37] At the opposite end of town, Luke Potter was one of a number of men owning a few acres in a spruce swamp in Watertown Corner, a good four miles from his house in the village as the crow flies, but a miserable five miles as the oxen walk. There may well have been white cedar (*Chamaecyparis thyoides*) in this swamp, along with the spruce. Cedar was highly prized for making clapboards and shingles and for moth-proofing chests. Like black ash, this species is unfortunately now all but extirpated from Concord.[38]

Such precise subdivisions of land so far from the village make it clear that the settlers of Concord were gaining a feel for their forest. They were no longer stumbling through the swamps but were dividing them into one-acre shares. They had come to appreciate the distinct uses of different species of trees, so that every man wanted a piece of the places where each kind grew among his holdings. Even the ubiquitous pitch pine (*Pinus rigida*), although often cursed for indicating mean, poverty-stricken grazing, was not entirely despised. It made a hot fire, was set aside near bridges to provide stringers, and probably served for house flooring as well. A tar kiln was built on the pine plain beyond South Brook, in the far west corner of Concord, and trees on the pine plains there were bled for their sap. Pitch pine was utilized in colonial New England for tar and other naval stores, including, of course, pitch.[39] There was probably much less white pine (*Pinus strobus*) to be found in early Concord, but there was some because the species appears as a boundary marker in several early deeds.[40] In brief, part of the puzzle of the

Second Division, with its small, irregular lots, can be explained by the desire of Concord's proprietors to secure an ample supply of very particular trees standing in their forest.

Meadows

A few of Concord's husbandmen added to their meadow holdings in the Second Division. More picked up swamp, again scattered far and wide wherever the land dipped and held water. Many of these swamps would be cleared and converted to meadow in the course of time. Luke Potter acquired twenty acres of meadow in three lots in the South Quarter, along with a piece of Virginia Swamp in the East Quarter. Neither William Hartwell nor Humphrey Barrett received any parcels designated meadow or swamp in the Second Division, but one of Barrett's one-hundred-acre blocks was later described as upland and meadow.[41] Other East and North Quarter men did add to their meadow, but not much. The choicest pieces of meadow had already been distributed in the First Division.

Houselots and Tillage

Undoubtedly the driving purpose behind the Second Division was to provide land for Concord's proprietors to set up new, more consolidated homesteads outside the village — or to afford that opportunity to their offspring. Some made the move at once, others took their time. Some never moved at all. In the South Quarter, George Hayward jumped from the center of the village to the far west end of Concord, where he established a mill. James Hosmer took up a consolidated holding by the North River beyond the New Field, two miles west of the meetinghouse. He settled his son James Hosmer Jr. on a similar holding nearby. George Wheeler remained in the village but established his sons Thomas and William on Second Division lands near the Hosmers. Wheeler later passed his village homestead and part of his remaining outlands on to his youngest son, John.[42]

Deacon Potter stayed put. Not only did the deacon stay where he was, but his son Judah remained in the village to the end of his days. Luke Potter did acquire eight additional acres of tillage land in the South Field during the Second Division, within a reasonable half-mile carting distance of his door. He split his far-flung collection of small holdings with his son, dividing many parcels neatly in half. The houselot by the Mill Brook and even the barn were split down the middle. Judah received his half-inheritance in 1695 at the tender age of thirty-eight. The deed of gift carried this preamble, so typical of that world:"To manifest my true love & fatherly care to my son Judah Potter, as also to return some requitall unto him for his faithfullness and obedience to his mother & my selfe, and to further his settlement in ye Town wth a competency of housing & lands for his comfort."[43] Judah's competency included fifteen parcels of land scattered across Concord, the largest of them twenty-five acres and most of them clustered between four and eight. The remainder of the land came to Judah in his father's will, freighted with still more filial obligations to his mother and sisters, all spelled out in painstaking detail.[44] Like his father, Luke, Judah Potter remained a common-field yeoman.

The East Quarter showed a similar mixed pattern of lumpers and splitters, movers and stickers. William Hartwell's sons John and Samuel both remained by him in the village—it was not until the third generation that the family's Second Division outlands were settled. Thomas Brooks, by contrast, sold his house by the millpond and moved with his three sons Joshua, Caleb, and Gershom to a large block of Second Division by Elm Brook Meadow, two miles east of the meetinghouse. Here the family proceeded, over several generations, to fashion a whole new jumble of intermixed plowlands and mowing lots. These East Quarter families will reappear in the next chapter.[45]

In the North Quarter the story was the same again. Humphrey Barrett Jr. inherited his father's lands during the course of the Second Division and remained on the homestead at the north end of the village. He married Deacon Potter's daughter Mary and had two sons, Joseph and Benjamin. The young men received their inheritances on July 13, 1702, and the move to the Second Division lands finally began a half century after they had been granted. Eldest son Joseph, "for his encouragement in order to settlement, together in consideration of several duties and obligations," received the homestead in the village, along with the well-worked cluster of tillage land, pasture, and meadows just across the road.[46] Youngest son Benjamin, "in order to his settlement," received forty acres two miles from the center that his father had purchased of John Smedley by the mill on Spencer Brook, along with the nearby one-hundred-acre block of Second Division upland and meadow at Annursnack. This formed the nucleus of the Barrett's Mill farm of Revolutionary War fame. But Benjamin also received two mowing lots in the Great Meadow.[47]

In just this deliberate manner were Concord's Second Division lands settled. It was a process that took three or even four generations to complete. The old pattern of general field husbandry in the center was largely maintained *at the same time* that new farms were being carved out two and three miles from the meetinghouse. These two sectors of the town, the center and the periphery, remained tightly bound not only by family ties but by practical ones as well. For generations Concord would remain a community of cousins, some of whom continued to live in the village and work dispersed tillage lands within nearby fields but also drove their carts to holdings on the outskirts to find hay and timber; others of whom worked more consolidated farms at the outskirts but also carted home the superior hay from their lots in the Great Meadow. As Joseph Wood has pointed out, the fundamental ecological and social principles of New England communities remained intact whether settlement took nucleated or dispersed form, and in some towns it took both at once. The move toward dispersed farmsteads by some of the first few generations of Concord townspeople need not be taken as a sign of a breakdown or decline of communal feeling or function.[48]

Concord's Second Division was directed by two impulses. The first was to secure greater individual control over particularly valuable resources, such as white oak, white cedar, and the town's remaining wetlands. The second was to provide land for more consolidated farms outside the village, whether at once or in the future. Both of these movements ultimately marked a withdrawal from the commons system. But abandonment of the commons and dispersal to the outlands was not uniformly abrupt. Many families continued to work the old, scattered general field holdings for generations. Many of the more remote Second Division lands went unsettled for generations as well. In the meantime, these backlands continued to offer Concord's husbandmen a resource they had been providing before being divided: common grazing.

Persistence of the Commons

Elements of the commons system that persisted in Concord can be conveniently divided, for purposes of discussion, into the Great Field, the Great Meadow, and common grazing. Needless to say, given the nature of mixed husbandry, there was a good deal of overlap among these tillage, mowing, and grazing categories in practice. Perhaps most surprising of all, at first glance, is the endurance of the Great Field. We have seen how village dwellers like Potter, Barrett, and Hartwell continued to work small private plowlands of several acres alongside their neighbors, within general fields. Still, one might expect, given the strong tide toward enclosure, that after the Second Division any common management of these general fields soon faded away. But that is not what happened at all.

In 1672, Concord town meeting instructed the selectmen to "take order that all Corne fields be sufficiently fenced in season, the Crane field and bricke kill field especially."[49] By then the Second Division was two decades along and well-nigh complete: hundreds of private lots had been granted and many were being enclosed throughout Concord— and here were the older general tillage fields still being ordered by the selectmen just as they always had been. How long this state of affairs continued is difficult to tell for most of the fields, but we do have records concerning the largest, the Great Field. The answer there is, for another century and thus for the entire colonial period.

On March 1, 1691, forty-one proprietors of the Great Field (together with the proprietors of the adjoining Great Meadow) signed a covenant, binding themselves "that the . . . medow & field shal ly within one entire field fence." And what a fence it was (fig. 5.4). The area enclosed was some two square miles, running from the Burying Place at the center of the village to what would become the Bedford line, a good two miles northeast. This field (including the Great Meadow) was in places a mile wide. A long stretch of the way to Billerica (later Bedford) lay within the field, with gates at either end. Perhaps most perplexing to modern eyes, even the village houselots along the north side of the Bay Road, some two dozen of them, also lay within the field.[50]

This should serve to remind us that the Great Field was not common land. It was a collection of privately owned lots over which limited common rights and management were retained, first by the town and later by agreement among the proprietors. The purpose of joint management in the case of the Great Field was precisely to construct a

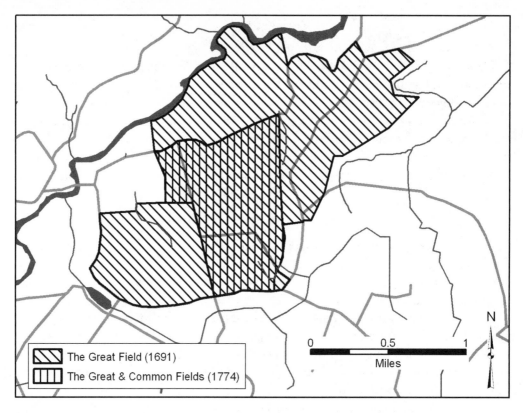

N

0   0.5   1
Miles

FIGURE 5.4.

The Great Field took shape from the time of Concord's founding in 1635. It was formally incorporated by its proprietors in 1691. For a time it included the broad upper end of the Great Meadow as well. The field shrank to a much smaller size by 1774 and was dissolved in 1778.

single fence around the whole, which was considerably less work than fencing all the individual parcels within. Most of the records of the field's proprietors concerned allotting this fence. Committees were periodically chosen to assess how many acres each proprietor owned within the field and to specify which section of fence each was to maintain, proportionate to his interest. Livestock were then excluded from the field during the growing season, "Excepting Laboring creatures and they to be Tended on our own lands."[51]

Once a year the cattle were turned into the field at the rate of two creatures to the acre, but only for two weeks in October, after the corn had been gathered. This limited grazing period may have provided a useful bite toward the end of the grazing season but could not have been the major reason for maintaining the common fence. The major reason was to keep marauding livestock out. Of course, sheep and cattle were not supposed to be roaming loose about Concord, but given the constant movement of herds and flocks to various pastures about town and to market, some inevitably were. And swine did have liberty to run at large. In 1672, the selectmen were instructed to take order "that swine, sheep & lambs be kept from doing damage in corn fields and meadows."[52] They did not discharge this duty solely by looking under haystacks for the napping rapscallion who had allowed the sheep and cows to stray. Instead, they saw to the construction of a good and sufficient fence.

The fencing that enclosed the Great Field and other general fields like it was no ramshackle split-rail affair. By law, it must have been a post and rail fence five rails high. We know such fences were made in Concord—in 1672, John Hoar contracted to deliver 50 five-hole posts along with 250 rails to a neighboring property.[53] The wood used was probably oak or, better yet, rot-resistant chestnut. These fences stood about chest high, bringing the five rails close enough to make a barrier that was tight against smaller stock as well as cattle. Building such fencing represented an immense labor, and it is easy to see why maintaining a Great Field with a single fence around it was an attractive proposition and why it persisted for generations.[54]

The covenant establishing the Great Field was periodically renewed through most of the eighteenth century, but the field itself gradually diminished in size, and its internal restrictions were apparently relaxed. Maintaining a general field did have some drawbacks. For one thing, it made it difficult for any land within the fence to be used as pasture—though in time, some lots within the field were pastured.[55] These lots were presumably fenced but were still a potential nuisance to other proprietors should the stock escape. The Great Field fence was also vulnerable at several points, exposing all the crops within to damage from wandering livestock. Once the stock were in anywhere, they were in everywhere. Over time, the most problematic areas were split away from the field. During the 1730s, the proprietors voted to "leave out the Road Leading to Bedford through sd Fields." Clearly, stock moving along the road within the field, with gates opening and (usually) closing behind them, presented a constant hazard. About the same time, the river meadow, which posed distinct management challenges of its own, was also left to its own governance.[56]

In 1738, the company assembled at Jonathan Ball's inn to consider "whether they will leave out the hils so called it being thought by many that in so doing the proprietors may be generally comoded and freed from Great Spoyl and Damages which heretofore many have sustained and as they now Lye are not likely to avoid for the futer."[57] The despoilers at this end of the field were doubtless livestock escaping from the houselots which lay within the official field fence. What the proprietors concluded that day is not recorded, but by the end of the colonial period the hills behind the village had indeed been largely excluded from the field, and only a few houselots remained within the fence.

By 1774 the field encompassed thirty proprietors but was down to only 237 acres (see fig. 5.4).[58] Simple geometry tells us that the smaller the field grew, the less advantage remained in having a common fence. On October 26, 1778, it was finally voted to dissolve the proprietorship of the Great and Common Fields. Disputes over boundaries had arisen, and the minutes of the final meetings are replete with motions to reconsider votes already taken, with individuals being added and subtracted from the fence walking committee, and with other signs of rancor and contention.[59] The Great Field had plainly seen its day. Even the gradual change in its name from simply the great field betrays a change in attitude. The Great and Common Fields, as it became known, reflects a shift in how the field was viewed—from a singular entity to a plural collection. The field had acquired the ceremonial grandeur of a barely breathing anachronism. As its name waxed great, the field itself shrunk, until at length it became a title without a function.

The Great Meadow was for a time included within the same fence as the adjoining Great Field. Yet the river meadow retained its identity and by about 1740 had been left out of the field. From that time the official field fence ran between the meadow and the upland, instead of along the river (see fig. 5.4).[60] Like the tilled plowland, the meadow lots needed to be protected from livestock damage until the hay was mowed, so there was some advantage in enclosing them all within a single fence. But the opportunities for supplemental grazing were much greater in the meadow than in the field, and so the meadow proprietors had a stronger incentive to find controlled ways of allowing cattle onto the meadow for a greater part of the year. Some wished to graze their meadow lots instead of mowing them during the summer, and all wished to take advantage of the considerable afterfeed that grew up in the fall, once the hay had been cut.

Something of the intricacy of managing the meadows for both hay and grazing can be seen in an earlier agreement concluded among several meadow owners in Concord in 1669 and recorded in the Town Book:

> Where as there hath been summe differance amongst the proprietors of the Great meadow with respect to a highway from Humphrey Barats Loest meadow unto the head of the Great Meadow, it is now loveingly agreed by us whose names are underwritten that from hence forth or from this time every Proprietor shall have the liberty without Purchace to Cart over every of our meadows as his need may Require, Every one helping from time to time to make and maintayne the Passages for Carts, according to his propriety: & no person to go out of the way whereby Dammage may be don to any parson, and if any Proprietor shall leave downe any mans shutt Rayles or Gates, he shall pay five shillings.[61]

The problem here was that in order to cart hay from their meadows, some of the mowers had to cross adjacent lots where livestock were enclosed—and might escape. A cartway was established along the slightly higher, harder ground at the riverbank for this purpose, and the proprietors agreed to hire two of their number, John Smedley and Humphrey Barrett, to "make or mend" this way. A hefty fine was imposed to discourage carelessness in closing gates.

But what about the opposite problem faced by those who wanted to get livestock in and out of the meadow to graze? The agreement went on to stipulate that anyone needing "to drive their cattle at any time to feed upon their meadow shall get a way laid out upon the upland for their particular use." In other words, private ways were needed to lead cattle into individual meadow lots from the upland side of the Great Meadow, rather than driving them along the river bank, where they would barge across lots that had been shut up for hay. The agreement no doubt resolved a tangle of long-established and evidently conflicting practices so as to ensure that every proprietor was obliged to do his fair share and to follow the same rules.[62]

After the hay on the Great Meadow was cut every summer, the situation changed. The agreement continued, "In the latter end of the year when the hay is gotten out of the meadow any proprietor hath liberty to drive his cattle down the river bank to feed, but not to do any proprietor damage without due satisfaction, and every person is to put in cattle according to his propriety, and the cattle to be put into the meado no longer than the major part of the proprietors conclude." In other words, after the hay was harvested the meadow was opened for common grazing, saving the individual owners the trouble of keeping their stock confined on their own lots, which no doubt would have required vigilance even with ditches and fences in place between some of the lots. The number of cattle each proprietor could graze was calculated according to how much of the Great Meadow he owned (his "propriety"), in the customary way of stinting commons. When the aftermath was nibbled down, the cattle were herded from the meadow.

This agreement was signed by fourteen proprietors, including Humphrey Barrett. William Hartwell's meadow lay farther down the river—whether this lower stretch of the river meadow required a similar agreement is not recorded. Perhaps few other meadows had acquired such an unruly snarl of customary ways of passing and repassing to be untangled. It is likely that the 1669 covenant among the proprietors of the "broad" of the Great Meadow provides a formalized example of dozens of simpler verbal understandings among the joint owners of smaller meadows concerning where the carts and creatures were to go and how the meadow was to be opened for feeding. It is unclear whether the 1669 covenant was completely superseded by the Great Field covenant in 1691. The Great Meadow was formally disassociated from the Great Field fence again by 1740, but there is no further evidence of regulations governing it.

The convenient practice of commonly grazing the afterfeed on larger meadows may have continued in Concord for many decades after 1740; perhaps for another entire century. Meadows concentrated a vital resource, required by all, into a few particular places in the landscape. Thus they lent themselves to both minute subdivision and a degree of continued common management. We so often think of common and private property rights as opposing principles, but they frequently overlapped in functioning commons systems. Meadows called for close cooperation in how they were drained, fenced, accessed, and grazed. Given its numerous small lots upon which virtually the entire community turned out to mow the hay at the same time every summer and given the ever-present danger of rain and flood driving the work forward, the Great Meadow retained something of the feel of a commons well into the nineteenth century.[63]

The Great Field provided two weeks of grazing for the proprietors' cattle in October, and the afterfeed of the meadows perhaps a month or two starting around September, but where were the stock to graze during the summer? This had been one of the central ecological problems facing Concord from the start, and simply dividing the land did not automatically solve it. A gradual improvement of pasture on enclosed lands was indeed under way, as will be discussed in chapter 7. But for the remainder of the seventeenth century and into the eighteenth, a significant part of Concord's stock continued to graze in common herds. The question that naturally follows is, since the great bulk of the town's land had been privatized in the Second Division, where were these common herds to find grass? The answer is, in three places: on a new grant of common land, on smaller commons retained during the Second Division, and on private lands that had not yet been enclosed.

The drive to acquire more common grazing land predated the Second Division. As the people of Concord grew frustrated with their situation in the 1640s, they began to petition the General Court for more land. First a group of disaffected Concordians requested land for a new plantation northwest of Concord in 1643, and although this request was endorsed by the Court, nothing seems to have come of it.[64] Another request for land in the same area went down to Boston in 1651, but this time it was a request not for a new plantation, but for an enlargement of Concord itself. These stockmen were looking for more pasturage, the pine barrens of Concord having proven woefully deficient in supplying adequate grazing for the common herd. Perhaps they had their eye on several long stretches of meadow along the brooks that ran from the uplands west of the town into the North River. In any case, by 1650 this was the one remaining direction in which Concord could still expand. The region surrounding Concord had largely filled with new towns and private grants since the town was pioneered in 1635.

In 1655, word came up that "the Court doth grant them five thousand acres of land for feeding."[65] The new grant was a strip of land along Concord's northwest boundary, extending as far as the Indian praying village of Nashoba which had been established in 1654.[66] In fact, it appears that some of Concord's grant ran over land that the people of Nashoba already claimed, either by virtue of their own recent grant from the General Court or by their presence in the region since time immemorial. In 1660, the inhabitants of Concord agreed to pay the "Ingenes of Nashoba" £15 to satisfy their right in the new grant. Lt. Joseph Wheeler picked up the tab and was granted a tract of 610 acres in return.[67] An additional 5,000 acres were granted Concord in 1665. When the entire tract was laid out in 1666 it was found to contain 9,800 acres. Later surveys put the actual size of the grant closer to 13,000 acres.[68] A nice size for another New England town, but it took most of a century for Concord's New Grant to be set off as Acton.

The New Grant was known as Concord Village, or just the Village. This is confusing because for several generations very few people apparently lived in this so-called Village. Like many villages being set up during the early decades of settlement in Massachusetts, this land may have been granted to Concord by the General Court partly for the purpose of establishing a daughter community made up of people who could not find adequate resources in Concord.[69] However, this was not the initial use to which the New Grant was put. Instead, for several generations it was devoted primarily (although certainly not exclusively) to *common grazing*—just at the time when the existing common land in Concord proper was being broken up into private holdings. In January 1668, the town voted that Concord Village "shall ley for a free Common to the present householders of Concord and such as shall hereafter be approved & allowed; except such parcel or parcels of it as shall be thought mete to make farmes for the use and benifit of the Towne."[70]

On March 20 of that year, a complicated twenty-one-year lease was arranged with Capt. Thomas Wheeler to keep the town's dry cattle on the New Grant. Dry cattle usually included barren or unbred cows, together perhaps with fattening steers and retired oxen. Wheeler was to supply a herdsman to put the cattle to pasture and return them to a pen every day. In return, Thomas was to be paid two shillings per head, and he was to have the use of a farm including a house and barn (which he was to build to precise specifications), along with 30 acres of tillage land sufficiently fenced. Wheeler also had liberty to take timber within the New Grant.[71] This was the "Towne's Farme" referred to above, created as a means to compensate someone to manage the dry herd. Concord had set up a remote grazing range upon which to run the stock that did not have to come home to be milked every night. The practice of sending dry stock to summer pasture up-country would persist into the twentieth century. In this earliest version of sending the cattle away to the hills, Concordians chose a commons approach which was to last through the next three generations—it was the 1730s before Concord Village was surveyed and heavily settled. In 1735, the Town of Acton was finally incorporated.

In the meantime, a large part of this tract of well over 10,000 acres (nearly half the size of the original town) was owned as commons by the proprietors of Concord and their heirs. There was some contention about exactly who qualified as a proprietor, and a list was drawn up in 1697. In 1698, the Village began holding a special session following Concord town meeting to manage their exclusive affairs. The Village meadows were leased, timber cutting was regulated, and the land was used for common grazing. Thus, at the same moment that Concordians divided most of their original commons among themselves, they set up another commons on a similar scale and retained it for the better part of a century.[72]

The New Grant was not the only common grazing land in Concord. When the town met on January 2, 1653, to make their sweeping Second Division, the first thing they did was to set the cow commons at one beast per 20 acres of land (of all the land men held).[73] A typical First Division holding amounted to 100 acres or thereabouts, so the average allowance for pasturing on commons would have been about five beasts.[74] This matches reasonably well with Edward Johnson's 1654 estimate of fifty families and three hundred head of cattle in Concord.[75] One grown cow was reckoned equal to two yearlings, one horse, or four sheep. The provision was amended two years later to allow "poor men" who lacked sufficient commons (that is, common rights) of their own to pasture at least four cows, at the town's discretion. This measure, designed to protect small householders from being dispossessed by the division of the commons, gives us an idea of what was considered a bare minimum number of stock necessary for subsistence. In January 1667 it was decided that men who lacked commons for their cattle could pay six pence to those who had commons to let them into their quarter.[76]

Where were these creatures to find commons to graze? Concord had acquired more land from the General Court for the dry stock, but what of the milk herd, the calf herd, and the sheep flock? The clearest answer is provided by the North Quarter, which created a well-defined common called the Twenty Score. This was an undivided tract of land just north of the houselots beyond the river, in which twenty North Quarter proprietors owned rights that amounted to about 20 acres each (fig. 5.5).[77] The Twenty Score included upland, meadow, and swamp and had embedded within it several privately owned meadows and uplands. We have no record of its precise regulation and use, but it seems to have provided common grazing to the quarter for almost another half century after the Second Division. In 1697, the bulk of the Twenty Score was divided into twenty private lots of 13¾ acres each, but some 65 acres at the southern end were retained as a sheep common for several more years after that. Thirty trees were to be left on this sheep common and not felled, presumably to provide shade. This suggests that the men of the North Quarter had been systematically clearing their commons to improve the grazing.[78]

A few other parcels in Concord were described as common or undivided in the land division records. It is difficult to tell whether these had been deliberately set aside as commons or simply had not yet been divided when abutting landowners recorded their holdings. Areas that remained undivided when land divisions were recorded in the late 1660s and early 1670s amounted to several hundred acres in each quarter. Many of these, however, were whittled away as men gradually took up the last of their Second Division grants over the ensuing decades. Finally, in 1733, a committee was appointed to search out the remaining common land in Concord. Six parcels amounting to 226 acres were returned and were either divided among those who had inherited unfulfilled claims or auctioned off.[79]

In 1672, Concord's selectmen were instructed by town meeting to take care that cattle be herded and that persons not overcharge the commons with cattle.[80] This must have been a difficult charge. Even with the acquisition of the New Grant to pasture the dry herd, the scraps of commons remaining within Concord proper could hardly have been sufficient to accommodate the growing stock of the town. Clearly, after the Second Division a large part of the available grazing was to be found on private land, and many husbandmen undoubtedly took to keeping stock on their own enclosed pastures. But there were plainly still common herds. Where were they feeding?

These livestock grazed private but as yet unenclosed land at the outskirts of town. The commons had been divided on paper and private parcels even surveyed on the ground, but for a period much of this land was still *managed* as commons, at least for purposes of grazing. A South Quarter meeting in December 1666 decided that "the drift way for the herd to fair haven should be to turn out of Marlborough way between brook meadow and nut meadow, passing over the brook at the usual place, the drift way to lie on the northeast side of the spruce swamp so to pass into the cart way to fair haven between the two spruce swamps."[81] The route that this common herd took is revealing, both in where it ended and in how it got there. Instead of going directly south on the existing cartway

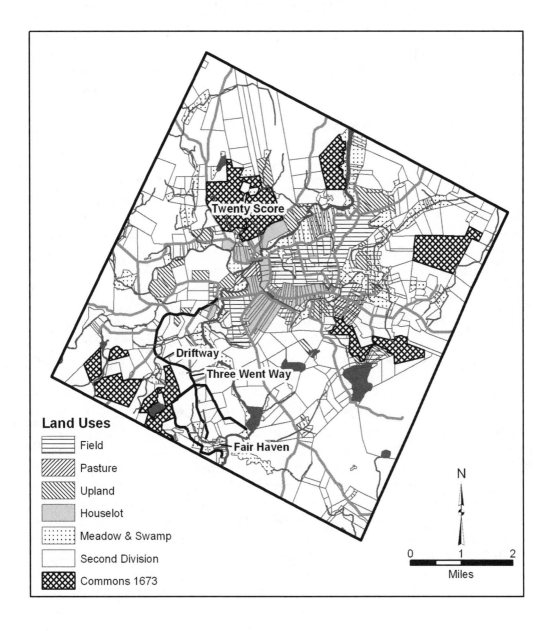

FIGURE 5.5.
Commons, 1673. At the conclusion of the Second Division several parcels were left as undivided commons. Best documented was the "Twenty Score" in the North Quarter. The map also highlights the westerly "Driftway" for the herd to Fair Haven, laid out in 1666, the "Three Went Way," and the three wents, or ways.

from the South Bridge, by 1666 the herd was being driven half a mile farther west along Marlborough Road. It was then brought back east to rejoin the old route at the Three Went Way (see fig. 5.5). Why make this long detour? Presumably so that the herdsman could pick up cows belonging to the Hosmers, Wheelers, George Hayward, and Michael Wood. These men had all recently moved to their Second Division grants, necessitating a new route for the common herd on its way to and from the village every day. In other words, even men who had left the village for enclosed holdings on the outskirts had not entirely withdrawn from older common practices.

This is further confirmed by *where* the cows grazed. The herd was on its way to Fair Haven. But by 1666 Fair Haven was in private hands—largely the hands of these same men, together with others such as John Miles, who still lived in the center. These land-owners evidently pooled their remote "wild lands" to provide common grazing for the cow herd. Undoubtedly the herd was also sometimes driven to true common land lying west of Fair Haven around White Pond, but this was mostly scrubby pine plain. By contrast, some of the best grazing in the quarter was to be found in the extensive meadows at Fair Haven. This corner of Concord was not settled until the 1720s, and many of the remote meadow lots there were doubtless among the last to be improved for mowing by their owners. In the meantime, they were grazed by the common herd. We have no record of the agreements by which this grazing was regulated in Concord. There is a record of very similar common grazing of privately owned land in the west end of neighboring Watertown continuing until almost the close of the seventeenth century. Until the outskirts of early towns were finally settled and fields and pastures enclosed, it was simply most sensible and efficient to graze them commonly.[82]

The foundation of private, enclosed land tenure had been laid down by Concord's Second Division, but it was not immediately built upon in every corner of Concord's husbandry. Common and private management coexisted for several generations as Concord's landscape evolved. The creation of new general tillage fields ceased, and new homesteads on the edge of the village tended to have their holdings more (although seldom entirely) consolidated. At the same time, near the village center the older general fields with their small, intermixed lots lived on and, at least in the case of the Great Field, retained a degree of collective management right through the colonial period. The pattern of ownership and management of the meadows hardly changed at all—most farmers' mowing lots remained fragmented and scattered into the nineteenth century. Even common herding persisted, partly on remaining commons and partly on unimproved private holdings in the far corners of Concord. A large New Grant was acquired for common grazing on the town's northwest border. At the same time private pastures were steadily being enclosed and improved as well. Concord's best woodlands were decisively privatized—although for some time the joint owners of the larger swamps apparently held undivided rights rather than distinct parcels. Unfortunately there would be no real effort to manage forests for the common good again for another three hundred years, until our own time.[83]

The persistence of a degree of common management covering some privately held lands followed an established tradition of English common field systems. In Concord, several of these practices continued to make ecological sense and so continued in a modified form long after the Second Division. The transition to private, enclosed farming was more nearly completed by the 1730s, when Concord had been settled to its borders and almost all land and livestock had come under individual management as well as ownership. But this did not mean that farmers even then became single-mindedly devoted to maximizing individual profits from their land. Throughout the colonial period, they remained bound one and all by family and community obligations and expectations and by the limitations of their environment and their markets into a system that was oriented primarily toward yielding a comfortable way of life directly from the diverse elements of Concord's landscape.

CHAPTER 6

Settling the East Quarter

In order to his settlement in the town of Concord.
—CONCORD TOWN RECORDS

LET us imagine that one cold morning in March of 1653 William Hartwell set out from his house in the village toward his mowing lot in Rocky Meadow nearly three miles to the east. The winter's snow was mostly gone, but the Bay Road was frozen and passable by cart, and in the raw northwest wind that was blowing seemed likely to remain firm throughout the day. Hartwell may have needed to draw home a load of hay from a large stack that had been standing all winter on the hard ground beside his meadow. By March he might have all but exhausted the hay carted from his mowing lots in the Great Meadow and along Elm Brook the summer before.[1]

Hartwell also needed to take a closer look at the upland lying about Rocky Meadow. Town meeting had passed the Second Division a few months earlier, and the East Quarter would soon begin dividing its commons. Hartwell had to consider what parcels he would propose for. Along with Hartwell might have gone his two oldest sons, twelve-year-old John and eight-year-old Samuel, to help at loading the hay. Young Samuel would at least relish making the trip with the men, and John was reaching the age when he should learn to judge the quality of land upon which his future comfort would depend.

Also accompanying Hartwell this morning perhaps was young Joshua Brooks, son of Thomas Brooks, who lived on the other side of the brook against the millpond.[2] Capt. Thomas Brooks owned a mowing lot alongside Hartwell's in Rocky Meadow and may also have stacked hay there for spring feeding. Joshua Brooks had reached his majority, so he no doubt had a keen interest in the lay of the land surrounding his father's meadows. If all went well, the two men would cart home two loads of hay—one in the forenoon to Hartwell's yard and another to Brooks's after dinner.

From the yard where they tackled the oxen and secured the yoke to the cart pole, the men and boys stepped off eastward into the sharp morning light of late winter. For the first half mile they passed before the houselots of Hartwell's neighbors under the shelter of the Hill on their left, with the Mill Brook and its flooded, frozen meadow lots below the road on their right. Then the Hill abruptly ended, and they were in open country. Here Billerica Way ran toward the northeast, crossing nearly a mile of the Great Field before exiting again at the gate past Virginia Swamp. Had Hartwell been carting manure they would have turned here, through Cranefield Gate and then left again down the lane that led to his plowland behind the Hill. Instead, they continued east along the Bay Road. From the corner onward, they passed no more houses.

The Bay Road ran across an extensive level of meadow, marsh, and swamp studded with a few upland islands. The road sought the shortest links between the islands, crossing the intervening low places to reach the stony uplands a mile to the east, beyond Elm Brook.[3] With one notable gap, this one-mile stretch across the lowlands to the base of Elm Brook Hill would soon be settled by Concord's second generation. The uplands beyond, reaching to the Cambridge line, would not be fairly settled until the third generation at the turn of the century. These patterns, which will be explored in this chapter, help reveal the minds of the settlers.

The land at the fork of the road would be improved for a houselot by John Meriam about 1663 and become known as Meriam's Corner. On this March morning a decade earlier there would probably have been no house standing there yet. John Meriam was still a twelve-year-old boy living by the millpond with his stepfather, Joseph Wheeler, next door to Thomas Brooks. Were nostalgia prospective Joshua Brooks might have felt a tug as he passed by the corner because his descendants would be intimately connected to that place, through many ties of blood with the Meriam family. Joshua's son Daniel was destined to marry John Meriam's daughter Anna, and his granddaughter Dorothy would one day marry John's son Joseph and become the mother of many generations of Meriams to come. The Meriams would form a tight-knit family neighborhood at the corner.

From the corner the men followed a causeway (or "cassey," as they would have said) over the lowland, crossing the north branch of the Mill Brook. To their right, mowing lots ran down along the carefully ditched brook through Bridge Meadow and on toward Town Meadow before the village houselots. To their left was a long stretch of swampy land curling away from the road toward the northeast, probably for the most part not yet improved for mowing. Hartwell may have given some thought to proposing for part of this swamp—a bit more meadow so close to home would have been convenient for him. In the event, the wetland would be claimed by several of his neighbors and in time become known as the Dam Meadow.[4]

The road gained the hard ground once more and climbed a gentle slope. Here was good land, both sides of the way, although wooded and stony in places. It would be claimed by Capt. Timothy Wheeler, one of Concord's leading men. But Wheeler would never live there, and neither would his equally prominent son-in-law, Capt. James Minot, to whom the land eventually passed. Both men would reside in the village and lease this outlying farm to tenants. A house would be constructed on the south side of the road sometime in the late seventeenth century. Not until well into the eighteenth century would a branch of the Minot family finally take up residence and build a new house on higher ground north of the road.[5]

The cart (empty but for the two boys) passed over the wooded rise and once more into open country. On the right, a gate blocked the lane into the Ox Pasture, a fenced common of about one hundred acres, much of it lowland along the southern branch of the Mill Brook. There the men of the East Quarter pastured their oxen during the growing season, turning them loose in the evening after plowing in the nearby general fields or carting hay and corn. This was a convenience, but there were those among Hartwell's neighbors who believed the land would be better improved if it were subdivided among them. And so it was that in the Second Division William Hartwell would receive five-and-a-half acres of the Ox Pasture bounding on the Mill Brook.[6]

Along the north side of the road ran a stout five-rail fence, and just beyond the crest of the hill the men passed by Brickiln gate. Within the gate lay the Brickiln Field, some fifty acres of fine plowland where eight or twelve men of the quarter owned tillage lots. The ownership of many of these lots had already passed from hand to hand several times as families left Concord during the difficult years of the previous decade, and a few had been consolidated into larger holdings. Hartwell himself owned a lot at the back of the Brickiln Field, with adjoining meadow.[7] But this meadowland was probably still ill-drained and unimproved in 1653, so let us assume that Hartwell had cut only a small amount of hay against the upland while reaping grain at his plowland there the previous summer, cocked it, and carted it home a few days later along with the sheaves of barley. The soil in the field was of excellent quality, strong enough for the small amount of beverage grain that Hartwell grew.

Although the land was good and the village nearby—or perhaps because of that— no farms would be settled at the Brickiln Field by Concord's second generation. Instead, the land would be managed as a general tillage field for some decades more, as the slow consolidation of holdings within it continued. Finally, in the last years of the seventeenth century, members of the third generation would establish houselots along the Bay Road where their grandsires had first worked lots in the Brickiln Field. They would cobble together farms from disconnected parcels stretching northeast from the old Ox Pasture across the Brickiln Field to Elm Brook Meadow and Brickiln Island. But try as they might, neither they nor their descendants would ever manage to fully consolidate these farms. The indelible stamp of subdivision and scattering would endure until far into the *nineteenth* century, long after the families who dominated the neighborhood during the colonial era had passed from the scene.

The team walked on past the field and crossed another bit of low ground, where bright green skunk cabbage was just emerging from the black muck under maple and elm trees. The maples would have been showing a distinct tinge of red in the swelling buds, their tops catching the early morning light. Farmers whose fodder was running dangerously low after a long, hard winter must have been cheered by these promising signs of spring. Now the road rose once more onto a long, nearly level bench, covered with a fine growth of white oak, hickory (or walnut, as these men called it), chestnut, and white ash. They passed by a rough private lane that led south over common land, winding a mile up the hill through the woods to the Chestnut Field, where Hartwell owned two lots.

As the men strode along beside the team, young Joshua Brooks might have asked the older man his opinion of the land thereabouts. Would it make good plowland? Hartwell might have allowed that it was likely enough looking land, not so wet and stony as the uplands beyond, yet not so droughty as a pine plain. He knew very well that Joshua's father, Thomas, owned large tracts of First Division mowing in Elm Brook Meadow a few score rods to the north. Captain Brooks had spoken of his desire to come in with the East Quarter for the major portion of his Second Division (even though he lived in the South Quarter). This was indeed fine land and would have offered a good start for either of William Hartwell's sons, but it was out of his reach.

It was certainly not, however, beyond the grasp of Joshua Brooks. In fact, Joshua would be married the coming October to Hannah Mason of Watertown, and in the next few years he would build a house along the Bay Road there above Elm Brook Meadow, on Second Division land granted to his father and passed along to him. Close beside the house, hard by the brook, he would establish a tannery which would preserve the prosperity of his descendants for a century and a half. Joshua's brother Caleb would set a farm next door, and close beside *him* would move brother-in-law Timothy Wheeler. An older man, Captain Wheeler was a widower at the time of the Second Division, but he would shortly take Joshua's sister Mary Brooks as his second wife and come to live alongside his new brothers. But not for long—in 1663, Wheeler would buy the extensive holdings of the Reverend Peter Bulkeley from the minister's widow and move back into the village. Joshua's brother Gershom Brooks would purchase brother-in-law Timothy Wheeler's homestead in 1665 and join his older brothers at Elm Brook, bringing their revered father, Thomas, along with him. For most of the next two centuries that stretch of the Bay Road would be inhabited almost entirely by people named Brooks, most of them descended from Joshua and Hannah.

Crossing the sharp dip in the road at Elm Brook, the men started the long, steep climb over Elm Brook Hill. Halfway up, the track bent sharply to the north, still climbing, skirting a frog pond on the upland ahead. These uplands were high and rocky, but by no means dry—they were full of springs, sumps, and swamps. The woods on the Hill were heavy—oaks of all kinds with great white oaks predominating, chestnuts, a few soft maples, and here and there a solitary white pine soaring above the rest.[8] The ground was thick with stones, and patches of grainy snow lingered in the deep shade of the trees and the dips behind larger boulders. This ground would never amount to much as plowland, but the timber was good, and it caught William Hartwell's eye—he would be granted eight acres there southeast of the road in the Second Division. Joshua Brooks would receive land on the opposite, northwest side of the road from his father.

At the top of the hill the road forked, the way to the bay swinging back to the right, while a rougher track continued ahead to the northeast. The men followed this narrow cartway, which would later be extended across country to reach the new town of Woburn—William Hartwell, who knew the way, would be one of the two men from Concord appointed to lay it out.[9] They bumped along for another half mile over the wooded upland and descended a gentle slope to Rocky Meadow, a few dozen acres of rough mowing containing lots belonging to half a dozen Concord men. A ditch angled across the remote meadow and then ran north along its east side, this being the headwaters of the Shawsheen River. Stones had been sledded from the meadow and piled into a straggling wall along the south end, and a rough rail fence—sufficient, at least, to turn away wandering cattle—surrounded the entire meadow, also encompassing a few acres of hard ground at the west side, along the cartway. It would have to be set right before summer—here and there the frost had heaved the posts and tumbled the rails. The boys leaped from the cart and ran ahead to take down the bars, and the men drove the team along a track to the edge of the meadow. They turned the cart there by Hartwell's haystack. A score or more rods to the north stood a second stack against Brooks's meadow—these two were the last of the hay at Rocky Meadow that spring.[10]

The boys scrambled up the pole onto the top of the stack and removed the coarse thatch that had protected the hay from the weather. The haystack had been built upon a wooden staddle to save it spoiling on the moist ground. The men saw, to their annoyance, that deer had nosed through the thatch and been feeding at the hay during the past weeks. There was little fence could do against these creatures—at least the bulk of the hay in a large stack was beyond their reach. The boys and Joshua Brooks began forking down the hay in bundles to the older man, who carefully built the load on the cart. As they worked, Joshua Brooks might have remarked that it was good hay and had kept well through the winter. And Hartwell may have replied that it was serviceable enough for dry cattle, and not to be despised at this late season, but that it could hardly compare to what he had cut on the meadows of Bedfordshire as a boy, before coming over into this country at about the age Joshua was now. Joshua had been born in England but could scarcely remember anything of his homeland, although his mother spoke of it often. At which Hartwell might have observed that his wife, Jasan, was the same and that leaving home so far behind was surely hardest on the womenfolk.

From where he stood in the bed of the cart, William Hartwell may have surveyed the upland round about Rocky Meadow. Farther east along the Bay Road, he knew, were a few stretches of well-drained, sandy land suitable for tillage that might make fine places for homesteads one day. The upland here by the meadow, by contrast, was rock-strewn and broken, with many an intervening brook and swamp—not the best land in Concord by any means, but not without its virtues. It was well watered and held good timber, and the swampy places could in time provide additional mowing. The Lord made nothing without some purpose.

However he may have come to his decision, William Hartwell would soon propose for, and be granted, two tracts of Second Division upland at Rocky Meadow—one running northeast to the Cambridge line, the other southwest to the Bay Road. He would not be granted any of the better-drained land to the southeast. That would go instead to a man named Richard Rice, who would move there in the coming decades and survive until 1709, being accounted over one hundred years old at his death. Rice was one of a handful of Concord's earliest settlers who would remove to the far fringes of the town himself, as the proprietors and their progeny moved deliberately to take up their divisions.

William Hartwell would represent the opposite extreme. His Second Division lots would be divided between the two boys John and Samuel—but, though he lived to age seventy-seven, Hartwell would never see his upland at Rocky Meadow inhabited. Instead (doubtless to his great satisfaction), his sons would remain close by their father in the village, working dispersed holdings alongside his own. It would remain for William Hartwell's grandson, Samuel's eldest son, Samuel, to move to the rocks and swamps by Rocky Meadow about the time of his marriage in 1692, two years after his grandfather's death and some four decades after the Second Division. Samuel Hartwell Jr. would assemble a prosperous farm in part by purchasing a sufficient parcel of well-drained plowland from his elderly neighbor Richard Rice.

Young Joshua Brooks may have been surveying the countryside as well, as he pitched hay down into the cart. The meadow lot where the second haystack stood would soon be divided between Joshua and his brother Caleb and would thereafter be divided among Joshua's sons—but would one day be acquired from them by Hartwell's descendants.[11] To the west were woods adjoining Joshua's father's rough, unimproved meadow at the Suburbs, woods full of black oak rich in tanbark and white oak suitable for ship timber.[12] There Thomas Brooks would propose for and be granted upland, which would be passed along to Joshua by deed of gift in 1667. In time Joshua would acquire adjoining meadow and upland, giving him a large tract to divide among his heirs some forty years hence. Hartwells and Brookses would own and work almost all of the land along both sides of Rocky Meadow Way well into the nineteenth century.

Their load made, the men started the team and completed the long, slow trip back to the village. There they forked the hay into the nearly empty mow above the cattle bay in Hartwell's barn, baited the oxen with a bundle of the same dry stuff, and went in for their own dinner. Those sitting at table that day might have included the hay party, mother Jasan Hartwell, the boys' two younger brothers and three younger sisters, and a neighbor girl—perhaps Priscilla Wright—whom the Hartwell's might have engaged as a servant to help with the heavy domestic chores.[13] John Hartwell and Priscilla Wright were to be married in 1662. After William Hartwell gave thanks to God, they dined together on plain fare: pease porridge with a bit of salt pork, Indian corn and rye flatbread baked before the fire, and maybe a little cheese—all washed down with beer. Like many of his neighbors, Hartwell had planted an orchard, but throughout his life he would keep the habit of drinking beer. His sons would come to drink chiefly cider instead, especially during the winter months.[14]

After dinner the men and boys may have put on their greatcoats and set out again for Rocky Meadow, making again the trip they had made in the morning. The Hartwell family would make this trip in reality hundreds of times—to mow and cart their hay, to pasture their cattle, and in time to visit and change work and visit with Samuel's son Samuel after he moved to the upland near Rocky Meadow in 1692—William for nearly another forty, John for another fifty, and Samuel for another seventy-two years.

In 1653, as the Second Division got under way, Concord still resembled an English common field village. The great majority of houselots were tightly clustered within a mile of the meetinghouse, most men cultivated tillage lots scattered within the general fields, and some three-quarters of the town was a grazing commons thick with trees. With the privatization of almost all of the commons and the coming of age of Concord's second generation, the slow dispersal of homesteads to the town bounds began. A few members of the aging but still vigorous first generation moved to new holdings themselves; most remained in the village and continued to work their First Division lots. They passed along Second Division lands to their sons—often their older sons—some of whom established new farms on the outskirts. The youngest son frequently stayed with his parents and inherited the homeplace in the village, along with additional Second Division land to pass along to the third generation.

By and large, the second generation settled lands covering the second mile from the center (see fig. 6.1). It remained for the third and fourth generations to settle the bulk of the third and fourth miles and to fill in the gaps. By the middle of the eighteenth century, after four fruitful generations, Concord had finally grown thick with farms.[15] What did those farms look like? What shape did they take, and how were they arranged on the ground alongside those of their neighbors?

The new farms created from Second Division grants were somewhat more consolidated than the dispersed holdings of the first generation. Building on what they had received from their fathers, many Concord husbandmen labored to assemble convenient, contiguous homesteads by buying land or exchanging with their neighbors. But equally powerful forces worked against consolidation. For one, the ecological necessity of possessing a balance of diverse resources enforced a wide scattering of parcels, particularly meadow lots and woodlots (I will return to this subject in chapter 7). For another, given the wealth of land available during the early generations, partible inheritance quickly became the rule in New England. Large families (which also became the rule) often prompted intricate subdivision when land was passed along, in order to provide each heir

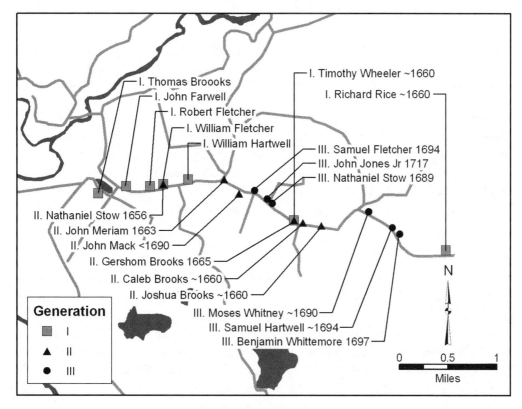

FIGURE 6.1.

East Quarter settlement. The first generation lived mostly within a mile of the meetinghouse. Members of the second generation moved out another mile or so (skipping past the Brickiln Field). Members of the third generation homesteaded at the Brickiln Field and settled the third mile to the town bounds.

with a living. As each generation came of age, wastelands lying farther from the village were parceled out for new homesteads, but improved lands closer to home were frequently subdivided among sons as well. Daughters were sometimes endowed with land at their marriage, as sons-in-law joined these tight-knit families. Consolidation was never complete, and if anything by the end of the colonial period was on the retreat.

The result was a pattern of fragmented farms, often knitted together into complex kin neighborhoods, across the landscape of Concord. A family trade such as blacksmithing or tanning (also passing from generation to generation) was often central to the family economic and social fabric. Close bonds of cooperation but also the chafing rub of competition for the ever-tightening supply of land marked this system.[16] The process of land inheritance and farm formation for the first four generations, down to about 1750, can be traced through a series of families and neighborhoods along the Bay Road from the village center to the uplands near the Cambridge line.

The Hartwells

The Hartwell family illustrates a pattern of inheritance in which older sons dispersed to new lands and new towns, while the youngest son stayed with the home place, enjoying its improvements and advantages but also shouldering its obligations (see appendix 1: Genealogies). This pattern was by no means universal but was common enough to be called the rule in Concord. With plenty of unsettled land to be improved, the logic of sending forth the elder sons is plain. William Hartwell helped set the pattern in the way he divided his lands between his two sons, John and Samuel. But as the generations passed, things did not always work out as the father planned.

The Second Generation

When the Concord settler William Hartwell died in 1690 he meticulously divided his land between his two remaining sons, taking care also to provide for the "convenient maintenance" of his wife, Jasan, who survived him by five years.[17] Neither his eldest son, John, nor his younger son Samuel left home to establish a farm on William Hartwell's Second Division grants. Each had long since settled in the village, married, and begun raising a large family. Samuel lived with his parents at the homeplace, while John moved only as far as the house next door. William had already provided each son with 100 acres by deed of gift, and the East Quarter had also granted each a small amount of land in his own right in the Second Division.

Thus, both John and Samuel Hartwell passed their lives as husbandmen and yeomen within the old common field pattern in which they had been raised. Each possessed tillage and meadow lots scattered from the Great Field to the Chestnut Field, from the Great River to Rocky Meadow. In 1672, the quarter even granted Samuel Hartwell a little pond in the midst of the Great Field, lying adjacent to a swamp belonging to his father. This area was subsequently drained by a long ditch to the Great Meadow and was broken up for plowland, becoming known as Polden.[18] Both brothers signed the covenant governing the Great Field in 1691, and John Hartwell was appointed to the committee that apportioned the fence.[19] In addition to these common field–style holdings, each brother inherited a share of William Hartwell's Second Division grants to dispose of as he chose—John was left 50 acres more than Samuel, who accordingly received his father's right in the Village, or New Grant (later Acton). To complete his holding John received 1¾ acres in the Cedar Swamp, four miles away on the Billerica line. In that same swamp, Samuel received 1 acre.[20]

When William Hartwell willed his residual property to his sons he gave his wife the power to sell land if she had need, taking equally from each. He also left his livestock to Jasan for her life, then to be divided equally between the sons. And he concluded his will as follows: "I also desire my son Samuel, according to his promise made before several witnesses, to take good care of his aged mother, and to be helpful to her in ye management of her concerns, for her comfort, he living most conveniently for it. Also I warmly desire my said two sons John and Samuel & as their father charge them to maintain brotherly love and unity between themselves, living as becometh brethren in mutual helpfulness each to other."[21] When they received this fatherly injunction John was forty-nine, Samuel forty-five. Both were on the verge of becoming grandfathers. How well they lived up to their father's warm charge is not recorded. John Hartwell died of smallpox in 1703, while his brother Samuel lived on in the homestead until 1725, where he died in the eightieth year of his age.

The Third Generation

By the time Samuel Hartwell died in 1725, none of his sons remained at home to carry on the home place. He left it instead to his unmarried daughter Ruth. Not one of his brother John's children was living next door any longer. The third generation saw the end of the Hartwell line under the Hill in Concord center. Except for Ruth, the numerous Hartwell offspring had all dispersed to the outskirts of Concord, neighboring towns, and neighboring colonies.[22]

Samuel Hartwell had four surviving sons at his death in 1725, yet none was living at home. His eldest son, Samuel, had taken up Second Division land south of Rocky Meadow on the Bay Road, to which I shall return to close this chapter. His next son, William, had been given Second Division land a bit farther north and would soon be among the leaders in the formation of the town of Bedford. Samuel's youngest son, Jonathan Hartwell Sr. had recently moved to Littleton.[23]

This was surely not what his father had intended. When Jonathan came of age in 1709, Samuel Hartwell deeded his youngest son half of his property "in order to his settlement in the town of Concord." The gift included 5 acres of the homelot, the south end of the barn as far as the threshing floor, and twelve other far-flung pieces of land. The largest parcel was 12 acres of woodland—most of the others were less than 4 acres, including 2¾ acres in the Cedar Swamp.[24] It was a half-interest in an old common field-style holding of about 100 acres composed of small, scattered lots, and the other half was surely coming to him. Jonathan (who also worked as a weaver) bought and sold a few parcels over the next decade, but apparently he did not like his prospects well enough to settle in Concord after all. In 1720, he bought a 104-acre farm made up of only three parcels (the houselot alone was 88 acres) and removed to Littleton.[25]

Consequently, when Samuel Hartwell died he left the residue of his estate—his houselot, three tillage lots, two meadow lots, and a woodlot—to his spinster daughter Ruth, then fifty-six years old.[26] Ruth lived on in the old Hartwell homestead for another quarter century until 1756, with a cow or two, one swine, and a dwindling legacy of lands in and about the Great Field. About 1748, she sold 18 acres of pasture at the old Polding lot to her neighbor Nathan Meriam.[27] She was the last of the home branch of the Hartwell family.

The Meriams

The Meriam family lived half a mile east of the Hartwells at the end of the village, at Meriam's Corner. The Meriams divided and passed on their land in a different manner, by which offspring from several branches of the family were accommodated within a close-knit neighborhood of kin. At the same time, like the Hartwell offspring, many Meriam children left home to settle new lands and new towns. Those who stayed subdivided both old family land and new acquisitions in a way that reproduced the fragmented, intermixed pattern of the common fields. The Meriam family also demonstrated the importance of a hereditary family trade in holding together an economically viable cluster of small landholdings.

The Second Generation

The first generation Concord settler Joseph Meriam, a clothier from Hadlow in Kent, most likely lived south of the millpond between his brother George Meriam and Thomas Brooks. He died young in 1641, only a few years after the town was founded.[28] John Meriam, Joseph's youngest, was the only son to spend his life in Concord. John Meriam married Mary Cooper in 1663 and probably built his house at the corner about that time. By 1672 his land amounted to 171½ acres, a middling estate for his generation (plate 5A).[29]

The 1675 map shows the core of John Meriam's land grants, together with parcels he purchased nearby. These homelands included tillage, pasture, orchard, and meadow.[30] Meriam continued to accumulate land throughout his life, both close to home and as far away as a lot in neighboring Cambridge Farms (later Lexington).[31] John Meriam's farm was typical of a second-generation holding on the edge of the village. He filled out a reasonably compact homestead close to his house but also continued to rely on scattered outlands (not shown on the map). He had a sizable parcel of plowland across the road from his house and additional tillage lots elsewhere within the Cranefield. He had meadow lots by the Mill Brook on both sides of the Bay Road at his door, but along with these went 3 acres in the Great Meadow and larger, probably rougher lots in Elm Brook Meadow and Virginia Swamp. Together with his neighbor Jacob Taylor, he acquired woodland at the Chestnut Field. Finally, Meriam had 91 acres of unimproved land on the northeast border of Concord in Shawshine Corner to distribute to his sons.[32]

The Third Generation

John Meriam divided his land among five sons as they came of age around the turn of the century. Two of them established farms on the fringe of Concord, while the other three subdivided their father's homeplace by the village. Through their lives these three brothers purchased additional parcels, enlarging their small inheritances into more substantial farms. This third generation also saw the emergence of the family trades of black-smithing and locksmithing that would allow several generations of Meriams to maintain themselves on small holdings.

The two sons who were given John Meriam's Second Division land at Shawshine Corner were Nathaniel (in 1696), and Samuel (in 1702).[33] Samuel went on to become a deacon and selectman in the new parish and town of Bedford, incorporated in 1729 from parts of Concord and Billerica.[34]

The three sons who settled in the immediate vicinity of Meriam's Corner were John, Ebenezer, and Joseph Meriam.[35] Eldest son John married in 1691, Ebenezer and Joseph not until 1705.[36] It appears that John built a new house across Billerica Way about the time of his marriage; Joseph built a dwelling just north of his father to house his wife, Dorothy, and their rapidly growing family; while Ebenezer and his wife, Elizabeth, stayed on with his parents in the ancestral home.[37] The 1710 map shows how the family plowlands and meadows about Meriam's Corners were split among the three brothers (plate 5A). Clearly, equity and family feeling confounded consolidation here—if consolidation was even an animating desire for these people. The Meriams chose to cluster their houses and scramble their land, rather than blocking it into three neat homesteads. Joseph and Ebenezer remained closely intertwined, often buying and selling property together.[38] It is possible that until the need arose for each to accumulate more land to divide among his own offspring, the two brothers did not even formally partition much of their property but held it in common.[39]

The 1735 map gives a more complete picture of the landholdings of the three Meriam brothers at the height of their careers in Concord (fig. 6.2, table 6.1). By this time parents John and Mary Meriam had died, and Ebenezer Meriam had purchased the homestead of one Jacob Taylor and moved a few doors west of the ancestral hearth.[40] This property added still more pease porridge to the Meriam's pot. Here we see thirty-seven parcels (the largest 17 acres) distributed among three men and spread over three or four miles of countryside. In addition to their homelands, the brothers owned both pasture and tillage lots within the Great Field, several mowing lots in the Great Meadow, more meadow and swamp in Virginia and in Elm Brook Meadow, upland and swamp at the Suburbs, woodland at the Chestnut Field, and a little piece of the Ox Pasture. Many of these lots adjoined one another. John and Joseph also owned rights in the New Grant, and Joseph held the family's hereditary 2 acres in the Cedar Swamp.[41]

How did these men feel about this commingled way of holding and working land? Did they regard it as a comfortable web of convenience, an awkward nuisance, or a little of both? About 1735, at the age of seventy-one, after more than forty years of farming at the corner alongside his father and brothers, Lt. John Meriam Jr. sold his 54 acres in Concord and moved to Littleton.[42] But the bulk of John's land stayed firmly in the Meriam family, absorbed by brother Joseph, who passed it along to his own son Samuel.[43] It remained an integral part of a working family unit. Whatever prompted John Meriam to remove to Littleton did not infect Joseph and Ebenezer. As they passed land to the

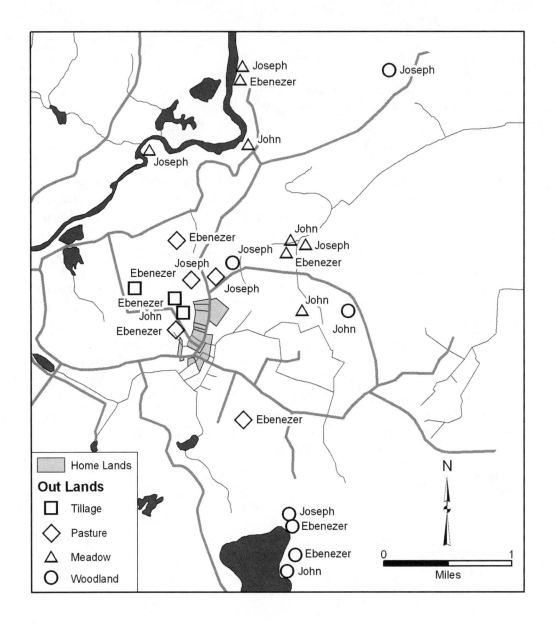

FIGURE 6.2.

Meriam landholdings, 1735. Between them, the three Meriam brothers owned some three
dozen parcels of land scattered over three miles (see table 6.1). This was no longer
a land bank for their offspring—it was their working landscape.

TABLE 6.1   Meriam: Third Generation Landholdings

Lt. John Meriam, Jr.—1737

	No. of Acres	Description
1.	6	houselot with house, barn, bounded E by highway to Billerica
2.	2	plowland bounded E on said highway, S on Ebenezer Meriam
3.	8	pasture bounded W on said highway, N on Joseph Meriam
4.	5	plowland in the plain
5.	6	meadow with upland adjoining, bounded S on Ebenezer Meriam
6.	16	woodland near Chestnut Field bounded W on Flint's Pond
7.	4	meadow in Island Meadow
8.	3	river meadow
9.	4	swamp and upland at Suburbs bounded on Joseph Meriam
10.		rights in New Grant

Ebenezer Meriam, Sr. ~1733

11.	2	houselot w/ house on N side road
12.	0.25	barnlot w/ barn S of road
13.	2	home meadow
14.	6.5	old houselot and meadow
15.	2	plowland in the plain
16.	2	plowland and swamp in polden in great field
17.	1.5	plowland below John Meriam Jr
18.	3	plowland bounded E on Billerica Way
19.	4	pastureland in ox pasture
20.	8	pastureland in Cranefield
21.	1.5	pasture called Horse Pasture
22.	3	meadow in Virginia Meadow
23.	5	river meadow in Bedford
24.	4	meadow called Potter's meadow (may be same as #7)
25.	5	woodland in Chestnut Field
26.	5.5	woodland in Chestnut Field

Joseph Meriam ~1735

27.	1.5	houselot, with house and barn
28.	3	orchard bounded E on Billerica Road
29.	7.5	plowland bounded E on Bilerica Road
30.	17	pasture and swamp in "house swamp"
31.	9	pasture bounded W on Billerica Road
32.	3	meadow and upland bounded N on Bay Road
33.	16	wood and meadow in Virginia Meadow—part of Elm Brook Meadow
34.	3	river meadow in the Holt
35.	5	river meadow in Bedford
36.	11	woodland in Chestnut Field
37.	15	woodland and meadow in Virginia Swamp
38.	2	cedar swamp in Bedford
39.		land in Acton

Note: The date at which landholdings could most accurately be determined varied slightly for each brother.

next generation of Meriams they proceeded to replicate, even intensify, the pattern of interlocking family holdings.

Beyond these intricate landholdings within their own family, the Meriams were also closely enmeshed with their neighbors in the Great Field. All three brothers had substantial interests in the field and did their part to maintain the fence and the gate by Lt. John Meriam's house. Ebenezer Meriam served on the fence allotment committee in 1731 and again in 1757 (at age eighty-two), as clerk of the "Proprietors of the Common or great fields and River meadows" in 1738, and as moderator of the proprietors' meeting in 1756.[44] For this branch of the Meriam family, husbandry during the first half of the eighteenth century was becoming if anything more complex and communal, rather than less—and it was about to become more intricate yet.

The Fourth Generation

During the 1740s, Joseph and Ebenezer Meriam completed the task of providing land to their children of the fourth generation—or, in Ebenezer's case, thought they had. The two brothers succeeded in settling four of their six sons about them at the corner. The family accomplished this by acquiring more land contiguous to their own (including their departed brother John's land) and by further subdividing what they already owned. The 1749 map shows the distribution of the Meriam homelands among these four new households (plate 5A).

This map would be even more complicated—in fact, incomprehensible to any but a Meriam—without treating the two Ebenezer Meriams as a single household. That is apparently how they lived. In 1743, Ebenezer Sr. divided his land with his son Ebenezer Jr., who then moved back home with his father.[45] They had been living separately, and both had recently lost their wives (and thus Ebenezer Jr. his mother as well). Father and son both remarried on December 9, 1742, to Elizabeth and Sarah Davis of Bedford.[46] Ebenezer Meriam Sr. and Jr.'s subsequent land dealings grew as jumbled as their domestic arrangements must have been, but it appears that between them they owned little more than fifty acres of land spread over three towns (Concord, Lincoln, and Bedford) in at least fifteen parcels, the largest of which was eight acres. In addition to the homelot on the Bay Road and the small barn lot and orchard across the road from it, they shared two lots of plowland, three lots of pasture, five meadow lots, and three woodlots. For a time, father and son were assessed for one ox apiece—even their team was neatly split.[47]

Ebenezer Meriam Jr. died young (at 34) in 1751, while his father lived on to 102 years of age, surviving his two grown sons by a quarter century and burying all five of his wives.[48] Ebenezer Meriam Sr. never became a great patriarch in the sense of seeing his offspring settled about him on substantial farms of their own, but he does seem to have been a quintessential eighteenth-century family man—and one who realized the inherent sorrows of an extraordinarily long life in full measure.[49] He farmed successively in collaboration with his father, his brother, and his son, cared for his father and mother in their old age, married five times himself, and ended his days being cared for by his granddaughter Sarah and her husband, a cooper named John Champney.[50] Through his life he shared ownership of property with four generations of Meriams. That is the way family ties and farming were intertwined in eighteenth-century Concord.

Ebenezer's brother Joseph Meriam, by contrast, did succeed in establishing three of his four sons at the corner.[51] Joseph's eldest son, Joseph Meriam Jr., moved to Grafton. His next son, Samuel Meriam, purchased uncle John's house and six acres of tillage land across Bedford Road in 1737, although he did not marry until 1741.[52] Probably his father helped Samuel acquire land before he was ready to set up housekeeping because the nearby family holding was going begging. Third son Nathan Meriam was given the greatest part of his father's homestead in 1747 and added to it with several substantial purchases about the same time. In making this gift, Joseph Meriam reserved half the dwelling and garden to himself and his wife for their natural lives, together with a few acres of plowland and meadow "to dispose of in case we should be brought to want." Josiah Meriam, Joseph's youngest son, was also established in life by his father in 1747, a year after his marriage. Joseph provided Josiah with the ancestral family house (just across the garden) at the corner of Bedford Road and the Bay Road, a modest selection of land for husbandry, and the tools of the locksmith trade.[53]

The resulting crazy quilt appears in the 1749 map. Once again, we see no effort on the Meriams' part to create compact individual holdings—if anything, just the opposite. Each brother's share of homeland was utterly disjointed, with uncle Ebenezer's lots interspersed for good measure. If this was an economic hardship there is no evidence of it, and we can only conclude that this was the way they wanted it. The new generation of Meriam brothers continued to acquire additional outlands throughout their lives, and by all appearances prospered. Nathan Meriam built his portion up to at least ninety-five acres and carried enough weight in Concord to serve five years as selectman during the War for Independence. All three brothers had smithing skills, and this must have been crucial to the family's modest success. Samuel Meriam was a blacksmith. He may have worked in the same shop as his father and his brother Josiah, who carried on the locksmith trade.[54] Josiah received the smallest landholding and must have concentrated primarily on his trade—the gunlock business was grimly brisk during Josiah Meriam's prime.[55] Nathan Meriam was primarily a yeoman farmer, but he apparently also possessed some of the family skills at working metal and may have helped in that business because he was given first refusal of his father's tools and shop should his brother Josiah decide to sell.[56]

The establishment of the fourth generation of Meriams in Concord in the 1740s took place at nearly the end of the period during which the landscape of the town was filling with farms. The Meriams demonstrate quite vividly that while (as we shall see) some of the outer parts of Concord were being settled with more consolidated farms, the older pattern of small, intermixed holdings was alive and well nearer the village. In fact, partible inheritance and the tightness of family ties, together with the practice of trades that diversified the family economic base and made very small holdings viable, was busily creating a *more* fragmented landscape as the generations passed. This was an echo of the process that had given birth to common field farming in the first place half a millennium earlier, back in England. By the time of his grandsons, John Meriam's original twenty-acre strip of plowland across the road from his house had been subdivided into six separate pieces. This, not consolidation, was the prevailing pattern in the central part of Concord throughout the colonial period.

The Brickiln Field

On the high ground a quarter mile east of Meriam's Corner lay the Minot farm and beyond it the Brickiln Field. Here the opposing forces of consolidation and fragmentation played themselves out in contrasting ways. The Minot farm belonged to a branch of one of Concord's wealthiest families. It remained integral and grew steadily throughout the colonial period. The Brickiln Field, by contrast, began as a general field with many small plowlands which were eventually gathered into three homesteads. Consolidation was never complete, however, and by the end of the colonial period the old Brickiln Field neighborhood had acquired many of the same features of intermixed holdings and intermarried families as Meriam's Corner.

The Minot farm originally belonged to Capt. Timothy Wheeler, one of Concord's most prominent settlers. The land passed to his son-in-law Capt. James Minot, who in turn became one of Concord's leading citizens: a justice of the peace, physician, and the owner of the corn mill in the center.[57] The farm at the east end of town by the Brickiln Field was tenanted from 1684 (if not earlier) until as late as 1730.[58] James Minot granted the place to his youngest son, Samuel, in 1726, along with a 40-acre woodlot lying on the hill east of the Ox Pasture.[59] Deacon Samuel Minot added steadily to his property, both adjacent to the home place and as far afield as Townsend, Massachusetts. He acquired 120 acres of grazing land in Princeton in common with his neighbor, Nathan Meriam.[60] He eventually passed the home farm to his son George Minot and supplied his other heirs with land elsewhere. With sufficient family wealth, consolidation overcame fragmentation. The ample Minot farm remained intact, and in fact grew, through the colonial period and into the nineteenth century.

The neighboring Brickiln Field was another story. At the end of the Second Division, the rolling landscape stretching north from the Ox Pasture across the Brickiln Field and over Island Meadow to Brickiln Island was a patchwork of forty or more small lots belonging to a dozen or more village men.[61] The Brickiln Field was still being managed as a general field at least as late as 1672, when the selectmen were instructed to look to its fencing.[62] No homesteads were settled there by the second generation (plate 5B).

By the early eighteenth century the picture had changed dramatically. Three families had accumulated the majority of the land in the neighborhood, forming larger (but still discontinuous) holdings. By 1710 two members of the third generation, Samuel Fletcher[63] and Nathaniel Stow,[64] had already set up homesteads at the old Brickiln Field, and John Jones[65] would be arriving shortly. These three families would remain on their homesteads for another century, through several more generations. They all worked at augmenting and consolidating their farms but none completely succeeded—they remained, for better or worse, in each others' way.

The Fletcher family, over the generations, achieved a modest increase in their holdings in the Brickiln Field and across the road in the Ox Pasture. Sgt. Nathaniel Stow next door certainly was able to accumulate land, but it was then split up among four of his six sons. In the end only one son, Joseph, stayed in Concord—but on a much-diminished holding.[66] John Jones was able to acquire some of the pieces from the breakup of his neighbor Stow's estate, but his farm still remained disjointed into the following generations. At the Brickiln Field we can see the drive for consolidation being thwarted by partible inheritance and by competition among neighbors for the same pieces of land, seesawing back and forth. But there were other factors. As will be detailed in the next chapter, the resources that each family required were themselves scattered over the undulating countryside: woodland at Pine Hill, pasture and mowing in the Ox Pasture, prime plowland in the old Brickiln Field behind the houselots, more mowing in Island Meadow, more pasture and woodland at Brickiln Island. All these farms had parcels distributed among these distinct parts of the local landscape.

We may presume that these families frequently exchanged work across their intermingled property. Thus they were all forever journeying to all these different bits of the countryside on their entangled business. The efficiencies of individual farm consolidation were not so compelling within a diversified system of husbandry that benefited from cooperation among neighbors and kin. By the end of the period a widowed Joseph Stow had remarried John Jones's daughter Olive, and it is possible that in practice these two neighboring farms became yoked together very closely indeed.[67] In short, from an unrelated starting point, the three families at the Brickiln Field converged on a pattern of intermixed landholding quite similar to the tight-knit Meriams back at the corner.

Brooks

With the Brooks family we see a pattern of land inheritance like that of the Meriams, only on a larger scale. Here developed a neighborhood composed of the commingled farms of brothers and cousins who possessed significantly more robust parcels than the Meriams. Many Brooks fathers succeeded at acquiring land from neighbors to accommodate their numerous offspring, and so Brooks farms spread over the countryside like puckerbrush.[68] As with the Meriams, the prosperity of the Brooks family revolved in part around a trade, in this case a tannery.

The Second Generation

Capt. Thomas Brooks came to Concord from Watertown (and before that from London) and made his home south of the millpond.[69] In the First and Second Divisions he acquired a nearly solid three hundred acres stretching south and east from Elm Book Meadow to the top of Elm Brook Hill, together with a far-flung assortment of other meadows and upland.[70] By 1665 his three sons Joshua, Caleb, and Gershom had settled on homesteads along the Bay Road overlooking Elm Brook Meadow, as shown on the 1675 Brooks map (plate 6A).[71]

For the next decade the three Brooks brothers lived side by side at Elm Brook, although not always in perfect love and unity as becometh brethren. Caleb Brooks was unhappy about the disparity between his inheritance and that of his brother Joshua. In 1673, "after long and serious debate and by advice of friends," they agreed that Joshua was to quietly enjoy the lands given him by his father as recorded in his deed and in the town book, and Caleb was to receive £4 from Joshua to end the dispute.[72] Whether this settlement left Caleb satisfied or still stewing we cannot be certain, but a few years later, in 1679, he sold all his Concord holdings to brother Joshua and moved to Medford.[73] Thus by the late seventeenth century, Thomas Brooks's legacy at Elm Brook had been concentrated on two of his sons, Joshua and Gershom—and they were adding more land.

The Third Generation

With the passing of land to the third generation, the Brooks farms at Elm Brook Meadow took the shapes they would hold—with some slight rearrangement, both from fragmentation and consolidation—through the colonial period and well into the nineteenth century. As appears on the 1710 map, what had been three (and then two) homesteads were split into five, belonging to one son of Gershom and four sons of Joshua (plate 6A). The departure of one brother made room for the remaining brothers to settle more offspring on their patrimony.

Gershom Brooks died in 1686 at age fifty-two, and his farm passed intact to eldest son Daniel Brooks Jr. once the boy came of age.[74] A few years later Joshua Brooks passed on the bulk of his portion of the family upland and meadow at Elm Brook, including what he had acquired from brother Caleb. On November 11, 1695, he split the remaining land between four of his sons, Noah, Daniel, Joseph, and Job.[75] Joshua died shortly thereafter, in 1696.

The 1710 map shows the resulting homesteads of the four brothers and their cousin. Daniel Brooks Sr.'s farm had uncle Caleb Brooks's old homeplace as its nucleus. By 1710 the farm included a stretch of upland on both sides of the Bay Road and attached mowing in Elm Brook Meadow, along with detached parcels in the Pine Plain and beyond Chestnut Field, upland and meadow at the Suburbs, and an acre in the Cedar Swamp.[76] Joseph Brooks received a slice of upland and meadow to the east of his brother Daniel.

He eventually chose to make his home not on the Bay Road, however, but on a parcel farther south along the way leading to the Chestnut Field—a way that came to be called Brooks Road. Joseph was also granted outlands and an acre in the Cedar Swamp.[77] Noah Brooks's portion centered on the upland south of the Bay Road. Noah also received meadow lying along the course of Elm Brook north of the road, an assortment of outlands similar to those of his brothers, and (of course) his acre in the Cedar Swamp.[78]

Finally, Joshua Brooks left his homeplace, together with substantial outlands north of the Bay Road, to his son Job Brooks. Probably this is where Joshua planned to live his remaining seasons, but he died the next year, and young Job died the year after that. Although not an old man (he was about sixty-six) Joshua died in the fullness of his years, his sons well established around him; but Job died with his life just beginning. Job had not yet married, and (aside from the property his father had just provided) his estate amounted to only two cows, a bed and two chests, five chairs, thirteen beaver skins, a "sabel and a forme," his clothes, a razor, and a belt.[79] Job's siblings quitclaimed the homestead to their next brother Hugh Brooks, two years younger than Job.[80]

Such were the ample portions with which the third generation of the Brooks family set about providing a comfortable subsistence for themselves and, if they could, furnishing their offspring with the means for the same. By then they had the family tannery to assist them.[81] It is likely Joshua Brooks founded the tannery and passed it on to his son Noah in an unrecorded deed—it was located adjacent to Joshua's house.[82] In any event, the tannery passed down through Noah Brooks's line for five more generations, and members of other branches of the family were also involved in the business. The consistent prosperity of those who owned the tannery during the colonial period speaks for itself.

Like the Meriams, the Brooks family chose not to partition their land into solid blocks and disperse their houses, but to keep their homes within sight of one another along a quarter mile of road and make a hash of their outlands. The exception was Joseph Brooks. It appears that Joseph had a homelot alongside his brothers granted him by his father in 1695, but when he finally came to marry in 1704, he decided instead to sell a piece of this to his brother Hugh and to live off by himself. We cannot tell what motivated Joseph Brooks. We can only notice that, coincidentally or not, things never did seem to go well for him or his progeny.

The Fourth Generation

Three things are evident from the map of Brooks family land in 1749, which mostly reflects members of the fourth generation in their prime (plate 6A). First, with further subdivision the landholdings had grown even more complicated. Second, the family had succeeded in accumulating surrounding land to help accommodate its growth (extending far to the south and east beyond what is shown on the map). Third, some branches of the Brooks family had been successful at holding and bequeathing land, while others grew pinched or vanished entirely—often selling their Concord land back to Brooks kin when they left.[83]

Daniel Brooks Jr. was the last of the Gershom Brooks line at Elm Brook. In 1746, at age seventy, Daniel moved to Westford.[84] His cousin Ensign Daniel Brooks Sr., who lived next door, acquired land throughout his life and seems to have been able to give all his children a good start. In 1725, Daniel provided his son Job Brooks Sr. with a twenty-acre houselot at the south end of his upland, along with two pieces of meadow and a small woodlot. Job immediately began to purchase more land from neighbors.[85] Next, Daniel split his homeplace with his son Samuel, and the remainder passed upon his death.[86] Samuel added meadowland in Elm Brook Meadow to the north and upland and meadow at the Suburbs a mile to the east, making this a farm rich in grass. One farm had become two farms, and the placement of Job's house just north of his uncle Joseph's added to a growing cluster of Brooks homesteads there.

Hugh Brooks left the old Joshua Brooks homeplace north of the Bay Road undivided to his younger son Job Jr. in 1740.[87] The property included the houselot and substantial meadow, upland, and woodland in the East Quarter, fifty acres in the North Quarter, rights in land in the New Grant (by then the town of Acton), and rights in the Cedar Swamp in Bedford.[88] Job Brooks Jr., who worked as a currier in his cousin Joshua's tannery next door, continued accumulating land—including, one by one, the broken shards of his uncle Joseph Brooks's lot to the west.[89]

Noah Brooks stands as an archetypal Concord patriarch. Noah was a tanner as well as a farmer, and he prospered: like his father, Joshua, he acquired enough land to see four sons settled in Concord.[90] His eldest son, Joshua Brooks, followed in the tanning trade, and upon his marriage in 1713 was provided with a houselot next door to Noah's own, with adjoining plowland and other pieces of a disjointed farm.[91] Joshua added steadily to these forty-six acres and became a wealthy man.[92] His property lay interspersed like a checkerboard with that of his uncles and cousins and particularly his younger brother Thomas, who was deeded the homeplace and substantial outlands in 1726.[93]

Noah's deed of gift to Thomas was another classic of New England yeomanry, whereby the elders of the community retained tight control of their children well into adult life. In exchange for his generous endowment of improved property, this youngest son agreed to "help his father and mother with husbandry and other necessaries to a comfortable subsistence during their natural lives." In the event of Noah's death, Thomas was to be allowed "to sell land if necessary to provide for his mother while she is a widow," but only "with the consent of the pastor of the church in Concord."[94] But selling land was the last thing on Thomas's mind. His own children were arriving thick and fast, and he would soon face the problem of providing them with farms and dowries, like his father before him.

Noah provided his two middle sons with yet more nearby land. Ebenezer and Benjamin Brooks joined other third- and fourth-generation Concordians who were pushing up onto the stony, heavily wooded glacial till uplands at the eastern fringe of town, carving out homesteads all the way to the old Cambridge line. Ebenezer Brooks moved to a farm on rocky Elm Brook Hill by 1716.[95] Benjamin Brooks was provided land lying south of uncle Joseph Brooks with a dwelling already on it, adding to the Brooks cluster there. The farm ran mostly eastward, over the hills toward Flint's Great Meadow.[96]

Noah Brooks lived with his son Thomas until he died in 1739, and his wife, Dorothy, lived another thirteen years, to age ninety. We may guess that they subsisted comfortably and died happy in a virtuous posterity. Besides seeing their four sons settled around them, they saw their three daughters married to solid Concord men.[97] But one child of Noah and Dorothy Brooks ultimately did leave Concord. Their second son, Ebenezer Brooks, was a worsted weaver by trade, dwelling on Elm Brook Hill half a mile east of his parents' house. Ebenezer added scraps of property until his ungainly farm stretched over the rocky surface of the earth for a mile and a half; long, narrow arms reaching down into Elm Brook Meadow to the west and into the swamps near Flint's Great Meadow to the southeast (fig. 6.3).[98] But the weaver lagged behind his brothers in the increase in value of his property.[99] Through the 1730s he sold pieces of his land. In 1742, three years after his father's death, Ebenezer Brooks finally sold his houselot to a cooper named Timothy Cook and moved to Grafton.[100]

When Noah Brooks constrained his son Thomas from too easily selling his patrimony, granting the minister veto power, he may have had in mind the downward course of his brother Joseph Brooks. Joseph had six sons to launch, along with three daughters. But all the while his brothers were buying land to prepare for this reckoning, Joseph was selling—often to his siblings.[101] By the 1730s, as he began passing land to his sons, Joseph Brooks styled himself a mere husbandman. Many of these gifts were painfully small—they were meant to be liquidated, not held.[102] The sons scattered to the hills, to Acton, Harvard, Bolton, and to Temple, New Hampshire. Joseph settled the home place on his youngest son, James, in 1734, adding a small augmentation from his dwindling store of land in 1743.[103] In 1751, five years after his father's death, James Brooks sold the shrunken homestead to his brother Amos, who returned from Harvard to take over care of the home place.[104]

On the whole, the Brooks fathers of the third generation did a remarkable job of settling still more sons of the fourth generation within a growing complex of intermixed family lands in their corner of Concord. But, as in previous generations, about the fringes of this family continuity we can see a few brothers leaving town, which helped provide land for those who stayed behind. Had it not been for this, the land would have grown even more densely settled and intricately subdivided than it was already becoming.

Hartwell

The uplands where Ebenezer Brooks made his home about 1715 stretch from Elm Brook Hill for a mile eastward within Concord and continue into neighboring Cambridge Farms, which became the town of Lexington in 1713. Ebenezer Brooks was by no means

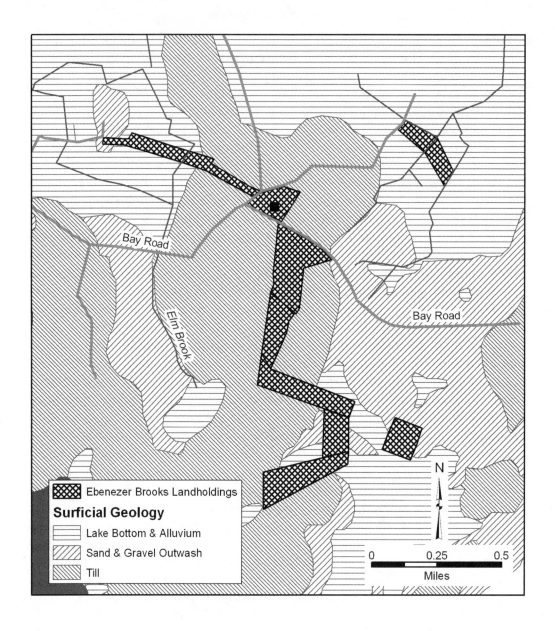

FIGURE 6.3.
Ebenezer Brooks landholdings. A marginal farm encompassing rocky till, scraps of meadow, and little else. A weaver by trade, Ebenezer Brooks did not achieve the prosperity of his brothers and moved west to Grafton in 1742.

the first to live there: these outskirts of Concord had been settled primarily by the third generation, around the turn of the century. Most of those who settled these heights were better able than Brooks to pull together substantial, contiguous farms from pieces of Second Division grants, both inherited and purchased. There was more elbow room this far from the meetinghouse, which may have appealed to those with a desire to assemble a consolidated farm. There was also less opportunity in the generations remaining before the end of the colonial era for fragmentation to set in. But even when these farms were divided among heirs we do not see the deliberate intermixing of holdings found among the Meriams or even among the Brookses. This suggests a stronger preference for keeping farms intact.

Second Generation

The first to dwell along the Bay Road east of Elm Brook appears to have been Sgt. Richard Rice. Rice was an early Concord settler who chose to receive a solid block of nearly two hundred acres of Second Division adjoining his First Division lot in Rocky Meadow. It seems safe to assume that Rice and his family had moved to the Cambridge line by 1666, because he left only a trace of his original First Division holdings in the form of "a little orchard" by the millpond, where he must have returned every fall to collect his apples.[105] During the last decades of the seventeenth century several of Richard Rice's children were living on portions of his Second Division land.[106] No other first- or second-generation settlers ventured out to this section of Concord. The land still lay largely in unimproved Second Division lots belonging to village men such as William Hartwell until nearly the end of the seventeenth century (plate 6B).

The Third Generation

The two men who assembled large farms on these uplands were Samuel Hartwell and Benjamin Whittemore, both of whom arrived in the 1690s. Lt. Benjamin Whittemore came from Charlestown in 1692, married Joshua Brooks's daughter Esther, and bought a "messuage or small tenement" on Elm Brook Hill at the corner of the Bay Road and Rocky Meadow Way.[107] Whittemore soon moved to better land down the road (the rocky place at Rocky Meadow Way eventually became the home of his nephew, Ebenezer Brooks). He purchased a larger houselot from Peter Rice in 1697 and steadily added adjoining land on both sides of the road. Whittemore eventually acquired 100 acres or so north of the Bay Road stretching from his houselot back to Rocky Meadow and some 150 acres south of the road, extending to Flint's Great Meadow.[108] A lowly weaver in his early years, Benjamin Whittemore went on to hold high office in Concord and to style himself a gentleman by the end of his days.

Samuel Hartwell Jr. grew up in the village, the eldest son of Samuel Hartwell and grandson of William Hartwell. He moved out to family Second Division land, probably about the time of his marriage in 1692.[109] In January 1694, Hartwell bought of Richard Rice eighteen acres "for the most part sowed in."[110] This field of prime arable soil south of the Bay Road at the old Cambridge line remained the family's principal plowland for generations. But Samuel Hartwell made his home half a mile west, building the first of two neighboring houses that would belong to the Hartwell family for the next century and a half.

Hartwell put together a large farm that stretched north from the Bay Road to Rocky Meadow and westward to the Suburbs. He and his neighbor Benjamin Whittemore frequently sold land to one another, seemingly very willing to accommodate each other's ambitions as they took shape. One parcel of plowland even moved from one family to the other and then back again, thirty-four years later.[111] By the time he began passing land to his son Ephraim, Samuel Hartwell had already consolidated a farm of over one hundred acres.

The Fourth Generation

With the fourth generation on Elm Brook Hill we see ongoing efforts to consolidate farms, and a determination that the land be passed along intact—or at least in large, solid divisions, not piecemeal. When Benjamin Whittemore died in 1735, he had already settled his two sons Nathaniel and Benjamin Jr., giving them each forty acres to get started.[112] The regular way in which the division was completed can be seen on the 1749 map (plate 6B).[113] Nathaniel Whittemore continued to buy and sell land to solidify his farm, including purchasing several parcels from the departing Ebenezer Brooks. In 1758, at age sixty, Nathaniel sold his substantial property to Elizabeth and William Dodge and moved to a farm in Harvard.[114]

Samuel Hartwell was also firm about not spoiling the whole. He settled his entire Concord farm on his youngest son, Ephraim—three older sons had moved to other towns.[115] In 1733, Samuel gave twenty-seven-year-old cordwainer Ephraim Hartwell 40 acres of his home farm.[116] It was enough for Ephraim to build a new house and enter the ranks of the yeomanry. He continued buying and selling pieces of land as his father had done, expanding and consolidating the farm. In 1745, Samuel died and Ephraim inherited the remainder of his land.[117] By 1749 the whole encompassed at least 141 acres (all but a small piece in Elm Brook Meadow are shown on plate 6B), and Ephraim Hartwell was a gentleman.

By 1750, Concord had been settled to its borders and filled with farms. The process of pushing out three to four miles took three to four generations of Concord yeomen about one century to complete. The Lexington line along the well-traveled Bay Road in the East Quarter was reached a generation sooner than the Sudbury line, lying over the river and beyond remote Fair Haven Meadow in the South Quarter. As in the East Quarter, both dispersed and consolidated patterns of farming were in evidence on this side of town. Near the village Deacon Luke Potter had passed along a widely dispersed, common field–style holding to his youngest son, Judah, in 1695, who passed it along only slightly more consolidated to his youngest son, Samuel, in 1731.[118] At distant Fair Haven, by contrast, the third and fourth generations of Hosmers and Wheelers settled their ancestor's Second Division lots in more solid blocks, resembling those of Samuel Hartwell and Benjamin Whittemore.[119]

There was a complex pattern to the way land was settled and farms took shape over these first four generations in Concord. The older ways were concentrated and intensified nearest to the village. About the original common fields landholdings remained dispersed, and more husbandmen were being incorporated at each generation. On Second Division lands farther out, farms often began in a more consolidated form—but, as the Brooks family shows, they tended to grow more jumbled as the generations passed. Many contradictory forces were at work, but fragmentation had the upper hand. Some yeomen worked to acquire contiguous holdings nearer home, but at the end of their lives they often took pains to divide land among their sons in an equitable way. In some places, for example, the Brickiln Field, it appears that persistent efforts to consolidate farms were just as persistently frustrated. But in other cases—most notably among the Meriams— it is equally clear that farm holdings were being *deliberately* scattered among kin, even when they could have been divided in a more compact manner. Even at the outskirts of town, where the farms were youngest and the owners displayed the strongest determination to keep them intact, subdivision was beginning to appear. The driving social imperative behind these patterns—and perhaps it was economic as well—was to cluster related households as closely together as possible. There were also overriding ecological reasons for the way the diverse pieces of a working farm *had to be* distributed across the landscape (see chapter 7).

From the very beginnings of settlement in Concord, the departure of some offspring to take up new holdings worked closely in tandem with increased density and fragmentation in the older neighborhoods. There was no apparent contradiction between these two movements, one outward and one inward—both were ways of accommodating a rapidly growing population, and both operated within the same families. The fecundity of colonial New England families is legendary, almost all women marrying young and bearing large families. Gloria Main has called the New England town a "land-gobbling engine of growth . . . fueled by an enormously successful family regime that produced hordes of healthy, long-lived, hard-working children."[120] There was more land to be had—at the expense of the Indians, to be sure—but having it required moving to it. In every family, in every generation, we see some sons and daughters going to new farms, either on Second Division land within Concord or increasingly to new towns beyond Concord. The more distant corners of Concord, once settled, themselves joined with corners of neighboring communities to form new towns: Bedford in 1729, Lincoln in 1754, Carlisle in 1780 (after several decades of trying).[121] Some of those who took up distant opportunities were not just starting out in life but already owned substantial farms in Concord. These they sold to neighbors or kin, enabling other family members to settle close to home—more family members in each generation. Even as some Concord natives left home, the number who stayed behind inhabiting the mother town increased.

Concord was not simply filling out to its borders, it was filling in, too. Up to some point, Concord was able to expand to accommodate more people on the same land—either as farmers themselves or as part of a diversifying local economy. How was this possible? What were its limits? Were Concord's forests and soils being steadily depleted, necessitating either constant outward movement or an eventual, painful decline in productivity? Were large families necessary to Concord's agrarian economy, leading toward an inevitable demographic crunch? Were the economic aspirations and opportunities of these yeomen and goodwives changing as well? We have seen the interesting way in which Concord's farms took shape upon the landscape. Let us now look more closely at how those lands were farmed.

CHAPTER 7

The Ecological Structure of Colonial Farming

Husbandry and other necessaries to a comfortable subsistence.

BY the middle of the eighteenth century, the yeomen of Concord had adapted the mixed husbandry system of their English ancestors to the soils and climate of New England. This was no longer a frontier community struggling with an unfamiliar environment. The frame of Concord's husbandry had been raised, sheathed, and shingled, and stood out clearly in the way the tax collector assessed landholdings and on the face of the land itself. Farms grew fragmented or consolidated (sometimes by turns) with the pull of competing social forces, but at bottom conformed to a consistent ecological pattern. Agriculture was given order by an underlying mosaic of rocky, sandy, and moist soils suitable for distinct purposes, and by the way these elements were organized into a working system of husbandry.

A particularly thorough enumeration of farm property, livestock, and products was taken by the Concord tax assessor in 1749.[1] Categories of land in this valuation included tillage, orchard, fresh meadow (coastal towns also reported salt marsh), English mowing, and pasture.[2] Woodland, although never considered improved land, was also necessary to husbandry and was added to the valuation list in 1784. These productive elements required particular soils and were by no means perfectly interchangeable on the landscape. The challenge to the husbandman, revisited generation after generation, was to possess workable proportions of tillage, orchard, mowing, pasture, and woodland, on the proper ground for each, in the most convenient arrangement (fig. 7.1).

Given these ecological requirements, a man's plowlands, meadows, and woodlots frequently lay mingled with those of his neighbors—with whom he often changed work as well. Every farmer knew the character of the soil field by field, sometimes stone by stone,

FIGURE 7.1.
The working landscape of Concord: dairy cows on a rocky hillside, an orchard on the knoll
below the road, and the river meadows in benign spring flood. Herbert Gleason, 1901.

on every farm for miles around—by hearsay if not by hard experience. The formal commons system had all but disappeared from Concord by the late colonial period, but in practice husbandry remained a collaborative undertaking. Farming involved clusters of farm families making comprehensive use of the land within their immediate neighborhood and ultimately—especially in the management of water, which remained a common resource—across the community as a whole.

Each kind of land had a special role in the agricultural economy. In practice, of course, these categories overlapped—for example, pastureland was occasionally plowed for tillage, orchards were often mowed for hay, meadows were grazed after mowing, and woodlands were sometimes foraged by livestock. Nevertheless, each piece of land had a primary function which in many cases remained fixed for long periods of time—years, decades, even generations. These elements were connected by the movement of livestock, the transfer of nutrients, and the flow of water. They had to be balanced not only in their places on the landscape, but also in their seasonal demands for labor. They supplied a wide range of complementary resources to the farm and household economy and to the local exchange economy of the community. Some of them also yielded commodities to be marketed beyond Concord. This all went to make up a complex, fluid system of mixed husbandry, but one that was a long way from an extensive system of shifting cultivation. By any reasonable definition, this was intensive farming.

The Elements of Husbandry

The Land Use map offers an overview of the landscape of part of Concord's East Quarter, along the Bay Road, in 1749 (plate 7). Most types of land use were tightly clustered. Tillage lots were concentrated upon patches of prime plowland, often divided among several owners. Orchards were strung along the road near each homestead. Meadow mowing lots collected (naturally) in wetlands and along brooks and were much subdivided. Pastureland was the most widespread element in the landscape, reflecting the broad range in the kinds of land that were grazed. In general, pastures overlapped with the tilled land and ran out from there onto rougher ground. Woodlands were located mostly at the back fringes, on steep, rocky slopes and in swamps.[3]

How were individual farms laid out across this mosaic of soils and uses? Few Concord farms were all in one piece; most were spread far and wide. But some were more fragmented than others, and there was a pattern to that as well. As one traveled away from the village through four neighborhoods along the road—from Meriam's Corner past the old Brickiln Field, up Brooks Hill to the Hartwell Tavern—the farms grew progressively more consolidated. But even farms like Ephraim Hartwell's, whose fields, meadows, pastures, and woodlots were mostly contiguous, often took elongated, rambling shapes, reaching from the farmhouse by the road far back into the countryside. The shape of farms was the outcome of a history of give and take within families and among neighbors.[4] The working components of these farms meshed in operation.

Tillage

Plowland in Concord tended to be tightly clustered for two reasons. The first was the yeoman's continuing preference for growing grain on light, well-drained soils. The second was the convenience of having the bulk of one's tillage close to the barn. Both factors were at work, usually in concert but occasionally at odds, in the neighborhoods along the Bay Road.

The three brothers Meriam, together with the two Meriams Ebenezer, held most of their plowland in intermixed lots in their home field north of Samuel Meriam's house, within the Great Field fence (plate 8A). Although the Meriams did till a few of the other parcels they owned deeper in the Great Field, these more distant holdings were frequently identified as pasture instead.[5] Much the same pattern prevailed half a mile down the road, where the primary plowland still lay within what had been the Brickiln Field, just behind the houselots (plate 8B). George Minot and Samuel Fletcher also cultivated more distant lots in the Great Field, while Joseph Stow had extra plowland on Brickiln Island.[6]

The situation was similar another half mile to the east, on the sandy glacial outwash terrace known as Brooks Hill (plate 8C). Job Brooks Jr.'s plowland lay within his homelot north of the road, while south of the road Joshua Brooks Jr. divided the "Great Field" behind his house with his uncle Thomas Brooks next door. Adjoining this field to the west was the home field of cousin Samuel Brooks before his house, and to the south, the tillage of Samuel's brother Job Brooks Sr. behind *his* house. So again, we see a single patch of arable soil carefully parceled into the home fields of four close relatives and neighbors, with a fifth parcel lying just across the road. Thomas and Joshua possessed smaller tilled fields east and south of their barns, while Samuel held additional plowland at Brickiln Island to the north.[7]

The picture changed among the boulders atop Elm Brook Hill, a mile farther to the east (plate 8D). Here tillable soils were hard to find, and so hard choices had to be made. The small homestead where Ebenezer Brooks had once lived apparently included plowland, but it must have been rough and rocky land to plow. The place appears to have been vacant in 1749, and no wonder.[8] The more prosperous Ephraim Hartwell next door didn't even try to break up the ground on his homestead—at least, not very often. His detached tillage lay half a mile from his door, on a prime piece of well-drained silt loam. Inconvenient, perhaps, yet this parcel of excellent soil was cultivated by the Hartwell family from 1694 until 1903.[9] The same stretch of arable soil supplied plowland to the Benjamin Whittemore family on its south and west ends.[10] The Nathaniel Whittemore farm north of the road was located adjacent to an even smaller island of sandy loam that comprised its home field.[11] In short, some husbandmen on these stony uplands were able to shoehorn their homesteads in close to patches of arable soil, while others had to choose between convenience and quality. Some, like those living at the Ebenezer Brooks homestead, chose (or had no choice but) to run a plow among the glacial boulders at their doorstep. Others, like the Hartwells, made the unusual but prudent decision to cart manure a fair distance in order to plow a better class of tillage.

The precision with which Concord's yeomen placed their plowlands is unmistakable. But why did they prefer these sandy, sometimes droughty, often not particularly rich soils? Because they were easy to plow—the same reason that Native women had preferred many of the same soils for hoe culture. The lowland meadows, rich with muck, may have been potentially more productive but were too wet to allow tillage. Besides, they were already devoted to growing hay, a resource in shorter supply than plowland. The upland tills were often simply too stony to plow. The best of them, with the stones removed, could become productive arable soils. But they were still hard to turn. Even with modern tillage equipment they can be disheartening, as close encounters with boulders continually trip the plow. In the eighteenth century such soils routinely snapped coulters, dulled shares, and sometimes shattered the wooden moldboard altogether.[12] Sledding rocks to the field edges to make walls did not end the problem: the polylithic fecundity of these soils has no bottom. Though the worst stones be removed, cultivation inspires the winter frost to heave a fresh supply up into the plow zone. Farming in New England, first, last, and always, raises rocks. Consequently, Concord husbandmen plowed mostly the easier sands and gravels laid down by glacial meltwater.

Sandy outwash soils were far from perfect—they had limitations of their own. Some were simply too steep or excessively drained to cultivate, and these were also avoided. Almost all suffered to some degree from droughtiness. Hence the principal bread grains these husbandmen settled upon, Indian corn and English rye, were those best suited to warm, dry soils. Another shortcoming of these soils was that they were not overly fertile. Strongly leached and short of organic matter and clay, they tended not only to lose water but to lack nutrients. They may have been easier to plow than the boulder clay, but they were still New England spodosols, hardly the rich molisols of the North American heartland. None of the arable land in Concord was intrinsically high yielding, come to that. In truth, except in some upland outskirts such as Elm Brook Hill, most farmers had more than enough land to plow. The great problem was how to make it grow corn.

The answer to that problem was to manure it, and the husbandmen of Concord did so. Of course they could have adopted the shifting, forest-fallow cultivation style of their Native predecessors, just as they had adopted maize itself. But there is no indication that they did, to any great extent—on the contrary, their home fields were fixed in place. In any event, shifting cultivation alone could never have accommodated the increase of four generations of fruitful Concord husbandmen and their wives.[13] They may have employed long fallow rotations on some of their outlying plowland and pastureland, but this was not the bedrock of their grain husbandry. Some may have continued to manure their corn with alewives, as they had in the early years of the settlement.[14] But this would not have answered for new farms at a distance from the river, especially as the fish runs began to falter in the eighteenth century. Instead, these farmers turned naturally to what their English ancestors had known, time out of mind, and to what still made sense in Concord: they relied on the dung of their stock.

There is documentary evidence of the importance of manure in colonial Concord, although one has to dig to find it. This is surely not because the use of manure was remarkable, but because it was routine. In March 1672, John Hoar exchanged his two-hundred-acre frontier farm in the North Quarter for Edward Wright's houselot and landholdings in the village. As part of this unusual agreement, Hoar was allowed to keep an unspecified number of cattle, sheep, and swine at his old place through the coming season, to be looked after by Wright "as if they were his owne." "Likewise," the agreement continued, "each p^{r}ty is to have w^t doung is made, and lying on ye ground, with ye land as p^{r}t of ye privileges."[15] In other words, in this odd case where there might have been some room for doubt, it was thought worthwhile to specify that the coveted manure belonged to the ground it fell upon, rather than to the stock that dropped it. Things were no different a century later, in 1774. A cordwainer named Jacob Walker sold Job Brooks and five other men stalls in his horse stable in the village, "reserving yearly the dung made in each stable."[16]

Clearly, manure was not lightly cast aside in Concord! Dung forks were common in eighteenth-century estate inventories, and they did not stand in the corner unused. There is no reason to doubt that dressing the corn was routine practice in New England—Howard Russell thought so and cited many accounts of colonial farmers carting dung.[17] From Bedford, New Hampshire, Matthew Patten's diary contains numerous entries about dunging the corn as it was planted—and this was in a recently settled upcountry town.[18] But the most convincing evidence of all comes not from the archives, but from the land itself: the sandy Concord plowlands simply could not have grown corn for so long without manure.

How was the manure delivered to the fields? There lay the heart of the matter, and the reason for drawing the tillage land close to the barn—or, more precisely, for placing the house and barn close to good plowland to begin with, if possible. The normal way, and the hard way, was to cart the winter's accumulation from barn, yard, and pen, then either spread it or lay it within the furrows as the corn was planted. Some farmers preferred better-composted "summer manure" from the previous year for placing directly in the hole with the seed. Carting manure was slow, heavy labor for men and oxen, and one good reason for keeping the haul as short as possible. An easier way was to let the livestock dung the ground themselves—depositing the end product of feed brought from another part of the farm, from the grasslands. This could be accomplished by yearly fencing of part of the plowland upon which to turn out the cows—for the evening in summer and for the day in winter. In New Hampshire, Matthew Patten appears to have fenced a new area for his cowyards each spring and to have plowed and planted these yards in following years.[19] There are no direct accounts of this practice in Concord—but again, there isn't anything remarkable about it. In this way, nutrients from distant pastures (in summer) and meadows (as hay, in winter) were concentrated on the nearby planting ground. It was a form of stock and manure management that made sense with tillage land clustered close to the barn, in a pastoral economy dominated by cattle rather than sheep. In the new soil and social environment of New England it logically replaced the ancient English practice of summer folding on more distant common fields.

Concord farmers spread and harrowed manure before planting corn and sometimes put additional manure in the holes.[20] Indications from Patten's diary and elsewhere confirm that in New England the heavy-feeding corn crop normally received the bulk of the manure.[21] The other grain that appeared most prominently in Concord inventories and tax valuations was rye. In rough terms, about two-thirds of the grain harvested was corn, one-quarter rye, and the rest mostly oats.[22] Rye, however, yielded only about half as many bushels per acre as corn (at best), so the acreage sown was probably nearly the same. In Concord, winter rye was seeded in September on land broken out of pasture, although

FIGURE 7.2.
A field of rye, covering a level sand plain. Herbert Gleason, 1916.

some was probably also undersown to follow corn. Customarily, little or no manure was spared for this hardy, light-feeding, low-yielding crop (fig. 7.2).[23]

Corn and rye provided the daily bread called rye 'n' Injun, along with a variety of biscuits, puddings, and mush. Corn was also an important fodder crop, beginning in September when the stalks were topped before harvesting and fed to the cattle—so the bulk of the vegetative growth was directly recycled into manure.[24] Rye straw could supplement hay for winter fodder and also served as bedding which absorbed urine, thus binding nitrogen along the metabolic pathways of the microbes working away on the cellulose. Rye straw was also preferred for making the "cheese" of pomace in the cider press. Oats, the next most common grain, were grown primarily for stock feed, usually in the year following corn before the field returned to grass.[25] More demanding wheat and barley went virtually uncultivated in Concord—although a few inventories did include malt. Concord families may have purchased their malt from other Middlesex County towns where more barley was grown.[26] Similarly, peas and beans cut no figure in Concord valuations, but beans were mentioned in some inventories and were certainly widely consumed. Presumably they were planted as a companion crop in the corn, as were pumpkins, in the Native way.[27] Probably very few potatoes were grown in Concord in 1749, since they had been introduced to New Hampshire by the Scots-Irish only a few decades earlier—although they were spreading rapidly throughout New England. Like corn, they needed dung.[28] Finally, many Concord farmers grew a few pounds of flax for linen.

The prevailing tillage pattern in Concord is now discernible. Farmers typically cultivated from half a dozen to a dozen acres, depending upon their family needs and labor supply—the average along the Bay Road was about seven (table 7.1). But deeds and inventories leave little doubt that by the mid–eighteenth century most Concord farmers had more tillable land at their disposal than they actually tilled in any given year (see appendix 2, East Quarter Land Use, by Deeds and Valuations—1749). This land lay along a spectrum which may conveniently be divided into two sorts: homelands and outlands. The home fields close to the barn received most of the manure and grew most of the Indian corn. They might also be planted to small patches of flax or potatoes, and a year or two of corn might be followed by oats or rye. But the section of the home fields which was planted and sown moved about from year to year, alternating with cowyards and with land that had been temporarily grassed down for pasture or mowing.[29]

Outlands farther away from the barn were less frequently broken up for tillage. Many such parcels were designated "plowland and pastureland" in deeds. When cultivated, these fields were usually sown to rye and were rarely manured. In fact, Middlesex County farmers *preferred* the flour made from rye sown on light pine plain land.[30] In brief, Concord's tillage pattern closely resembled the infield-outfield systems long used in pastoral regions of Britain, where the arable land was limited chiefly to providing for subsistence and the bulk of the land was devoted primarily to livestock.

Gardens and Orchards

All Concord homesteads had gardens. They were not considered an important enough part of the yeoman's world of husbandry and real property to be distinguished in tax valuations, but they did appear now and then in deeds and inventories—though even there they were often subsumed in the houselot or barn lot. Nathan Meriam's garden was before the house, where it adjoined his brother Josiah's garden.[31] Joshua Brooks's garden lay within the small lot called the Hog Pasture just east of his house. Husbandmen may have legally owned these gardens, like everything else, but gardening itself fell within the world of work and exchange that belonged to their wives. For example, when Edward Wright disposed of his land to his sons Edward and Samuel, the heirs obliged themselves not only "to till six acres for the comfort of our ffather and mother where he shall appoint it in our land yearly," but also that "our mother may have what ground *she* will to garden."[32] Through a painstaking examination of inventories (many of them

TABLE 7.1   East Quarter Tax Valuation—1749
Selected Farmers

Name	Ac Orch	Brl Cidr	Ac Till	Bu Grain	Ac Mow	Tons Eng	Tons FM	Ac Pas	Horse	Oxen	Cows	Swine	Sheep	livestock grazing units	bu. grain/ ac. tillage	tons hay/ ac. tillage
Thomas Brooks Jr	1	40	7	150	27	1	26	32	1	4	7	0	0	12.33	21.43	3.86
Samuel Brooks	1	25	7	140	40	1	14	14	1	4	8	3	0	13.33	20.00	2.14
Joshua Brooks Jr	1	15	7	130	11	2	12	24	2	3	8	2	0	13.67	18.57	2.00
Ephraim Hartwell	1	40	10	180	35	2	25	20	3	6	8	1	15	20.14	18.00	2.70
Ebenezer Meriam Sr & Jr	0	4	5	90	16	0	12	10	1	2	5	2	0	8.33	18.00	2.40
Nathaniel Whittemore	1	40	12	200	40	6	30	10	2	4	5	2	13	13.52	16.67	3.00
Samuel Minot	1	20	8	120	23	1	18	14	1	4	6	2	1	11.48	15.00	2.38
John Jones	1	10	10	150	24	1	20	13	1	4	9	2	0	14.33	15.00	2.10
Nathan Meriam	0	4	7	100	11	2	8	14	1	2	7	2	0	10.33	14.29	1.43
Samuel Fletcher	1	12	7	100	14	0	12	14	1	2	5	2	0	8.33	14.29	1.71
Josiah Meriam	0	0	3	40	7	0	4	7	1	2	3	1	6	7.19	13.33	1.33
Job Brooks Jr	0.5	10	8	100	20	1	18	7	1	2	7	0	0	10.33	12.50	2.38
Joseph Stow	0.5	4	4	50	10	1	5	4	1	2	3	1	0	6.33	12.50	1.50
Samuel Meriam Jr	1	5	5	60	7	1	2	8	1	4	5	1	0	10.33	12.00	0.60
Average	0.71	16.36	7.14	115.00	20.36	1.36	14.71	13.64	1.29	3.21	6.14	1.50	2.50	11.43	15.83	2.11

Source: Massachusetts Tax Valuation, 1749, Town of Concord Assessors Records, Microfilm Box 3 Roll 008, CFPL

Note: John Jones livestock from 1749 East Book, not Valuation

Note: tons hay/ac. tillage is a proxy for manure supply, not a yield.

from Concord), Sarah MacMahon has shown the importance of gardens to expanding and enlivening the diet of colonial New Englanders during the course of the eighteenth century.[33] As gardens were small and lay close to home it was a simple matter to bring plenty of organic matter to them, and we may assume they were well manured.

Gardening could also be an important part of the living of poor people who owned little or no land and who passed their lives without leaving a scratch in the hard records of tax valuations and deeds. Daniel Pellet swept the meetinghouse and rang the church bell in Concord during the early decades of the eighteenth century. In 1699, Pellet had been granted the liberty to improve a "little small corner" of the parcel in Concord center that Capt. Timothy Wheeler had given for a schoolhouse, which had been overrun with bushes. "Dan[l] doth engage to clear away y[e] brush & make it suitable for Inions to be sown on it," the selectmen's minutes recorded. Daniel's yearly quitrent was set at one onion. Partly by cultivating an unused corner of land intensively, perhaps with dung collected from the street, this man of small means endeavored to scrape together a living. The little market garden must have been at least marginally profitable because Pellet kept paying the rent (which rose from an onion to a shilling or two as the sharp-eyed selectmen observed that the enterprise went on) for decades. This tells us that even in the early eighteenth century there was a market for garden produce in Concord and perhaps beyond.[34]

Besides gardens, Concord was full of orchards, from the village center to the far sides of distant hills. Most Concord farmers had at least a small orchard, and these did make it into tax valuations—at least, some of them did. About two-thirds of Concord farmers returned orchards in the valuation of 1749.[35] Deeds and inventories reveal, however, that many men who reported no orchard actually did have one. In truth, it appears that almost all farmers along the Bay Road owned an orchard—if only a small one back of the garden. But many Concord husbandmen were beginning to make revealing choices in where they planted their apple trees. By the late colonial period, a few orchards were appearing that were not so close to home. Orchards were no longer a mere extension of the garden but had come into their own as an important element in the agricultural landscape. In fact, apple orchards could stand very well for the transformation of English husbandmen into New England farmers.

The Meriams apparently made their cider as a single extended family. Ebenezer Meriam and son had an orchard by their barn that was too small to trouble the assessor in 1749, yet they owned a cider mill.[36] Samuel Meriam reported an orchard, but deeds suggest his brother Nathan actually owned it. This three-acre family orchard had stood since grandfather John Meriam's day on a rough hillside across the road from Nathan's house and across the way from Samuel's. It was not tillable land—later deeds indicate it was sometimes pastured.[37]

It appears that most of the farmers in the Brickiln neighborhood deliberately placed their orchards on marginal land. Deacon Samuel Minot had orchards on both sides of the road, both before and behind his house.[38] Samuel Fletcher next door had his orchard below the road beside Minot's.[39] These two orchards stood on a rocky little knoll which one would not have wished to plow. John Jones had his orchard below the road in front of the house, probably at the rocky base of a till slope at the east end of a bit of meadow there.[40] In addition, both Jones and Fletcher owned small parcels of woodland half a mile away at the head of the Ox Pasture, on the stony slopes of Pine Hill. These were later identified as orchards and may well have already been so in 1749.[41]

All the Brooks homesteads along the Bay Road had orchards. Samuel and Job Brooks's orchards were most likely at their homelots.[42] Thomas Brooks's home orchard lay between his house and his nephew Joshua Brooks Jr., but by the 1740s Thomas had planted a second orchard a quarter mile to the east, where the road climbed Elm Brook Hill.[43] Joshua Brooks (whose house was newer) seems to have had no home orchard, but he, too, had one at the top of Elm Brook Hill, a good half mile east of his door.[44] By the mid–eighteenth century members of the Brooks clan were likewise finding new sites for orchards on the hill.

The farmers who lived on the uplands beyond the top of the hill did not have so far to search for orchard sites as for plowland, and for the same reason: the rocky lands at their door may have been hard to plow but were often first-rate for growing apples. Ephraim Hartwell and Nathaniel Whittemore both reported pressing forty barrels of cider in the 1749 valuation. Whittemore's orchard was on the downslope between the road and a piece of meadow just west of his house.[45] Hartwell's home orchard lay east of his house and had once been the orchard of his father, Samuel, but he seems to have owned a second orchard further down the road at the corner of his tillage field. Here also stood a cider mill. Hartwell was possessed of yet another "young" orchard on the pasture across the road from his house by the time of his death in 1793, but this may not yet have been planted in 1749.[46] Like their neighbors, these farmers were gradually moving their orchards away from good tillage and onto glacial till.

The changing position of orchards was one small but telling part of the adaptation of English husbandry to New England that took place in the early generations of settlement. This movement accompanied the shift from beer to cider as the workaday drink of the yeomanry. According to McMahon's analysis of Middlesex County inventories, cider largely replaced beer during the first half of the eighteenth century, as part of an ongoing intensification of food production.[47] In Concord, cider dominated the beverage scene almost from the beginning—there is only a small trace of barley or beer in seventeenth-century Concord records. When Gershom Brooks bought his brother-in-law Timothy Wheeler's houselot on the Bay Road near Elm Brook in 1665, by contrast, it already had an orchard "on the end of it."[48] Orchards were widespread by the end of the seventeenth century and ubiquitous by the middle of the eighteenth.

Cider's advantages were both topographic and temporal. The switch to cider released tillage land from beverage production—or, more to the point in Concord, it conserved manure for the corn crop since apple trees' deep foraging roots did not demand fertilizer, at least not in the immediate way barley did. Concord yeomen like Joshua Brooks soon discovered that orchards grew well on rocky slopes that could scarcely be plowed in any case. Stony till soils were not merely suitable for apples, they proved superior. The characteristic hardpan blocked drainage so that water seeped slowly downhill near the surface, ensuring good soil moisture all summer. In fact, the slopes of Elm Brook Hill where the Brooks and Hartwell families expanded their orchards became a center of commercial fruit growing a century later, lasting well into the twentieth century. If this rocky land is agriculturally marginal, we have only discovered it recently.[49]

Colonial orchards were composed largely of nameless seedling trees, which produced a random mix of sour, crabby fruit — perfect for cider. Perhaps a few favorites were grafted for winter keepers, but for the most part trees were grown from seeds in pomace, and the thriftiest whips set out in the orchard.[50] The fruits were small but offered ecological advantages. Being high in acid and tannin and having a thick skin, they possessed significant natural defenses against pests. Most important of all, by virtue of choosing healthy young trees from their seedbeds and nurseries to transplant, farmers were selecting trees with tolerance for prevalent strains of apple scab fungus and other diseases. They were creating a locally adapted population of cider apples. Colonial husbandmen were able to get reasonable production from these trees with little care, something next to impossible in modern orchards in which fruit varieties and pests alike have become adapted to a steady rain of pesticides.

Cider had one further advantage: it released labor from mowing barley during the summer reaping and haying season, a tight squeeze in the yearly round. Apples did not have to be gathered and pressed until fall, often after the corn had been harvested.[51] This made for a better seasonal distribution of labor, especially because with no further effort the cider reached its prime during the next haying season, exactly when it was most required. Trees could be pruned in March, another slack season, and planted in late fall or early spring. All in all, cider apples proved an excellent solution to the problem of making a fermented beverage from Concord's offering of soils, which were so backward at growing beer. Apple cider replaced barley malt for drink just as Indian corn and rye had replaced wheat and barley for meal — and for similar reasons. In doing so, the beverage fruit and the bread grains migrated to distinct parts of the landscape.

Meadow

The meadows lay at the bottom of husbandry in Concord. Without the meadows, agriculture would have required another footing and taken another form — as indeed it did in the nineteenth century. Fundamentally, the amount of hay that could be cut determined the number of cattle that could be kept, and cattle were the key to both subsistence and wealth. Meadows provided the great bulk of fodder for the stock, and thus manure for tillage, year in and year out. The adoption of sown grass and clover that drove the English agricultural revolution had barely touched eighteenth-century Concord. The primary source of grass, and hence of soil fertility, remained native meadow hay.[52]

In Concord, meadow lots were strung along nearly every stream, forming tracts that often ran for miles. By the middle of the eighteenth century, the process of converting swampy land into smooth, grassy meadows had transformed all but the remotest, boggiest wetlands in Concord. The acreage of meadowland would not change appreciably for the next hundred years.[53] Most of the meadows were composed of a succession of small lots divided by ditches, which dried them sufficiently to allow access and also served as fences—especially if a steep bank was thrown up alongside the ditch. Reported meadow holdings of yeomen along the Bay Road ranged from seven acres for blacksmith Samuel Meriam to forty acres for Samuel Brooks (see appendix 2).[54] Most farmers had mowing lots scattered among several meadows, in the way of their fathers since the time of the First Division. Typically they owned one or more pieces of home meadow within sight of the barn, along with several lots in various brook meadows and at least one parcel of river meadow. Once again, however, the farther from the village the homestead lay, the greater the consolidation of meadow holdings.

By 1749, the Meriam family's scattered mowing land was divided among four households (plate 8A). These men had home meadows at their doorsteps along the Mill Brook and in the nearby Dam Meadow. They had a cluster of mowing lots (along with some swampy woodland) in Virginia Meadow and farther down Elm Brook at Birch Island in Bedford. All four households also had lots in the Great Meadow in Concord and Bedford. This far-flung galaxy of small mowing lots did not add up to a very robust endowment of meadow, however—well under twenty acres average for each household within the family.[55]

The farmers in the Brickiln neighborhood also mowed scattered meadows, though not quite so dispersed as the Meriams (plate 8B). They averaged a bit more than twenty acres of mowing. The meadows belonging to the prosperous Brooks family were still more consolidated and more substantial. Thomas, Samuel, Joshua, and Job Brooks Jr. held the bulk of their mowing side by side in Elm Brook Meadow just below their houselots, along with a similar but smaller collection at the Suburbs a mile up the road (plate 8C). In addition, Thomas, Joshua, and Job each had a lot in the Great Meadow some three miles away, while Samuel had an extra lot in Virginia Meadow a bit closer to home. These meadow holdings ranged from just over twenty acres each for Job and Joshua to between fifty and seventy for Samuel.[56] Plentiful grass allowed the Brookses to keep more stock than most of their neighbors who lived toward the village.

On the uplands beyond Elm Brook Hill, the mowings were concentrated and large (plate 8D). This abundance may seem geographically odd at first glance, but it wasn't far from the high ground along the road to the headwater swamps of the Shawsheen River to the north and of Hobbs Brook (which ran to the Charles River) to the south. These farms had been assembled so that their back ends reached into these all-important wetlands. Ephraim Hartwell and his father, Samuel, before him had steadily accumulated parcels in Rocky Meadow and at the Suburbs, amounting by 1749 to over fifty contiguous acres of meadow. Hartwell also owned four acres in Elm Brook Meadow in Bedford, his only outlying parcel of land aside from his plowland. Nathaniel Whittemore next door (following in the footsteps of his father, Benjamin) had accomplished the same thing—he owned more than sixty acres of meadowland, strung along a brook that arose by his house and flowed north through his elongated homestead. Some of this meadow was surely grazed rather than mowed. Across the road, the children of Nathaniel's deceased brother Benjamin Whittemore Jr. also owned plentiful meadow at the lower end of their farm, lying at the head of Flint's Great Meadow. Both Hartwell and Whittemore kept substantial herds of cattle, horses, and sheep on the growth of these plentiful meadows, several miles upstream from any river.

Wide dispersal of meadow lots was not universal among Concord husbandmen, but it was so ordinary and so persistent that most of them must have found it either advantageous, unavoidable, or both. Many held tenaciously to their lots in the Great Meadow for generation after generation—even those like the Brookses who had abundant mowing land closer to home. This suggests that the river meadows grew the best hay. But there was not enough of the Great Meadow to go around, and besides, the lush growth alongside the river also carried a high risk of flooding. In rainy summers the river meadow hay may have been spoiled by flowage, but perhaps in such years the brook meadow hay grew all the better to compensate: spreading locations spread risk. Finally, the hay may have ripened at different times in these meadows, spreading out the labor of harvest.

By 1749 a sprawling network of meadows stretched from the headwater swamps among the remote hills, all along the wandering brook courses down to the broad river floodplain (figs. 7.3 and 7.4). Converting these wetlands from rough marshes and swamps into productive mowings had absorbed the labor of the first three generations of Concord husbandmen. The work of maintaining and improving the meadows was carried on by their sons and grandsons. In their native state many wetlands had not been dominated by grasses and had to be cleared to begin with. Swamp vegetation ranged from thickets of buttonbush and black willow to forests of elm, black ash, red maple, and swamp oak. After the trees and brush were cut, the new meadow had to be fenced. The brush was dried and burned, and in some instances grass seed was sown. Certain native grass species such as bluejoint and fowl meadow grass made better stock hay than the rest, and farmers did what they could to spread them. After a year or two, the grass was established and the meadow was ready to be cut for hay. Thereafter, annual mowing encouraged perennial grasses and forbs, while discouraging the regeneration of trees and shrubs.[57]

But simple clearing of a meadow and sowing of grass was seldom enough to allow it to be mowed. The meadows also needed to be systematically drained to support the desired vegetation and to support the weight of the mowers and their oxen at haying time. A mired ox was a great inconvenience and sometimes a fatal calamity.[58] The meadows in their native state had high water tables, and many of them lay flooded much of the year. The kinds of grasses that make tolerable stock hay are not emergents that flourish in standing water. With the odd exception of horsetail (*Equisetum fluviatile*), or "pipes," as it came to be called locally, such marsh emergents as sedges (*Carex* ssp), arrowhead

FIGURE 7.3.
Nut Meadow, a small headwater meadow and swamp in southwest Concord (see plate 3).
Pasture set off by rail fence, pine woods in background. Herbert Gleason, 1903.

(*Sagittaria* ssp), bulrushes (*Scirpus* ssp), and cattails (*Typha latifolia*) are unpalatable to cattle. Meadow grasses thrive when water covers the ground during the winter but drops well below the surface during the growing season.[59] Large stretches of Concord's lowlands were initially too poorly drained to grow decent hay, even after being cleared.

Draining the meadows first required ditching the brooks to ensure a regular flow. Laterals were then needed to cut off seepage from the uplands, to lead springs out to the main channel, and to get the water out of low places. Meadows sometimes had to be ditched all over again once the water's true inclination was revealed. Once put in place, these ditches had to be periodically cleaned of accumulated muck or they would cease to carry the water away. Drainage work was customarily done at low water during late summer and early fall with shovels and a special tool called a ditching knife.[60] Whether or not meadow owners traded work with their neighbors at this task, it was inescapably a collective enterprise: unless ditches throughout a meadow—or an entire watershed, really—were laid out coherently and faithfully kept clean, none of those upstream would drain properly. In most cases this coordination was no doubt informal, driven by both self-interest and social pressure, and left no record other than the networks of ditches themselves, connecting for miles. If now and then a more formal covenant was needed to

FIGURE 7.4.
Cows on Brook Meadow, at the outlet of Nut Meadow brook to the river.
Pines invading upland pasture behind. Herbert Gleason, 1903.

ensure that the work was properly done, the "major part of the proprietors of a Tract of Meadow Land" could petition the General Court to form a commission of sewers with power to compel the work and apportion the cost.[61]

Once a meadow had ditches so that the water could be gotten off in the summer, one more improvement was desirable: a dam, so that water could be held on during the winter. Most meadows flooded naturally during the spring freshet, of course. Flooding through the dormant season until April was beneficial, and farmers sought to augment and control it by deliberately flowing the meadows as their forefathers had done in England. Samuel Meriam and John Jones, among others, owned lots next to Daniel Hoar's in the Dam Meadow, which was "usually flowed in winter seasons by s^d Hoar's damm."[62] Road causeways also served as handy dikes. The object in flooding was to capture organic matter and nutrients. This practice was considered so valuable that disputes sometimes arose over who had rights to the water. One protracted legal struggle among meadow owners at Brickiln Island was settled with the finding that the defendant "hath no right to divert and spread the said water over his land in such a manner that the surplusage of water if any after fertilizing his land shall not return to ye sd watercourse."[63] That was an extreme case (to which I shall return) in which the flow had been turned from one watershed into another, depriving those downstream. More commonly farmers were able to part the water amicably and to share the cost of regulating its flow to their mutual advantage.[64]

Regular flooding deposited sediment on the meadows. Even in their native state, the wetlands were the only part of the country that built a rich, organic soil. As part of a working agrarian landscape, they served to recapture a good part of the accelerated loss brought on by cultivation of the uplands. Like meadows in England, Concord meadows received a steady augmentation of nutrients that leaked from pastures, fields, houses, farmyards, and shops higher in the watershed, and so could be cropped indefinitely. They were, in essence, inexhaustible. Although they were never manured, the meadows were still yielding hay year in and year out a century—in fact, two centuries—after they were first mowed. The yield was not huge, and the hay was not the best; but it did maintain stock, and it was dependable. In dry years when upland crops failed, the river meadows could still be counted upon. In wet years, when the hay on the lower meadows might be flooded, upper meadows along the brooks tended to yield heavily, upland pastures flourished and carried the stock into the winter in good condition, and lush corn stalks afforded extra fodder. The meadows did not require manure, they did not receive manure: they *made* manure. Up from the meadows came soil nutrients in the form of hay, to be returned to tillage land in the form of dung. Even the muck that had to be cleaned from the drainage ditches year after year could be dried and then carted back to sandy upland fields, enriching those coarse soils. In short, the meadows were like the bottom paddle of an undershot millwheel, recycling the flow of nutrients that drove the farm.[65]

English Hay

Very little English hay was grown in colonial Concord. English hay was a mix of cultivated forage species of European origin, including herdsgrass (or timothy), red top, and red clover, all of which spread through America in the course of the eighteenth century.[66] Domesticated hay supplied only 15 percent of Concord's hay crop in 1749. Few farmers along the Bay Road reported putting up more than one or two tons (see table 7.1). English grasses were often mowed in orchards and might be planted on higher meadow ground that wasn't flooded in the winter. English hay was a better quality (and potentially better yielding) fodder crop than native meadow grass, but at midcentury it had not yet assumed a central role in Concord's husbandry.[67]

The great migration of Puritans to Massachusetts Bay took place in the 1630s, several decades before the widespread adoption of domesticated grass and legumes in old England. Ordinary settlers brought with them the habit of relying on low-lying meadows of native grass for winter fodder and greatly expanded this practice in New England. They had little experience of systematically *planting* hay, let alone of the regular rotation of grass and clover with arable land that would drive the English agricultural revolution. Only as the supply of meadow hay grew pinched in the older towns in the mid–eighteenth century and as hill towns with few natural meadows began to be settled did farmers begin to plant much English hay. Concord, with its broad river meadows, was one of the last towns in Middlesex County to make this transition, although it was under way by the last decades of the colonial era. English hay would transform Concord farming in the early nineteenth century, but it was not an important feature of colonial agriculture.[68]

Pasture

Pasture was the most widespread category of improved land in eighteenth-century Concord, and the most obscure. It covered a broad ecological range, stretching from the cowyards behind the barn to the most remote woodlot in Concord and beyond. This range might be divided into four kinds of pasture. The first was warm, sandy land that would occasionally be plowed and sown, usually to rye. The next was cold, moist meadowland that was grazed in place of (or as well as) being mowed. A third was stony, glacial till upland, partly or wholly cleared of forest. The last sort of Concord pasture was not in Concord at all: it was backcountry pasture belonging to Concord men. These kinds of grazing lands had very different qualities and were surely employed accordingly, but unfortunately the system by which they were pastured by different stock at different seasons is largely unrecorded and unknown.[69]

What patterns can be discerned among the pasture holdings of husbandmen along the Bay Road? The Meriam family was not overly rich in pasture, either in quantity or quality. Most of their pastureland lay just beyond their tilled fields, either within the Great Field or to the east of Billerica road. These were light, sandy soils—most were tillable and probably were tilled from time to time. But in Concord, the best plowland did not usually make the best pasture. Light soils are too droughty for grass, especially in the late summer months. The Meriams must have been hard-pressed to find grazing for their cattle in a dry season such as 1749. It is not surprising to find that the Meriams owned fewer livestock than most of their neighbors down the road.[70]

Farmers in the Brickiln Field neighborhood were a bit better endowed with grazing than the Meriams. The Brooks clan possessed substantially more pasture and so carried more stock. They had also discovered the utility of stony glacial till for summer grass. The bulk of their grazing was located on high ground to the south and east of their houselots. Job's "home pasture" lay across the Tanyard meadow, on the same till slopes above Elm Brook as Joshua's "Schoolhouse pasture" and "Top of the Hill pasture" and all Thomas's home pastures. Both Thomas and Samuel Brooks had additional grazing a good mile away at the Suburbs on similar soils. Many of these hill pastures were only gradually emerging from the woods, as indicated by Joshua Brooks's "wood pasture" and Thomas's "Holt pasture." But the water-rich, rockbound till was proving as well suited to growing grass as it was to growing trees. The Brookses found additional grazing in their broad meadowlands.[71]

The upland farms located beyond Elm Brook Hill, in the neighborhood of Hartwell tavern, were abundantly endowed with pasture—indeed, were set in the midst of their pastures. Neither Ephraim Hartwell nor Nathaniel Whittemore reported an unusual acreage of pasture in the 1749 valuation, but their deeds tell a different story: Whittemore possessed nearly fifty acres and Hartwell well over that number of acres of pasture— significantly more than any of the farmers nearer the village. These pastures lay on the same bouldery slopes that were so desirable for orchards and so difficult for tillage. These farms were also blessed with a surplus of meadowland which could be grazed as well as mowed. The large supply of grass was reflected in the size of Whittemore's and especially Hartwell's herds and flocks. And, as with the Brookses, the abundance of manure from carrying large herds was manifested in superior grain harvests.

But the livestock reported in the valuation of 1749 don't give us the whole picture. The cows, oxen, and horses that dominate the lists represent the principal working and breeding stock of these husbandmen. Missing are the stock being reared and fattened, the beef cattle, the cash crop. What this meant can be seen in the inventory of Samuel Brooks, who died in 1758 at age sixty-two, before his sons had come of age. At his death Brooks owned a mare, a yoke of oxen, two swine, two milch cows, and five other cows— a fair approximation of the livestock shown in his 1749 valuation. But he also owned two two-year-old steers, two one-year-old steers, two calves, one colt, and a bull—all missing from the tax records of that year. The more complete valuations of 1784 and afterward more accurately reflected this heavier stocking of substantial farms with neat cattle.[72]

A few farmers such as Samuel Brooks and Ephraim Hartwell had sufficient grass to fatten beef at home. But there was another way to pasture those animals which, unlike milk cows and working oxen, did not need to be available everyday. Samuel Minot owned 40 acres of woodland and pasture "within fence" across the river in north Concord.[73] Similarly, Job Brooks owned 50 acres of pasture and woodland in north Concord called the "Moon lot."[74] But north Concord was hardly the end of it: Samuel Minot also owned half of 120 acres in the town of Princeton twenty-five miles to the west—the other half belonged to his neighbor, Nathan Meriam.[75] Ephraim Hartwell was another who owned land in Princeton.[76] Job Brooks owned 30 acres in Littleton, Thomas Brooks's son Noah picked up 50 acres of pasture in Princeton, and Joshua Brooks acquired 120 acres in Pepperell that stretched into New Hampshire.[77] As William Jones wrote of Concord in 1792, "The pasture land is not in proportion to the meadow land and other soil, but the principal farmers own pastures back in the country, where they fatten their beef, and pasture their young cattle."[78] This pattern was already well established by 1749 and in fact had appeared as early as the 1650s with Concord's acquisition of the New Grant (later Acton) to graze the dry stock of the town. By the mid–eighteenth century Concord yeomen had moved from that commons approach to privately owned backcountry summer pastures, but for the same dual purpose: to accommodate the offspring of their cows and eventually to accommodate some of their own offspring.

Concord's system of grazing can now be at least glimpsed. A yeoman required from ten to twenty acres of pasture within about a half mile of the barn to maintain his core stock. Some of this pasture, particularly among farmers who lived near the village, was found on light soils that might occasionally be tilled. That thirsty ground couldn't carry much stock past early summer, except in rainy years—but it greened up earliest in the spring. Better pasture, for those who had access to it, was to be found on the heavier glacial till uplands which were slowly emerging from the forest in the eighteenth century. But almost all farmers could utilize another late-summer grazing resource: the second growth of their well-watered meadows that came on in August and September, after the hay had been cut. Some combination of these three kinds of pasture proved adequate to support the basic herd of half a dozen cows, a yoke of oxen, a horse, and perhaps a few sheep.

Those farmers who had large supplies of meadow hay had a further opportunity to fatten a herd of beef cattle, reared by their cows—if they could also locate the summer grazing to match their hay "in proportion." They might clear additional pasture close to home for this purpose, if they had it; or acquire back pasture to the north and west, if they could afford it. These cattle were driven to the hills in the spring and brought down in the fall—an ancient tradition over much of the stock-raising world. The fattened beasts would be sold for slaughter in Concord or driven on to market in Boston; the rest would be fed through the winter on meadow hay and then sent up-country for another season. While at home they would convert that hay into manure for the cornfield. Those yeomen who lacked the grass resources to fatten beef themselves sometimes supplied calves to their neighbors' herds.

We can get a sense of the interlocking orbits of the cow herd and beef herd from the inventory of Job Brooks's son Asa, who died in 1816. On the home farm by the Brooks tanyard Asa Brooks kept a mare, a yoke of oxen, eight cows, a bull, two sheep, and four swine—substantially the same as his father kept in 1749. These were the working, breeding, and subsistence stock, the home herd. At his sixty-four-acre pasture in the hills of Princeton, meanwhile, Asa had twelve fattening oxen, two three-year-old steers, a beef cow, a yearling bull, a three-year-old heifer, a horse, and a colt—his beef herd and young replacement stock, representing, in part, several years' offspring from those eight cows at home. There is no reason to suppose that father Job Brooks had managed much differently with his back pastures in north Concord and Littleton half a century before.[79]

By the middle of the eighteenth century, the old practice of grazing livestock on commons or on unenclosed private lands had all but vanished from Concord—with the exception of swine, which most years were allowed to run at large on commons. The only legal commons left were along the highways, but in practice *at large* meant on any land that wasn't fenced. This posed a nuisance great enough that every so often town meeting reversed the usual policy and ordered the swine confined for the coming year.[80] The principal stock, the neat cattle, grazed on pastures that were enclosed and integrated into individual farms, although in a complex variety of ways. Concerning the condition of these pastures it is hard to generalize. Grazing capacity was on the rise until late in the colonial period, as new pastures were fenced and cleared. At the same time, older pastures were no doubt already declining in productivity (see chapter 8).

Pastures that rotated (if *rotated* is not too regular a word) with tilled land must have been of low quality at best. Those only occasionally broken up for rye were probably poorest, both because of the lightness of the soil and because little or no manure was carted on. On grassland that followed manured crops a better sward may have sometimes been obtained. If English grasses such as timothy and clover were sown, perhaps a few mowings of hay were even taken before the field was returned to grazing—the beginnings of convertible husbandry. However, this practice was still comparatively rare in mid-eighteenth-century Concord, as the scanty valuation returns for English mowing attested. In many cases, the pasture sward probably formed on its own from grasses that had survived while the field was in cultivation and from seeds that had been reintroduced in the dung of the livestock. Most of this rotational pasture doubtless suffered because of the undistinguished mix of forage plants that grew up, because of lack of lime and poor grazing practices, and because droughty arable soils simply made indifferent pastures, no matter how well managed.

But as Concord husbandmen pushed onto the upland outskirts of town such as Elm Brook Hill, better pasture ground steadily emerged from the woods. Such pastures were gradually cleared of hardwoods by the cutting of fuelwood and timber and by the suppressing of regeneration of trees through grazing and fire.[81] These pastures increased in grazing capacity as tree cover diminished and grass spread. Where the overgrowth was felled and some or all of the slash burned, a boost in soil pH and nutrients could give rise to good forage, especially on the moister till soils. New Englanders spoke of an initial burst of "white honeysuckle" (white clover) that appeared like magic when new ground in the woods was broken up.[82] Grazing livestock introduced the seeds of such European grasses as bluegrass (*Poa pratensis*), speargrass (*Poa annua*), witch grass (*Agropyron repens*—better known as an aggressive weed in cultivated ground), orchard grass (*Dactylis glomerata*), fescue (*Festuca* ssp), and white clover (*Trifolium repens*) to compete with the native shrub and herbaceous species.[83] In the words of William Wood, "In such places where the Cattle use to graze, the ground is much improved in the Woods, growing more grassie, and lesse weedy."[84] How long this ground would *remain* improved as the cattle grazed on through the years was another question, but during the colonial period, the initial gains in pasture acreage and forage species probably outstripped the gradual losses in quality.

The most reliable grazing of all lay in the meadows. After the hay was cut in midsummer, the grass renewed its growth. This second growth was anciently known in England and New England as the aftermath (which literally means "after mowing"), eddish, or rowan. The aftermath was customarily grazed—it could have been mowed a second time, but until the advent of mechanical mowers in the mid–nineteenth century the rowan was seldom considered worth haying. The second cutting produced far less hay than the first cut, although it was of better quality, containing more leaf than stem. Besides being unprofitable to mow, it was often desperately needed for grazing. Meadow hay was generally cut in Concord beginning in late July, by which time many upland pastures were nibbled to the roots, especially in a dry year. At this point part of the stock could be turned onto the meadows. During the 1770s young Jake Potter still drove his father Ephraim's cows from the village to graze at Fair Haven meadows in the fall, after they had been mowed. Because the meadows were fertilized by winter floods and because annual mowing discouraged woody invaders (the bane of pasture management), the productivity of this lowland grazing was maintained for generations.[85]

Woodland

One vital element in the Concord landscape that was not reported in the 1749 valuation was woodland. When woodland did begin to be reported in the late eighteenth century, Concord contained about four thousand acres, or approximately 35 percent of the reported total acreage of the town.[86] At midcentury forest covered at least one-third and probably closer to one-half of Concord. The 1749 Land Use map of the Bay Road gives the same picture (plate 7). Forest was still plentiful on hillsides and in swamps even near the roads, and deeds and probated estates indicate that yet more woodland was to be found farther back into the country, especially towards Flint's Pond. What forest resources did various farmers along the road possess?

The woodlands belonging to the Meriams were small, scattered, and remote. They ranged from two acres of woods on the steep hillside behind Ebenezer Meriam's house to Nathan Meriam's two acres in the Cedar Swamp three miles away; from Nathan's seventeen acres at the "house swamp" to several small woodlots at the Chestnut Field by Flint's Pond two miles in the other direction. The family also owned woodland adjoining their meadow lots in Virginia. Among them, the four households possessed some ninety-three wooded acres in thirteen separate lots (see appendix 2).

Those in the Brickiln neighborhood were better supplied with woodland. Samuel Fletcher owned only twenty-two acres, most of it a mile or so to the south on Pine Hill. Joseph Stow owned twenty-eight acres, about half at the Chestnut Field and the other half on the back side of Brickiln Island. John Jones was better endowed with about forty acres, mostly on Pine Hill and at the Chestnut Field. Deacon Samuel Minot owned a solid forty-acre block of woods on Pine Hill, in addition to a small woodlot in the back corner of his homelot—not to mention another forty acres of woodland and pasture in north Concord.

Members of the Brooks family had plenty of woodland to their name, on Elm Brook Hill, at the Suburbs, and at the Chestnut Field. Thomas Brooks owned thirty-one acres of woods, Samuel Brooks forty, and Job Brooks Jr. fifty-five—plus the fifty-acre "Moon lot" in north Concord, which was partially wooded. Joshua Brooks Jr., having just been given land by his father, owned only about fifteen acres of woods in 1749, but he was about to remedy that deficiency by purchasing twenty acres more at the Chestnut Field during the 1750s, much of it from the Meriams and Joseph Stow.[87] Meanwhile, grandsire Joshua Brooks's two lots in the Cedar Swamp, #7 and #10, which had gone to his four sons in 1695, had been scrupulously divided among them in 1731 by an agreement recorded in the Town Record book. These minuscule lots passed along to at least Job and Thomas of the fourth generation but finally disappeared from family estates by the latter part of the eighteenth century—perhaps the valuable cedar had been logged out by then.[88]

Ephraim Hartwell and Nathaniel Whittemore appear to have owned surprisingly little woodland within their homesteads, having improved all but the roughest slopes and swampiest bottoms for pasture and mowing. By 1749, each had only a bit more than twenty acres remaining in forest. Both men were well-off and could have acquired more woods had they felt the need. What we see along the Bay Road suggests that something like twenty or twenty-five acres of woodland was the minimum Concord farmers felt would comfortably satisfy their household needs. That would have generated some twenty cords of wood yearly, along with some timber—depending on the age and composition of the lot. Domestic wood use included timber for buildings and rail fences and above all fuel to burn. Concordians burned both oak and pine, valuing pine at about three-quarters the worth of oak.[89] Through the early eighteenth century families no doubt met most of their wood needs in the course of improving land, as pastures and meadows steadily emerged from forests and swamps. But some farmers also began to deliberately acquire woodlots in certain rough and remote areas such as Pine Hill and the Chestnut Field. These lots lay at the eastern end of a large tract of rocky hills and pine plains that had long been thought of primarily as woodland and was now being confirmed as an extensive collection of woodlots several miles long, stretching from Flint's Pond through Walden Woods west to the Sudbury River.

The local agrarian economy required many kinds of wood for many special purposes, and the proper trees were found in different parts of the landscape. But not every yeoman or artisan had to acquire all of these for himself. The Meriams, for example, appear to have possessed barely adequate woodland to meet their home needs and to have purchased distant woodlots at the Chestnut Field to assure themselves. Still, they did not attempt to produce all their own building material. On January 10, 1743, "Samuel Maraam" and his uncle Ebenezer visited the yard of the Concord merchant and wood dealer Ephraim Jones to buy timber. In typical Meriam fashion they split the expense down the middle, as Samuel purchased and hauled away thirty-one "feat" and Ebenezer thirty—a comfortable sled or cartload each.[90] The Meriams may not even have cut all their own fuel. Samuel was a blacksmith and his brother Josiah a locksmith, so they required large supplies of wood and charcoal for their shop fires. But some of the farmers for whom they shod oxen and sharpened plowshares would have paid in just those commodities—just as Calvin Wright in 1796 paid the Concord blacksmith Samuel Jones for his "showing" by "cutting, splitting, & carting wood & by lodes oake wood."[91]

The Brooks family made use of another important wood product in running their tannery: oak bark. The bark was ground in a mill located near the "hog pasture" up the brook and brought across the road to the tannery by the meadow.[92] The Brookses may have procured oak bark from their own extensive woodlands, but they doubtless obtained much of their supply from their customers. We can easily imagine a local yeoman arriving at the tannery with an ox hide to be tanned, along with a load of bark in payment. The Brooks tannery sat on a forty-rod (or quarter-acre) piece of ground just east of Job Brooks's houselot, on the western edge of the meadow there. It included a tan house, a tanyard, and "tann fatts," the vats or pits in which the skins were immersed in water made rich in tannin by soaking in oak bark. The tannery included a "currier shop" where the leather was further prepared. Like many tanners, the Brookses also ran a slaughterhouse on the premises.[93]

The Brooks tannery endured for six generations, from the late 1600s to the 1830s. Joshua Brooks evidently learned the family trade from his father-in-law, Hugh Mason, of Watertown in the mid–seventeenth century, and from Joshua it passed to his son Noah, and then through three more Joshuas until it ended with Isaac.[94] Other members of the family found employment at the tannery as well: Job Brooks Jr., who lived next door, called himself a currier into his thirties, even after he had inherited his father's substantial farm.[95] Neighbors also worked as curriers dressing leather or as cordwainers making shoes. Indeed, cordwainers proliferated in eighteenth-century Concord, as land became more scarce with the passing generations and young men looked for a start in life.[96]

Tanning was essential to the internal economy of a small farming community, joining two kinds of local resources to produce leather, a versatile and widely used material. Leather married the strengths found in the tough skins that two of the most prominent members of Concord's biota turned to the weather: the hides of cows and the bark of oaks. To make cowhides into durable leather required treating them and literally combining them, in close chemical bonds, with substances derived from the woods. First, the hides were soaked in lye leached from wood ash, to ease the scraping and removal of hair—the hair itself later served to mix with lime for plastering.[97] Then the hides were soaked in a series of vats containing tannin, derived from oak bark—black oak was reckoned best. Tannins bonded with proteins in the hides and filled physical pores, preserving and coloring the leather. Next came currying, in which the tanned leather was rubbed with a "dubbin," or mixture of tallow and oil, replacing the natural oils lost in the tanning process and rendering the leather strong and supple. The resulting product was used for many domestic and farm purposes, including britches, work aprons, saddles and harness, and above all for making shoes.[98]

The slaughterhouse and tannery was thus a critical node where the products of pasture and forest intersected on their way to supporting every aspect of life in Concord. The making of leather employed not only the cow's hide but the tallow as well; not just the oak's bark but the ashes. Even the hair found its use. The process was also no doubt a source of noxious pollution. Offal from slaughtered animals and effluent from the pits was probably sluiced directly into Elm Brook.[99] We can well imagine how pleasant it must have been mowing hay by the tanyard on a sultry day in early August, with the most sluggish of southwest breezes wafting over the meadow. But in the end, some part of the nutrients in tannery wastes made their way back to fertilizing the croplands in a round-about fashion, as anything organic that settled upon the lowly meadows was faithfully returned, in the form of muck and hay.

Woodlands were an integral part of Concord's economy, and by the middle of the eighteenth century were increasingly consigned to a designated part of the landscape, just like tillage or meadow. The forest held a diverse range of resources, far beyond the important and obvious fuel, fences, and timber: cooperage for barrels, bark for leather, even charcoal for gunpowder. The woods no doubt also continued to supply a wide range of foraged fare, from deer and turkeys to chestnuts and blueberries, although many of these resources grew scarce as the landscape was improved for cultivation. Little of this transformation was directly recorded, but it is telling that in 1745 the town forgot to elect its deer officers at the annual meeting in March and had to remedy the mistake at the next meeting—one senses that this always minor office had grown entirely ceremonial. Bounties for wolves and wildcats killed within the town were less frequently collected as the century wore on.[100] Flocks of migrating passenger pigeons still passed through Concord and were utilized even as they were depleted—for example, Samuel Meriam owned a pigeon net.[101] The landscape was being transformed to a settled, agrarian condition, at what may have seemed an incalculable loss to Henry Thoreau and to many of us today but was seen as precisely natural in its time. But although the forest was growing less extensive and less wild than it had been in the early days of the town, it was no less providential. The remaining woods were as essential to life in Concord, in direct and practical ways, as any other kind of land.

Water

It would not have been possible for English husbandmen to organize the land into a complex agrarian ecological system without exercising similar control over the water. In Concord, the newcomers encountered an erratic, unmanaged water system that immediately gave them trouble. The well-watered English countryside had long been carefully ditched and ponded, by furrow and leat, river and stream, from top to bottom. New England waterways seemed familiar in character to the English but raw and ill-mannered: meadows were swampy and unredeemed, rivers flooded at the wrong time. The waters of Concord, like the land, held promise but were not at first in good working order, as far as the English could see. As the farm economy of Concord grew and the landscape was improved, the water system too had to be brought under cultivation. A number of potentially conflicting uses had to be harmonized. Generations of effort were expended to reform the flow of water into a tractable English pattern.

The yeomen of Concord never quite succeeded in taming the "great and peevish" river to their complete satisfaction, but they made progress in that direction throughout the watershed.[102] By the mid–eighteenth century the farmers of Concord had brought a familiar order to their water as well as their land. As in England, they attempted to direct and utilize the flow of water for a variety of not always complementary purposes. They also had to deal with the obstacles that water placed in their path. The relationship of Concordians to water was intimate and complex. For the most part it was routine and unremarked, but it did have its share of drama. In fact, almost every Concord incident that attracted attention beyond the borders of the town seemed to have water at the bottom of it. It seems fitting that when Concord's moment of fame finally came, it took place at a troublesome bridge over its own rebellious river, just at the head of the Great Meadow.

The various uses of water that had to be integrated in Concord included watering stock, driving mills, accommodating fish, transporting people (upon or across), and, of course, flowing meadows. We can get an idea of how complicated this could all become and of some of the conflicts and compromises involved by following Elm Brook from its source down through the meadows surrounding Brickiln Island and on to . . . well, into which river does Elm Brook "naturally" flow, the Concord or the Shawsheen? By the middle of the eighteenth century, what seemed a simple question for the law of gravity to settle had been diverted into the courts.

Elm Brook—or Tanner's Brook, as the Lincoln branch is sometimes called—originates today in an ordinary maple swamp tucked away in the uplands near the Cambridge turnpike. In the eighteenth century this swamp was sufficiently drained not only to be mowed for hay, but to be tilled—part of it was the three-acre "further field" given to Thomas Brooks by his father, Noah, in 1726 and reached by a lane that passed between land given to Thomas's brothers Joshua and Benjamin.[103] The brook may have begun there, but the water was only passing through. Water that runs in streams does not really originate in springs or swamps, after all, but on the slopes above them and in the heavens above those, as part of an endless cycle of evaporation, precipitation, and runoff back to the sea. Rainfall percolates into the earth, where much of it moves downslope within the soil as what is called throughflow. Farther along it may gurgle forth as a spring or collect in a wetland and flow on in a stream.

This elementary hydrology is of surpassing practical importance to stockmen. Concord farmers needed watering places in every pasture.[104] Livestock herded on commons could be driven to water as needed during the course of the day, but as the land was enclosed, places for watering the cattle had to be built into the new pastoral infrastructure along with the walls and fences for confining them (fig. 7.5). Pastures had to be laid out so as to bound on a stream or ditch or to include a spring or small pond, if at all possible. The landscape of Concord is still dotted with these watering places, once neatly stoned and cleaned out but now overgrown and silted in and reduced to wet seepages.

Daniel Brooks Sr. was particular about water, and he carefully recorded what was doubtless taken for granted in hundreds of other transactions. When Daniel bought half of his brother Hugh's upland at the Suburbs in 1701, the deed specified that a great rock near the head of a spring was the bound between them and that Daniel was to have six rods to the water for a watering place. When Daniel bought fifteen acres of upland

FIGURE 7.5.
A small pasture watering place, with a great spreading white oak for summer shade.
Stony glacial till slopes have many springs. Herbert Gleason, 1901.

and swamp from Hezikiah Fletcher in 1713, it was duly noted that the east bound took in a "living spring of water." When Daniel passed part of his land along to his son Job Brooks Sr. in 1725, one piece of upland and meadow was conveyed "reserving to Daniel Brooks for his life a watering place to drive cattle to water upon s^d parcel from November 1 until March 31 annually." Water from this spring fed Tanners Brook just to the east. It evidently stayed open in cold weather, a desirable feature. Attention to water ran deep in this family—a century later, Daniel's great-grandson Eleazer Brooks Jr. included in a deed of sale the right to use a walled-in watering place lying across the lane in the next pasture. That spring, which ran north past Joseph Stow's house into Elm Brook meadow, remains open today.[105]

Tanners Brook flowed north through a narrow little valley between the pastures and wooded slopes that fed it. Arriving at the Brooks homestead, even so small a brook was put to work and detained by at least one dam to power a bark mill and, earlier, a sawmill. Such small ponds along streams had other uses as well—some irrigated grasslands, either by flooding or in the English manner of flowing hillsides. Both Joshua and Thomas Brooks mowed lots along the brook near the sawmill dam, on land that was probably simply flooded by the millpond in the winter and spring and mowed in summer once the water was drawn off.[106] But elsewhere in Concord can be found at least one instance of a more elaborate arrangement, by which water was diverted from a brook to run in a ditch and to flow over a hayfield on the slope below.[107] There is no evidence that tilled crops were irrigated in Concord (though gardens may have been), but a good part of the mowing land was deliberately watered.

The waters of Elm Brook crossed the road, supplied the tan vats, received the wastes of the tannery, and flowed on into Tanyard Meadow, the head of the sprawling reaches of Elm Brook Meadow. And here life grew unexpectedly complicated for the unassuming little brook. Its natural inclination (as far as one can judge) was to continue north past Brickiln Island along the eastern side of the meadow, and then past the Virginia Road causeway and so northeast into the Shawsheen River in Bedford—as indeed it does today. But even in nature, over geologic time, the headwaters of one stream will sometimes cut through a divide and steal the flow of a neighboring watercourse. On November 11, 1695, the Mill Brook stole the waters of Elm Brook and bore them away west to the Concord River—with a little help from the sons of Joshua Brooks. This feat of vernacular engineering was significant enough to be recorded in the Middlesex County Registry of Deeds.

The day Noah, Daniel, Joseph, and Job Brooks received their inheritances from their father, they also signed an agreement concerning the water that passed through their meadows. The indenture noted that "the water now goes through Noah Brooks's meadow and part of Job Brooks's meadow"—that is, Elm Brook ran through the Tanyard meadow and then touched the east side of Job's mowing lot on its way north between Brickiln Island and the upland to the east (fig. 7.6). But under the agreement, a new ditch was cut toward the west, through Job's meadow and passing south of a great rock in Joseph's

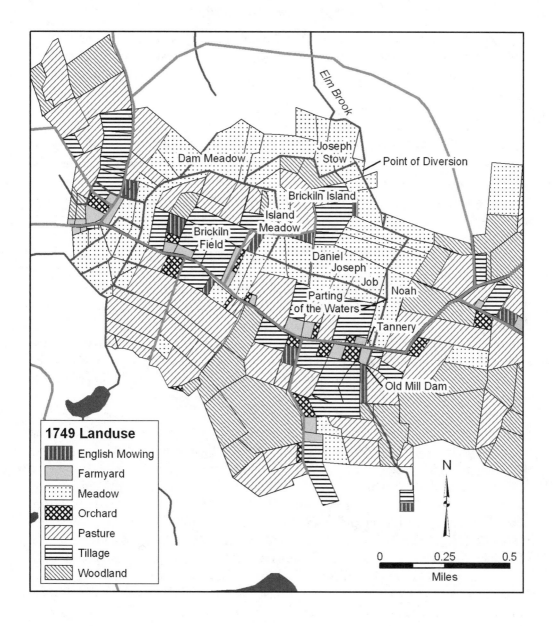

1749 Landuse

▥	English Mowing
▦	Farmyard
⣿	Meadow
▨	Orchard
▧	Pasture
▤	Tillage
▨	Woodland

N

0   0.25   0.5
Miles

FIGURE 7.6.

Elm Brook drainage, showing the "parting of the waters" where the four Brooks brothers
turned Elm Brook west into Island Meadow and the Mill Brook in 1695, and the contested
point where Nathaniel Stow and his son Joseph diverted it west again into the Dam Meadow.

meadow and then on through Daniel's meadow and into a meadow belonging to Francis
Fletcher to the northwest. This sent the water around to the west side of Brickiln Island
into the Dam Meadow, drained by the Mill Brook on its west end. To conclude the agreement, Noah and Job Brooks obliged themselves and their heirs not to make any ditch
or drain north of the new ditch and to fill the existing ditch that had carried the water
north; while Daniel and Joseph Brooks obliged themselves and their heirs to make and
maintain the new ditch in their meadows.[108] As the Dam and Island meadows below
them had been slowly drained and improved during the late seventeenth century, the
Brooks brothers had doubtless concluded that by cutting a ditch to the west they could
get a better flow. And so it remained: a 1733 sketch in the estate papers of Daniel Brooks
(passed down to his grandson Eleazer) showed Elm Brook flowing past the Tan House
to a point labeled "the parting of the water." From there one line marked "course of the
water" ran west, while a second line marked "course of part of the water" continued
north.[109]

But that was only the first diversion of Elm Brook. As numerous farmers later testified, except in times of drought some water still ran north past Brickiln Island, fed by
more springs. Then in 1715, Nathaniel Stow apparently dug a ditch "for a fence" along
his meadow on the *north* side of Brickiln Island, which turned that flow of water west as
well, into the Dam Meadow and Mill Brook again. Nathaniel's son Joseph Stow renewed
the ditch, but many times, "contrary to his allowance or approbation" (as he testified in
1756 and again in 1761), his ditch was stopped up and the water carried to the north in
another ditch that sent it toward the Virginia Road "causey" and Bedford, down Elm
Brook proper—presumably its original course. The perpetrators were John Wheat and
Jonas Wheeler of Virginia Road, who wanted to flow their own meadows.

This counterdiversion did not go unchallenged. The Virginia Road men were sued
for trespass by the Concord miller Timothy Minot, who laid claim to the "ancient and
normal course of water running into the Mill Brook," including all the water obtained
"by any means formerly attempted," under the 1668 mill charter from the Great and General Court. Minot's witnesses were many and willing. Joseph Stow, John Jones, Samuel
Fletcher, Samuel Meriam, Ebenezer Meriam, and Joshua Brooks all swore that as far
back as they could remember the water had always run west—except, they admitted, in
times of great freshet, when it recollected itself and ran north. This was hardly disinter-
ested testimony. The deponents all *benefited* by the water going the westerly course set
forth in their depositions, as Wheat and Wheeler objected; but eventually they gave in.
Wheat agreed in 1762 to stop diverting the water, and with that, Minot released his claim
of damages.[110]

One wonders whether Timothy Minot wasn't suing partly, or even primarily, on behalf of the farmers of the Brickiln neighborhood. His brother Deacon Samuel Minot
was not deposed, but he also owned meadows just below the Dam meadow; and Joshua
Brooks was Timothy's brother-in-law. But the miller certainly had his own interest in
protecting every drop of water that might augment the rather piddling Mill Brook. The

dispute had not gone dry, in any event—it bubbled up again a generation later. In September 1792, August 1793, and August 1794 (and at diverse times between), one David Wheeler "broke . . . down the bank of the watercourse & diverted & obstructed the water so that the mill did not . . . grind as commodiously as usual." Dr. Timothy Minot Jr., who had taken over from his father and rebuilt the Concord mill in 1763, took Wheeler to court. This Wheeler, however, was wealthy, had a distinguished Revolutionary War record, and was perhaps a more formidable opponent at law. This time the jury found against Minot—perhaps they had a hard time swallowing the claim that "a certain ancient . . . watercourse which comes from a spring . . . in *Elm* brook meadow so called *naturally* runs to" the Mill Brook. But Minot appealed to the Supreme Judicial Court and in 1798 a five-man committee found in his favor, awarding him five dollars damages, seventy dollars costs, and, of course, the precious water.[111]

The vigilance and tenacity of these men gives us an idea of how tightly controlled and sometimes contested the flow of water had become in Concord—even in the smallest of streams. An intricate system of interconnected ditches, elaborated by generations of husbandmen, served both to drain the meadows during the haying season and to direct the water to flow and fertilize the meadows afterward. Wheeler was eager to turn the water his way immediately after the hay had been cut in late summer, perhaps to recharge the soil and augment the growth of the aftermath before actually flooding the ground.[112] But this was a dead loss to the meadow owners lying to the west toward the Dam Meadow, who wanted the water for the same purpose; as it was to the Concord miller another two miles down the Mill Brook, who acted (at least in effect) on their behalf.

Thus every drop of water coming down from the uplands was spoken for, sometimes more than once, just as every piece of land had found at least one purpose. A network of dikes and ditches and a corresponding network of lanes and causeways had completely reordered the swampy landscape along the Mill Brook into a neatly defined patchwork of low meadows and upland islands. In 1720, at the request of Ensign John Jones, Samuel Fletcher, Nathaniel Stow, Judah Potter, Joseph Fletcher, Daniel Brooks Sr, Daniel Brooks Jr, John Jones Jr, and John Meriam, Concord had laid out a "free way" to Brickiln Island, "for those with land adjacent for carting and driving creatures to pasture."[113] A century after Concord was settled, this extensive lowland of swamps and marshes had been transformed into an intricately ordered landscape containing a myriad of small ditches, dams, culverts, causeways, cartways, and mowing lots surrounding higher islands of plowlands, pastures, fences, and gravel pits.

Below Hoar's dam the water flowed again in two channels past Meriam's Corner and across the Bay Road, after which it joined the main branch of the Mill Brook running west through the Town Meadow to the village mill. Here the water was again seasonally regulated for the benefit of the miller and neighboring farmers, this time more amicably, it appears. Minot had a right under his ancient mill charter to raise the water four feet ten inches above the bottom of the mill trough, without being liable for damages to those whose property was flowed at that head. But in 1738 he relinquished that right to several upstream neighbors during four summer months, agreeing to draw off the millpond and let the brook run from May 1 until September 1. This allowed much of the pond to be cleaned and the rich muck to be carted away, and for hay to be cut over most of the previously flowed meadow.[114] In this way, the millpond not only did not impede cultivation—it was turned to a positive agricultural purpose, augmenting the growth of the hay upstream.

From the mill waste, the Mill Brook flowed another half mile north into Concord River—arriving about a quarter mile below the confluence of the North and South rivers at Egg Rock and a quarter mile above the head of the Great Meadow. Over its winding course of four miles, the little stream that began as Elm Brook before its diversion west had driven several mills and passed via an intricate network of dams and ditches through hundreds of acres of meadows, draining them during the summer and flowing and fertilizing them during the winter. Now, as part of the river, the water would be called upon to perform another round of overlapping tasks, again by seasonal turns. Concord husbandmen faced much the same challenge of multipurpose engineering with their river as with their brooks, only on a larger and less tractable scale. They needed to drain the broad meadows and prevent the river from rising into them during the haying season, while allowing it to flow them beneficially during the winter. If the river was to be harnessed to drive mills, this too had to be harmonized with the all-important production of the hay. At the same time, the river's bountiful spring fish runs needed to be protected. Finally, the river posed a special challenge to transportation: it had to be safely bridged.

The river meadows required constant attention. The same geography that created the great meadows in Concord and upstream in Sudbury, that is, the drop of only two feet over twenty-five miles of channel winding through ancient glacial lake bottoms, left them vulnerable to flooding. A small, lethargic river with a rise of only a few feet could inundate an impressive floodplain and stay up for days or weeks on end, while the water passed indolently over the narrow bedrock bar downstream in Billerica. The problem of plumbing the river meadows was essentially the same as that of improving the smaller brook meadows lying above: conveying the flow from springs and brooks expeditiously through the meadow and out to the river channel. The difference was that following a summer rain the water would rise and then pass off the brook meadows quickly, and it was simple enough to dig a ditch sufficient to expedite the flow of water in the brook. But the river was another matter. After a rainy spell it would rise ponderously and might stay up long enough to spoil the hay. Trying to induce the river to drain faster was a real chore.

As early as 1636, their first summer in Concord, the inhabitants of the new town had complained to the General Court about the flowage of their meadows by water held back in Billerica and were granted liberty "to abate the Falls in the river upon which their towne standeth." In 1644, a commission of sewers made up of prominent Concord and Sudbury men was empowered "to set some order which may conduce to the better surveying, improving, and draining of the meadows, and saving and preserving of the hay there gotten, either by draining of the same, or otherwise." But whatever they may have tried to do about the falls was of no use. Edward Johnson reported in 1654 that "the Rocky falles which causeth their Meddowes to lie much covered with water . . . these people together with their Neighbour Towne, have severall times assayed to cut through but cannot, yet it may be turned another way with an hundred pound charge, as it appeared."[115] Perhaps having failed to cut through the bedrock in Billerica, the meadow proprietors proposed to dig a channel from the Great Meadow east to Elm Brook, sending excess flow down the Shawsheen.[116] This was not so far-fetched—something similar had already been accomplished by the town of Dedham in 1639 by diverting part of the Charles River through Mother Brook into the Neponset.[117]

Nothing so dramatic ever took place on the Concord River, and the commissioners of sewers soon fell into the routine business of cutting water weeds and dredging channels through all the sandy bars for twenty miles above the rocky Fordway bar, encouraging the water to move along as commodiously as possible. These measures apparently had some effect, for in 1702, soon after the province passed a new act authorizing the appointment of commissioners of sewers, Concord, Sudbury, and Billerica together petitioned for another commission to be appointed to carry on anew. "Something hath been done in order to the lowering of the water, by Removing Rocks and bars of Sand,"[118] the petitions ran "which hath effected so much good as doth encourage the proprietors. But some of the proprietors are unwilling to yield suitable and proportionable help or money to carry on the work whose neglect doth hinder that which may be of public good to the proprietors as also discourages those who have been and would be active in the matter."[119] A new commission was duly empowered, and the endless labor of improving the river went doggedly on. It was, however, the last time Billerica would join with its upstream neighbors.

An obstruction in the river which threatened real havoc with the meadows above was a milldam at the Billerica falls, constructed in 1710. The town of Billerica had granted a mill privilege to one Christopher Osgood in 1708 provided he would "secure and defend the town of Billerica from any trouble and charge that may arise for damage that may be done to the medows of the towns above us by said mill dam."[120] Clearly, the inhabitants of Billerica were expecting trouble, although they probably did not anticipate that the ensuing controversy would last for a century and a half. As long as the top of the dam remained below the level of the Fordway bar half a mile upstream, it posed no threat beyond that. Raised substantially above the bar, it could impede the flow of water for twenty-four miles to the northern border of Framingham, right through the Concord and Sudbury meadows.

The men of Sudbury and Concord cast a cold eye on the new dam, although their earliest objection was that it blocked the spring fish migrations. That was bad enough, and I shall return to that topic in a moment. But by 1721 the dam must have been raised above the bar because a new objection crept into the complaints being dispatched to the General Court in Boston. A petition from "Y^e Towns of Concord, Sudbury &C^A on Concord Riv^R" stated that

> they have formerly been much Benefited by the Fish Coming up Concord River, but Since the Erecting of a Mill Dam thereon within the Town of Bilrica By Christopher Osgood, The Fish have been almost wholly Obstructed from passing up, That they Conceive the Said Dam to be Erected Contrary to Law the River thereby Stopped and *a Great Quantity of their land Laid under Water*, that the Said Osgood being directed by the Commissioners of Sewers to remove the Said Dam utterly neglects and refuses it, and therefore praying that Nothing may be Revoked, of the power Given to the S^d Commission^{rs} But that Such Further Strength, & authority may be Given them as this Court in their Wisdom may think meet.

Laying the meadows under water was a far more serious offense than stopping the fish because it struck at the very heart of husbandry in the upstream towns. Fish they prized but could live without; without hay they would perish. The commission of sewers took its power to regulate the flow of the river for the common good seriously and expected to be backed up by higher authority when necessary—and so it was.[121] The commission was duly empowered to pull down the dam far enough to allow the river to discharge, and—after some further delay and the intercession of a special committee appointed by the Governor's Council—the dam (or at least the top part of it) was demolished in September 1722. A fracas ensued when a group of defiant Billerica men rebuilt the dam in November, but the matter must have finally been resolved by holding the head of the dam below the Fordway bar, as Concord and Sudbury demanded. Osgood's dam remained, but no further protests were heard from the upstream towns through the course of the eighteenth century—at least, none on account of the flowage of their meadows. Rebuilt during the 1790s and raised nearly three feet above the bar to supply water to the Middlesex Canal, the dam would once again become an object of bitter struggle for more than half of the nineteenth century.[122]

Meanwhile, the workaday effort to improve the river had to be renewed in each generation. In 1742, Concord and Sudbury farmers once again petitioned the General Court to empower a commission of sewers to remove bars and "stoppages" in the river, complaining as always that they often suffered great damage to their hay from floods.[123] But no matter what was done, the lowest meadows were never far from flooding in a wet summer. As the generations passed, these determined yeomen were to a certain extent working against themselves, by steadily clearing forests and draining wetlands higher in the watershed. This surely increased the summer flow of the river and especially the crest of floods. The effect was partially offset by the construction of millponds on every stream, which served as a large network of small reservoirs. On balance, it appears that once the controversy over the height of the Billerica dam was worked out by compromise, flowage of the river meadows was not the all-consuming problem that it would become again a century later. It was more a routine matter of constant vigilance, proper diligence, and fatalistic acceptance of the occasional loss by flooding of an otherwise productive resource. Yet another commission of sewers was formed in 1789, and one last generation of yeomen plunged into the river in their turn in 1816. Had the Industrial Revolution never occurred, their grandsons' grandsons' grandsons (and by now, perhaps their granddaughters, too) would no doubt be out there today, dredging mud and cutting weeds.[124]

But while the controversy over flowage of the meadows died down for a few decades after 1722, the same cannot be said for the struggle over fish. This dispute actually predated the construction of the dam by a few years and recurred periodically throughout the eighteenth century. The first sign of trouble came in 1709, when a petition arrived at the Governor's Council and General Court in Boston from the selectmen of Concord, on behalf of several towns along Concord River. The petition complained that "the fish that comonly runneth up sd Concord river in the springtime use to take their course from Merimack up by weymesit falls did formerly run up sd river by multitudes up so far as Concord & higher, & by that means the upper towns were accommodated (viz) Lancaster, Marlborough, Sudbury, Sherborn, Watertown Farms, Stow, Framingham, Medfield with some other towns & villages, until the fish were unseasonably stopped at Billerica by hedgewares that were made cross the sd river." This new weir in Billerica was too distant for inhabitants of the upper towns to go to catch fish, the petition went on, and even if they did the fish would taint on the way home. Billerica already had adequate fishing places, the upstream communities complained. The petition requested that no weir be allowed across the river below Concord and that a weir be allowed in Concord at a convenient place—but to be regularly opened to allow "fish to have a free passage to run their natural course."[125] The petition was granted, and at the Court of General Sessions the following spring the inhabitants of Concord were allowed to set a weir and spread a net over the river, but only during four days of the week. By 1721 the Concord fishermen had liberty to fish for five days, while the fish had liberty to run Tuesdays and (needless to say) on the Sabbath.[126]

Concord river and its tributaries once supported a large spring run of shad, alewives, and eels. The shad bred mainly in gravelly sections of the river channel itself, but alewives and eels also made their way up many of the smaller brooks. The anadromous alewives (a species of river herring) bred in ponds, the young returning to the ocean in late summer; whereas the catadromous eels lived within the inland waterways for several years until reaching maturity and then returned in August to the ocean to breed the following winter in the Sargasso Sea. Migratory fish had been an important element in the Native diet—indeed, the first mill in Concord (later Timothy Minot's) was sited at an Indian weir—and they remained attractive to the English, who had likewise husbanded fish within their home waterways for millennia. In New England, the English settlers had to become familiar with the ecological requirements of a new set of fish species. The newcomers were imposing complex changes on the rivers and brooks and so faced the challenge of either making their agrarian patterns compatible with the fish runs or accepting the depletion of a valuable common resource.

The changes made by the English to small streams like the Mill Brook *could* have proved beneficial to some fish species, if care were taken to accommodate them. Ditching up to the headwaters of these brooks in order to drain the meadows there may have opened some that were formerly impassable to fish, while millponds and farm ponds potentially constituted new spawning grounds—if the fish could get there. Seventeenth- and eighteenth-century farmers showed themselves capable of building fishways to allow fish to reach inaccessible lakes and ponds and of deliberately stocking those ponds with alewives—a practice which could soon build a substantial run. Milldams along the brooks did pose a hazard to the fish, but this could be mitigated by good management, which consisted of maintaining an adequate flow of negotiable water through the wasteway during the run in April and May and allowing the young fish down later in the season. Since the fish were regarded as common property of broad benefit to the public, towns imposed these standards on local mill owners throughout the colonial period.[127] Well into the eighteenth century, when milldams were small, mostly confined to brooks, and embedded in a legal tradition of tending and husbanding common resources, the volume of fish running in Concord River may even have been on the rise.[128]

Weirs and dams that blocked the river itself were a more difficult matter and did become a cause of controversy, as we have seen. The shad and alewife runs benefited many towns far upstream but were vulnerable to being monopolized or cut off by downstream interests. The objection of the upstream towns to the Billerica weir (and to the dam that soon followed) was not that Billerica was catching its share of the fish as they passed by, but that an unlicensed obstruction had been erected that caught too many fish, preventing them from running their natural course. As in the case of the diversion of the Mill Brook, the word *natural* here included a measure of customary cultural practice in that these men believed nature had been created by Providence for their use. The natural course of the fish thus comprehended both allowing enough fish to spawn to sustain the run and fairly apportioning the resulting resource among human communities. In fact, it could include the adopting of measures that encouraged more fish to spawn in every brook and pond. During the eighteenth century the runs were still extensive and attractive, pushing far into the headwaters. They were vigorously defended as a common resource.[129]

Responding to such concerns, the General Court in 1710 passed an act to prevent nuisances by hedges, weirs, and other encumbrances obstructing the passage of fish in rivers.[130] Unfortunately, Concord's troubles with nuisances in Billerica were just beginning—by then they faced not merely a weir, but a dam. Similar conflicts were occurring on other rivers, such as the Ipswich, where the builders of three dams were ordered to install wastes to allow the fish up.[131] In a pair of acts "to Prevent the Destruction of the Fish called Alewives" in 1736 and 1742, the General Court declared that milldams were required to leave a convenient sluice open from April 1 to May 30 to allow salmon, shad, alewives, and other fish to pass up. Town selectmen were allowed to appoint up to sixty days for the fish to pass down in the fall as well. Towns were also given the power to designate places for taking fish with scoop nets.[132] These acts were designed to give towns equitable access to the fish and to protect the valuable runs from overly destructive forms of exploitation. They must have been at least partially successful because fish continued to ascend at least as far as Concord through most of the eighteenth century.

Towns such as Concord took up the yearly business of seeing that the law was enforced, both designating who could fish and keeping a watchful eye on encroaching dams below. In 1741, Concord and Sudbury sent another committee to Billerica to treat with Mr. Osgood to remove his obstruction to the fish.[133] Every five years at town meeting, Concord auctioned the privilege of erecting the weir and fishing for shad and alewives.[134] In 1773, yet another committee was dispatched to apply for "removal of those obstructions in Concord River where by the fish that usualy come up . . . are prevented."[135] It may be that by this time the run was in decline, as dams began to be built across the upper reaches of the Sudbury and Assabet as well, truncating headwater spawning grounds. Concord continued auctioning the fishing privilege from 1727 until about 1800, when the run was cut off entirely at the old Billerica mill seat by the new Middlesex Canal dam.[136] By the early nineteenth century, according to the recollections of Edward Jarvis, consumption of fish from rivers and ponds was rare in most Concord households and only important for a few "thriftless" families.[137]

The waterways figured in transporting people as well as fish. Communication with coastal towns via the Merrimack River was blocked by Wamesit Falls near the mouth of the Concord, but the rivers were boated locally, for fishing and fowling at the very least. Travel across the water was a necessity—and a major headache. Bridges over the three rivers were costly to maintain and a source of interminable controversy between Concord and the General Court. Concord petitioned for tax relief in 1660, arguing that its ratepayers were at unfair expense in maintaining three bridges that were in general use by the whole country above.[138] By the early eighteenth century Concord lay along well-trafficked roads to the developing towns of northern Massachusetts and southern New Hampshire. An abatement was not granted until 1708, but still the colony had to constantly keep after the recalcitrant town to repair defects in its causeways and bridges.[139] Long before 1775, the North Bridge in Concord had become famous, but as a place where an unwary traveler might find himself hopelessly mired or even drowned. In 1738, Concord town meeting took up an article "to see if the Town will allow John Laughling of Stow anything for the Loss of his horse that was drowned with his Brother at the North Bridge Last spring." Unable to recover for his brother, Laughling at least hoped to be compensated for his horse. The inhabitants of Concord considered this request, but the motion "passed in the negative," as the clerk duly recorded.[140]

Commissions of sewers to relieve meadow flowage, petitions to remove blockages of fish by weirs, riots against forced dam removals, controversies over who paid for bridges—throughout the seventeenth and eighteenth centuries, Concord troubled the legislature repeatedly over matters pertaining to water. Aside from events during King Philip's War, nothing else that went on in Concord seems to have warranted much notice. Living and sometimes striving with the river, and with others over it, was a tradition that was already two centuries deep by the time the great "flowage" controversy arose in the nineteenth century. This litany of complaints reflected the turbulent spots in a long, difficult effort to remake the entire watershed to better serve an agrarian economy. In many ways, especially given the crucial role of meadow hay, good water management was the key to the whole enterprise. This long transformation of waterways undoubtedly diminished certain plant and animal species and was by no means free of human conflicts. But on the whole it appears to have been a remarkably successful effort to channel a range of ecological cross-currents—drainage and flowage, millponds and fish runs—into a single stream.

By the mid-eighteenth century, after four generations and a little more than a century, the English settlers of Concord had adapted a European mixed-husbandry system to their new country. They had done so both by modifying their husbandry to fit unfamiliar conditions and by transforming the environment into a more familiar pattern. They had kept some English crops, adopted a few American ones, and integrated them into a new working whole. In accomplishing this, they had also moved from a common toward a more enclosed form of husbandry—as was occurring at the same time back in England. But the outcome of this transition in Concord, in both its social and ecological dimensions, was strikingly different from that in England. In England, by the late eighteenth century, the agricultural revolution was moving rapidly toward so-called high farming by commercial farmers who rented large, consolidated acreages, hired the bulk of the rural population as laborers, and used new tools and crops for specialized market production. In Concord, by contrast, enclosure of the commons brought about the flowering of solid, independent yeomen whose husbandry was deeply embedded within a community of neighbors and kin with whom they traded labor and goods, with whom their scattered lands often lay intermixed, and with whom they shared in the regulation of many common resources through town meetings. Concord farming had come a long way since 1635 but was still at root a mixed pasture and arable system designed to yield a comfortable subsistence directly from family land and by exchange with neighbors, who all relied primarily upon the diverse resources within the bounds of the community.

The mixed husbandry that the men of Bedfordshire, Derbyshire, and Kent brought with them had been transformed in many ways. The ancient supremacy of wheat and barley had been cut down and only less demanding rye and oats among the English grains let stand. Native maize, well suited to sandy soils and wonderfully productive, had been adopted as the principal bread corn. Plowlands that lay scattered in the old general fields had been drawn home to the barn to facilitate manuring by cowyard and dung cart. Apple orchards, increasingly consigned to rocky uplands composed of boulder clay, had replaced barley fields as the primary source of drink. Through Herculean labors, far more meadowland than had been required in England had been drained and improved to meet the long New England winter months that left little or nothing to graze. As in England, many of these native meadows were systematically flowed to increase their productivity. Pastures had been steadily cleared from the forest and enclosed for cattle and sheep, although (most years) the swine still ran at large. Many yeomen had discovered that stony glacial till uplands, though hard to plow, proved best for pasture because they held water. European cool season grasses and legumes had been widely established in pastures, although very little English grass was yet cut for hay. Given the colder climate and the abundance of forest, woodlands had assumed an importance they lacked in England, and Concord farmers had gained new skills in utilizing the wood from a rich diversity of trees. The use of the forest for foraged fare such as chestnuts, pigeons, turkey, and deer had doubtless declined as the country was improved for cultivation, but it had by no means disappeared—and huckleberries, at least, were thriving as well as ever. Finally, the brooks of Concord had been disciplined to water English stock, drive English mills, and obey English laws, although the Great River sometimes reverted to its wilder American nature. And every spring, up through the neatly ditched waters of this reordered landscape, pushed teeming schools of adaptable American fish.

This system of husbandry was obviously quite different from the Native system that had preceded it, but it was also very different from the new system of convertible husbandry that had taken hold in England since the Puritans' departure. Most important, although white clover had been established in Concord pastures and some red clover was sown for hay, legume rotations had not assumed the crucial role in improving soil fertility that they were playing in the European agricultural revolution. But the private possession of substantial acreage of previously uncultivated forest land had *not* induced Concord yeomen to practice a less laborious, more extensive system of farming that rapidly used up the natural fertility of the virgin soil or led them to adopt slash-and-burn or brush fallow methods that resembled Native American horticulture. There is little evidence of that in Concord. Instead, the raw lands granted in the Second Division were slowly and deliberately filled with new farms as the generations passed; and once established, substantially the same set of plowlands remained in tillage on each farm for generations. In Concord, the soil had very little natural fertility to waste. To make it productive, Concord farmers elaborated a new version of an older pattern of mixed husbandry that revolved around the movement of stock and manure—a version of infield-outfield husbandry typical of pastoral regions. Instead of closely integrating grain and grass on the same soils through legume rotations, they continued to rely on maintaining a balance between grain and grass in largely separate parts of the landscape. The productivity of their principal corn land depended entirely on their meadows. That was the nub of mixed husbandry in colonial Concord.

English husbandmen had taken the world that the Indians had made, retained some of its features and added some of their own, and refashioned it into an entirely new (but similarly diverse) ecological system. There was more than one way to live sustainably in Musketaquid, or Concord. Taken as a whole, the new agroecological system made very complete, integrated use of local resources to meet local needs and in doing so was still

FIGURE 7.7.
Elements of mixed husbandry. The edge of a cornfield in the foreground, more plowland and mowing behind that, rocky pastures and woodlots in the background. Herbert Gleason, 1900.

largely bound by local limits. The system had a well-ordered structure, one in which nearly every part of the landscape was put to work producing something either directly for human use or supporting another part of the agricultural system. The landscape had been thoroughly and carefully subdivided into these elements, and an elaborate infrastructure of ways, walls, weirs, dams, and ditches had been built by which the system functioned (figs. 7.7 and 7.8). This is not to say that every last acre had been fenced and settled into a permanent task. Pastures were still emerging from woods, swamps were still being drained, and young husbandmen were still searching for more land that could be tilled as their fathers' farms were further subdivided. There were tensions in this world, but Concord had worked out its system of husbandry.

The orderly agrarian landscape was embedded in a similarly orderly social structure. Each yeoman's independently owned and operated family farm was deeply enmeshed in several concentric circles of community. First, the evidence suggests a high degree of economic coordination among families clustered in small neighborhoods, families such as the Meriams at Meriam's Corner and the Brookses at Elm Brook. Many families supplemented their farming with a trade that brought in extra income for themselves as well as furnished an important good to other members of the community, such as the Meriams

FIGURE 7.8.

A world of walls and ditches. A closer look into the background of figure 7.7, showing a piece
of meadowy ground set off by walls and crossed by a ditch. The massive wall at the back
of the meadow suggests how many stones had to be removed to improve it for mowing.
Beyond the meadow, successive walls delineate pastures and woodlots.
Herbert Gleason, no date (but probably the same day as fig. 7.7).

their smithing and the Brooks their tanning. By such means a wide range of resources
derived from the local landscape were given shape and distributed among the inhabitants
of Concord. Finally, in order for the system to function, a higher degree of coordination
was required at the community level. Much of this—for example, the maintenance of
networks of ditches to drain (and cartways to reach) intermixed holdings and the opening of meadows for fall grazing—was accomplished informally among neighbors. Some
ordering was accomplished more formally—through such covenants as the one signed by
the Proprietors of the Great Field or through such action at town meeting as the granting of a license to take fish. Every now and then what neighboring husbandmen and
neighboring towns did with their land and water caused conflict and had to be adjudi-
cated through the courts. No household was fully independent or self-sufficient, but the
community as a whole was closely tied to the land within its borders for the continued
production of the great bulk of its livelihood.

By 1749, the town of Concord was essentially full and a New England system of husbandry fully elaborated. The era of development, in which settled homesteads composed

of enclosed fields were gradually established from Concord's center out to its borders, was complete.[141] The old commons system had been transformed to one of independent farms embedded in community regulation, cooperation, and exchange. There was little slack left from which to create new farms in Concord—or indeed, sustain economic growth of any kind—unless by some fundamental intensification or reorganization of the agroecological system. This was a very difficult proposition because the existing system was already quite intensive and closely bound by limitations of soil, climate, and technology. Already, many members of the fourth generation were making their homes in new towns on land they or their fathers had acquired through grants, purchase, or military service; places like Acton, Grafton, and beyond. As the fifth generation came of age after 1750, the challenge of supporting more yeomen in Concord would only intensify.

CHAPTER 8

A Town of Limits

The necessary stock of the Country hath out-grown the meadows.
—JARED ELIOT, *Essays Upon Field Husbandry*

BY 1750, the town of Concord was full. Private enclosed farms had replaced the
commons and spread to the town's borders. The mixed husbandry by which the
inhabitants drew their living from the land had matured into a comprehensive
system that fit Concord's ecological conditions and had reached capacity. Population had
risen steadily through the first four generations; now further growth became difficult.[1]
But families were still large. More people could have subsisted in Concord but not without beginning to fray both the social and ecological fabric of the community. They could
not have enjoyed the comfortable subsistence to which they had become accustomed, let
alone the expanding material prosperity to which they had begun to aspire. Eighteenth-
century Concord had become a world of limits. It was not yet for the most part a "world
of scarcity" or of dramatic environmental decline.[2] It was, however, a world with little
further room, short of an ecological transformation, for demographic or economic growth
within its boundaries. Concordians would either have to adjust their way of life to stay
within those limits or find ways around them.

The challenge Concord faced was not one of soil exhaustion or environmental degradation. Concord's system of husbandry was fundamentally sustainable at the level of
population it had reached, although there was room for improvement. The problem was
less one of depleted resources than of swelling demand meeting fixed resources: more
mouths to feed, more children to settle within the town. Concord, like many New England towns, faced a demographic crunch. The mixed-husbandry system that had developed was labor-intensive and relied heavily on family: boys to help work the fields, girls

to help with the gardening, spinning, cheese making and child rearing. Given the initial abundance of land within Concord and given the social, religious, and military drive for a growing population of Christian souls in the new land, large families were desired. Given New England's healthy climate, large families were obtained—among the largest families ever known to human demography. These children needed farms and trades of their own, either within Concord or beyond.[3]

Within Concord was to be preferred but not always to be expected. The ambition of New England parents, the mark of their success (as well as their natural desire), was to settle their children close to home with a competency to take a respected place in the community. We have seen this wish expressed again and again in the deed of gift itself, with words such as "in order to his settlement in Concord." And in fact, through four generations many families had been able to establish several children within Concord. Family lands distributed in the Second Division were systematically filled with working farms until the edge of town was reached, existing farmlands were subdivided and redistributed as plowlands and meadows within them were steadily improved. But even in the early generations most families had also seen some sons and daughters move to frontier towns, as more territory was wrested from the Native people. Concord parents may have held an ideal of settling children at home, but they also accepted that circumstances might draw some of their offspring away.[4]

By the second half of the eighteenth century, as the fourth and then the fifth generations looked to give their children a start in life, the demographic dilemma had grown acute. Partible inheritance was still the rule, but increasingly the homestead was settled on only one son, "in order not to spoil the whole," and no further Second Division outlands remained to pioneer. Other sons had to be endowed in some other way. Only a few yeomen had sufficient family land in Concord to establish another young husbandman close at hand. In some cases a second heir received a small landholding, meant to supplement a trade—land to live on, but not land for a living. Only the wealthiest fathers had the means to help procure an existing farm on the land market within Concord. By the fifth generation in most families along the Bay Road only one son settled in Concord, while all the others left for the frontier.[5]

Meriam

The new pattern was evident at Meriam's Corner. Ebenezer Meriam's land passed first to his son Ebenezer Jr. (who died), then to his granddaughter Sarah and her husband, John Champney (who subsequently moved to New Hampshire), then into the possession of his daughter-in-law Sarah Blood (Ebenezer Jr.'s widow), and finally, after his death, was reabsorbed by his nephew Nathan Meriam and his great-nephew John Meriam. During the course of Ebenezer's life, these pieces of land had been passed about among five members of four generations of the Meriam family.[6]

Samuel Meriam died in 1767, leaving four sons and four daughters. True to the Meriam tradition, Samuel bequeathed one teenage son, Ebenezer, two steers and a colt, and another teenage son, John, two heifers and two calves, with instructions that they were to divide the land equally when they came of age—and if they could not agree, their uncles Nathan and Josiah Meriam were to make the division for them. How well John and Ebenezer agreed is not recorded, but Ebenezer died young in 1775, and two other brothers disappeared from Concord. This left John Meriam alone on his father Samuel's holdings at the corner.[7]

Nathan Meriam, solid yeoman and selectman, lived until 1782. He and his wife, Abigail, also had four daughters and four sons. The older boys moved to Mason and New Ipswich, New Hampshire. Nathan left the home farm to be divided between his younger sons Amos and Ephraim. But Amos and his wife, Deborah Brooks (Joshua's daughter), soon chose instead to move to Princeton, Massachusetts, where both families owned pasture. Once again, a single son, Ephraim, remained in Concord.[8]

Nathan and Samuel's youngest brother, the locksmith Josiah Meriam, left Meriam's Corner himself late in life, after his brothers had died. Josiah and his wife, Lydia, had seven daughters and four sons, and he appears to have provided for them well—but not at the old home place. One son became a physician in Framingham, another died young. When in his seventies, Josiah Meriam moved with his youngest son, Joseph, to a new homestead on Virginia Road, alongside some of the family meadows at Elm Brook. Josiah Jr., a carpenter, acquired a farm across the road from his father and brother. It was as though Josiah and two of his sons were bent on reprising the old Meriam pattern on the outskirts of Concord, where they had at least a little more elbow room.[9]

Most of Josiah's land about the corner was absorbed by his nephews John and Ephraim. The ancestral house, which had successively housed John, Ebenezer, Ebenezer Jr., and Josiah Meriam, disappeared about 1805. In the previous generation there had been four Meriam households on the old cluster of family lands. By the time the fifth generation was fully settled, only two remained.

Brickiln Field

At the Brickiln Field neighborhood, only the wealthy Minot farm was growing. Deacon Samuel Minot established his eldest son, Samuel, as a goldsmith in Boston and procured a farm on Virginia Road for his second son, Jonas.[10] The home farm on the Bay Road he deeded to his youngest son, George Minot, in 1765. Samuel and George after him both steadily accumulated land in the Ox Pasture across from the house and elsewhere in Concord (and in other towns), making this a solid and substantial farm.[11]

Meanwhile, yeoman Samuel Fletcher's more modest farm next door was deeded in 1770 to son-in-law Edward Flint, who had married Hepzibah, the youngest of six daughters.[12] Next door to Fletcher, John Jones died in 1762, leaving his equally modest farm to his youngest son, Farwell Jones—three older boys had already left home. Jones's daughter Olive married an elderly neighbor, Joseph Stow, in 1759. Stow died in the 1770s, leaving his still more modest farm to the widow Olive and two minor children. It is possible that for some time these last two tightly interwoven farms were worked as a single unit by Farwell Jones. The Brickiln neighborhood had all the farms it could hold, and even some of these were barely viable.

Brooks

The Brooks family was among the best endowed with land in Concord, but as the fifth generation came of age, even they were sending a majority of sons away from home. When Samuel Brooks died in 1758, eldest son Samuel took over the more-than-adequate home place. Two younger sons were left a few outlands in Concord (presumably to sell), along with more substantial property in other towns.[13] Samuel's younger cousin, Job Brooks Jr., passed his equally solid home place and all its outlying parcels intact to his youngest son, Asa Brooks, in 1794. Job's oldest son, Matthew, had been given land in Littleton and moved there.[14] Across the road, cousin Thomas Brooks farmed together with his youngest son, Noah, who died a year before his father and so finally inherited the farm from the grave. Thomas was able to settle his oldest son, Aaron, at the Benjamin Whittemore Jr. place a mile up the road. Middle son Luke moved to Stow.[15]

Lt. Joshua Brooks, tanner and gentleman, did about as well at establishing his sons in Concord—or in Lincoln, of which he was a leading founder in 1754—as any man of the fourth generation. Born in 1688, he was as old as some of the third generation. In 1745, Joshua passed the tannery and the home place on to his eldest son, Joshua Brooks Jr., and moved to a nearby farm that had belonged to his brother Benjamin. In 1760, Joshua supplied his middle son, Ephraim, a carpenter, with the makings of a farm on the uplands to the east lying toward Flint's Great Meadow—perhaps one of the last farms in Concord to be carved from raw Second Division land. Finally, at the end of his life, by deed in 1762 and by will 1768, Joshua Brooks passed his second home place along to his youngest son, Timothy. He was one of the few men of his time who were able to fulfill so completely the role of patriarch.[16]

But the same was not possible for Joshua Brooks Jr. of the fifth generation, himself a successful tanner, deacon of the church in Lincoln, and gentleman, who died in 1790. He passed the tannery and home place by increments to his eldest son, Joshua, beginning in 1781. Another son, Abel (who was apparently not fully competent), was established on a few scant acres nearby. Joshua's three other sons scattered to the winds— Jonas to Pepperell, Massachusetts; John across the border to Groton, New Hampshire; and William to Hollowell, in what would become Maine. To provide for most of his children this Joshua Brooks, unlike his father, had to convert his success at home into land on the frontier.[17]

Hartwell

The full range of responses to the tightening supply of land can be seen on the uplands east of Elm Brook. In 1754, a currier named Joseph Mason bought the nine-acre home-lot of what had been the Ebenezer Brooks place. Besides working leather, Mason tilled three of his rocky acres and pastured the other six with a single cow. He owned no mowing and no woodland. In his will of 1788, he left his currier tools to his eldest son, Jonas, and the rest of his real and personal estate to his wife, Grace, for life. When she died in 1802, her son and executor Jonas, now himself a currier, had the estate settled upon him to pay his siblings in equal proportion—that is, an estate worth $413 was to be divided among at least ten children. A few months after the estate was probated, Jonas sold the property to his neighbor John Hartwell and departed.[18]

The Hartwell family was a good deal more secure. Ephraim Hartwell owned a roomy tavern, his father Samuel's house next door, the large farm he and his father had once shared, and two pastures in Princeton. His eldest boy, Ephraim Hartwell Jr., moved to Princeton. The next son, Samuel Hartwell, moved into the neighboring house built by his grandfather and namesake. Ephraim gave Samuel a part of the farm for love in 1769, and he settled into the life of a clockmaker, blacksmith, farmer, and gentleman. The bulk of the home place passed upon Ephraim's death in 1793 to his youngest son, Deacon John Hartwell.[19] By the end of the eighteenth century the Hartwell land accommodated much the same two farms as when Ephraim Hartwell built his house alongside his father's in 1733.

The Whittemore family, equally well endowed with land, took a different course and chose to leave Concord. Nathaniel Whittemore of the fourth generation followed in the footsteps of his father, Benjamin, and continued to acquire land adjoining his own, but in the end he did not settle it upon any of his children. Instead, in 1758 Nathaniel sold 203 acres, his house, two barns, and his pew in Lincoln meetinghouse—all his real estate—to Elizabeth and William Dodge and moved his family to Harvard, Massachusetts.[20] On the south side of the Bay Road, Nathaniel's brother Benjamin Whittemore Jr had died young in 1735, leaving his children to inherit the remainder of a substantial farm his father had just begun passing to him. The farm was sold by the heirs in 1750 to Aaron Brooks, a second cousin.[21] The Whittemore land remained in the shape of the two farms it had taken by 1730, but the Whittemores themselves were gone.

Clearly, by 1750 Concord's system of husbandry could no longer provide for the continuing expansion of a population of simple yeoman farmers—no more new farms were being formed along the Bay Road, and no old farms were being further subdivided. Could Concord's economy grow in other ways, and thus support more people, at an equal or even higher standard of living? It could, and it had been, slowly but surely. If household labor was sufficient to take on other tasks beyond just running the farm during the course of the year, a larger economy of craft production, also based largely on local resources, could grow up alongside husbandry. This had been happening throughout New England towns in the eighteenth century as the population grew, and there were numerous tradesmen at work along the Bay Road: coopers, blacksmiths, locksmiths, housewrights, carpenters, tanners, curriers, cordwainers, weavers, and even a clockmaker, most of whom were also farmers.[22] So long as each farm was capable of producing (on average) at least a small surplus of food and wood beyond household needs, a larger population of tradesmen could be supported. Thus every new farm that was established from unimproved land allowed the growth of a more elaborate local artisan economy as well.

In Concord, a broad middling class of interdependent yeomen occupied intermixed lands, each producing much the same crops as his neighbors and aiming for a comfortable subsistence from his own soil, through family labor. To achieve that, most constantly exchanged work and traded farm goods with their neighbors, many were also artisans, and all traded with other local artisans as well. Like any ecological system, this one was highly competitive at the same time that it was cooperative.[23] It functioned through both male and female exchange networks in fantastically complex ways, as families passed through their life cycles, striving to maintain themselves at a respectable station and to enable their offspring to do the same. Concord's economy grew slowly in this way throughout the eighteenth century, by the steady improvement of farms and the surplus they generated. But for such internal growth to continue once the town had all the farms it could hold, those farms would have to not only sustain but *increase* their productivity.[24]

Concord's economy could also grow in another way, if farmers could reach "a market beyond themselves," as the nineteenth-century Concord native Edward Jarvis put it. That is, they could carry their produce to the ports and thus engage in the wider Atlantic economy. They could do this either by trading with local merchants such as Ephraim Jones or by driving their cattle and their carts to Boston and doing their marketing there. And increasingly, Concord farmers did just that. They had always imported some goods that were difficult to manufacture in the fledgling colony, so as to maintain what they considered a comfortable English way of life. From the beginning, Puritan settlers sought the "modest prosperity that they called a 'competency,'" and achieving that economic independence had always required some connection to the Atlantic trade world.[25] By the middle of the eighteenth century a "consumer revolution" was under way: farm families, even in country towns like Concord, desired and were able to obtain more imported goods, and cheaper: tea and sugar, rum and molasses, pots, pans and tableware, firearms, cloth, ribbons, and pins. To do this they had to market more of their own stuff, either directly by finding exportable commodities or indirectly by provisioning the port towns and shipping industry. Historians have argued that New England farm families increasingly reorganized their farm and household production to take advantage of these opportunities and to satisfy their own rising aspirations.[26]

This development increased the pressure on Concord's land: now it had to support not only more people, but people who wanted more. In part, these demands might be met by expanding production across the board to generate a larger marketable surplus—

not an easy proposition. In part, they might be met by reorganizing labor to concentrate more on those products that were the most profitable. Initially, such adjustments might fit comfortably into the existing rural economy of household production and exchange. But they had the potential, ultimately, to transform Concord's system of husbandry—to move it away from diverse production largely for local consumption and toward specialized production of a few marketed commodities. Such a shift might stimulate more efficient farm management, as economic historians have argued—it might even relieve pressure on Concord's land, if goods that could only be awkwardly produced at home could be imported instead. But farming that focused more narrowly on profit might also drive reckless, extractive use of the environment and of human lives—as was surely the case on the sugar plantations of the West Indies, where many of those cheap imported goods originated. So where was Concord in the final decades of the colonial era?

We have seen that by 1750 Concord's farmers and artisans were putting pressure on the land by their simple numbers; and that involvement with the market was adding to that pressure but might also be redirecting it. The town's population all but stopped growing for the next half century, as the inhabitants struggled to adjust to these new realities. The limits of what a certain way of living from the land could deliver had been reached. Can we delineate the precise agroecological nature of the tensions that had arisen within this intricately balanced mixed-husbandry system? What elements were showing strain in simply supporting the people of the town, what elements were being asked to generate a surplus, and what impact was this having on the land? Because Concord's husbandry was complex, relying on the integration of a diverse set of resources found in various parts of the landscape, it is worth taking a closer look to discover exactly where limits were being encountered. By 1771, something like one-third of Concord's land was still forested and might have been cleared if what was needed was simply more land to farm. But was it the right kind of land? Where was the rub in the harness, where was the hitch in the pull? Tax valuations reveal something about the state of husbandry in Concord.

Tillage

At the end of the colonial period, farmers were cultivating every acre of plowland they could manure, and then some. Farmers and artisans along the Bay Road were tilling eight acres on average in 1771, up from seven acres in 1749 (table 8.1). This land was sown with corn and rye to feed Concord's burgeoning population, which was still consuming bread made overwhelmingly from grain harvested on Concord's own soil. By 1771 the number of acres under tillage had been pushed to the limit—to the highest point ever recorded in Concord before or since, in spite of the loss of several hundred acres of plowland to Lincoln in 1754 (table 8.2).[27]

But though more acres were under the plow, no more grain was being harvested—in fact, slightly less. Average grain yields along the Bay Road stayed flat at just over 15 bushels per acre, or even got a bit worse (see tables 7.1 and 8.1). Across Concord, reported

TABLE 8.1 East Quarter Tax Valuation — 1771
Selected Farmers

Name	Ac Orch	Brl Cidr	Ac Till	Bu Grain	Ac Mow	Tons Eng	Tons FM	Ac Pas	Horse	Oxen	Cows	Swine	Sheep	livestock grazing units	bu. grain/ ac. tillage	tons hay/ ac. tillage
Job Brooks Jr	–	14	8	200	24	3	17	20	2	6	8	3	0	16.67	25.00	2.50
Samuel Brooks	–	25	9	160	35	3	12	30	2	2	6	3	0	10.67	17.78	1.67
Nathan Meriam	–	6	8	140	23	3	13	15	1	2	9	2	6	13.19	17.50	2.00
George Minot	–	8	10	160	30	10	14	24	2	4	11	2	2	17.95	16.00	2.40
Farwell Jones	–	20	10	150	24	5	10	6	1	0	4	1	0	5.33	15.00	1.50
Joseph Stow	–	3	3	40	12	2	4	10	1	0	4	0	0	5.33	13.33	2.00
Esther (Widow of Samuel) Meriam	–	4	8	100	30	4	9	30	1	2	7	2	0	10.33	12.50	1.63
Josiah Meriam	–	0	5	60	12	0	6	10	1	2	4	0	0	7.33	12.00	1.20
Samuel Fletcher	–	7	8	70	10	1	3	12	1	2	7	3	6	11.19	8.75	0.50
Thomas & Noah Brooks	0	–	10	–	28	–	–	32	2	4	15	6	0	21.67	–	–
Joshua Brooks	1	–	5	–	30	–	–	27	2	6	12	3	0	20.67	–	–
Ephraim & John Hartwell	2	–	20	–	37	–	–	25	2	7	10	4	5	20.38	–	–
Samuel Hartwell	0	–	5	–	5	–	–	5	1	0	3	2	6	5.19	–	–
Average	0.75	9.67	8.38	120.00	23.08	3.44	9.78	18.92	1.46	2.85	7.69	2.38	1.92	12.76	15.32	1.71

Source: Massachusetts Tax Valuation, 1771, Town of Concord and Town of Lincoln
Lincoln Tax Valuation, 1774

TABLE 8.2 Concord Land Use 1749–1850

	1749	1771	1784	1793	1801	1811	1821	1831	1840	1850
Total Improved Acres	6351	7475	7129	8011	7989	7261	8347	8547	8953	8593
Tillage Acres	1471	1487	1188	1064	1112	1156	1137	1098	1229	1068
% of Improved Acres	23	20	17	13	14	16	14	13	14	12
% of Total Acres					9	9	9	8	9	8
English Mowing Acres	445	908	753	722	841	992	1205	1279	1644	2206
% of Improved Acres	7	12	11	9	11	14	14	15	18	26
% of Total Acres					7	8	9	10	12	16
Fresh Meadow Acres	2581	2086	2089	1827	2236	2131	2153	2111	2088	1495
% of Improved Acres	40	28	29	23	28	29	26	25	23	17
% of Total Acres					17	17	17	16	15	11
Pasture Acres	1854	2994	3099	4398	3800	2982	3852	4059	3992	3824
% of Improved Acres	29	40	43	55	48	41	46	47	46	45
% of Total Acres					29	24	30	30	28	28
Woodland Acres			3878	4436	3635	3386	3262	2048	1994	1534
% of Total Acres					28	27	25	15	14	11
Unimproved Acres					1282	1732	1392	2833	3256	3744
% of Total Acres					10	14	11	21	23	27
Tons Hay/ Acre Mowing	0.82	0.71	—	—	0.70	0.74	0.64	0.65	0.72	0.89
Tons English Hay/Acre		0.69	—	—	0.87	0.85	0.73	0.73	0.82	0.93
Tons Meadow Hay/Acre		0.72	—	—	0.64	0.68	0.59	0.61	0.63	0.83
Bu. Grain/ Acre Tillage	13.2	12.2	—	—	15.1	12.5	14.9	16.4	17.4	16.7
Tons Hay/ Acre Tillage	1.69	1.43	—	—	1.95	1.99	1.89	2.01	2.17	3.07

Source: Massachusetts Tax Valuations, Town of Concord, (1811–1831 as reported in Lemuel Shattuck, *A History of the Town of Concord*)

Notes: 1749 "Mowing" = 3026 acres. English and Fresh acres are estimates derived from reported hay tonnage.

"Woodland" in 1784 and 1793 includes woodland and unimproved combined.

"Unimproved" also includes "Unimprovable"

Lincoln set off 1754, Carlisle set off 1780

grain yields went from 13.2 to 12.2 bushels per acre between 1749 and 1771 (see table 8.2).[28] Part of this apparent decline may have been caused by an expansion in land devoted to growing potatoes, whose production was not reported. But whether yields were actually decreasing or not, they were certainly stuck at very low levels—scarcely above 20 bushels per acre of corn, and 8 bushels of rye. That was but two-thirds of what they would be in 1801 and only about half of what would be considered normal in the mid–nineteenth century.[29] Given the husbandry of the day, Concord farmers were unable to squeeze more grain production from their land.

Why? The explanation advanced by many historians has been that New England's tillage land was getting worn out. Colonial settlers had farmed extensively, cropping their land without making adequate (or any) use of manure, moving on to fresh soil. Now only marginal land was available to replace the old, depleted ground, and fallow periods during which the land could rest and recover were growing short. "The general practice was to sow grain crops successively on the same land without manuring it until it was exhausted and then to leave it to fallow," said Percy Bidwell and John Falconer. Unfortunately, added William Cronon, "no manuring could be done to increase crop yields because cattle were rarely housed at night in a place where their dung could be gathered."[30] But this does not describe Concord. In Concord, as deeds and valuations make clear, all farmers had barns and cowyards to confine their cattle, all farmers cut hay and carted it home to feed their cattle, and all farmers relied on manure from their cattle to make their corn crop. Concord farmers lacked sufficient manure, but it was surely not primarily because of neglect.

One oft-cited critic who charged that New England farmers didn't house their stock and couldn't be bothered with manure was the anonymous author of *American Husbandry,* published in London in 1775. "The great want of the country," this man wrote, "is the want of dung." That much was true. What New Englanders needed, he continued, were great stocks of cattle, "not ranging through the woods, but confined to houses and warm places. This can only be done by providing plenty of winter food: at present, they keep no more than their hay will feed, and some they let into the woods to provide for themselves, not a few of which perish by severity of the cold."[31] This certainly seems a sweeping indictment of New England husbandry. However, the phrase "they keep no more than their hay will feed" simply states the obvious: the vast majority of New England's cattle were indeed kept on hay, at home, in the barn. If some stock were allowed to take their chances in the woods it was because fodder was in many places desperately scarce. The passage points to an acute shortage of winter feed and thus of manure, not to widespread ignorance and neglect. This author, who had only a secondhand acquaintance with New England husbandry at best, was largely ignorant of the nuances of his own subject.[32]

The problem faced by Concord's husbandmen in growing more corn was not that they were neglecting their manure, but that they were already using all the manure they could get. The same was no doubt true in many older New England towns. The bind was elucidated by Jared Eliot, a minister and physician from Killingworth, Connecticut, whose *Essays Upon Field Husbandry* appeared between 1748 and 1759. Eliot was a man speaking from within New England rural society, an acute observer and energetic experimenter whose recommendations were generally right on target. Indeed, a large part of what Eliot suggested in 1750 would become standard practice among New England farmers by 1850—not because they had read Eliot, necessarily, but because he was living among ordinary yeomen at a time when they were beginning to confront these problems and find solutions themselves.[33]

At the end of his first essay Eliot turned his attention to "our old Land which we have worn out. This is a difficult Article without Dung, which cannot be had for Love nor Money."[34] This did not refer to dung lying alone in the woods unloved, but to a critical resource in great demand but short supply, because it was already assiduously collected and stretched thin in application. As we have seen, in Concord even the droppings of horses stabled behind the meetinghouse during Sunday worship were spoken for, and matters undoubtedly fell out the same among Reverend Eliot's flock. Eliot wrote of the impracticality of raising hemp as a staple crop because "it would consume all our Dung to raise it in any great quantities; so that we should not be able to raise Bread Corn."[35] In Connecticut, as in Concord, manure raised the corn. In a later essay, Eliot recommended augmenting cow dung by mixing it with soil or other organic substances in the barnyard to better absorb and preserve the volatile nutrients. These worthy practices were standard on New England farms by the early nineteenth century and were doubtless common long before that. They had been practiced in England time out of mind, and Concord farmers never had any reason to scorn or neglect them.[36]

The shortage facing Concord farmers was less one of tillage land than of dung. There was more than enough land to plow, but there wasn't enough manure to adequately fertilize the required corn crop. The older plowland was undoubtedly worn out in the sense that it was suffering from the loss of some nutrients that manure alone could not well supply; in particular, calcium. The strongly acidic nature of New England soils was a handicap that would have been difficult to remedy even had it been fully understood, because there was no ready source of lime or marl in the neighborhood of Concord. Acid soils surely depressed crop yields and help explain why potatoes were such a godsend to New England: corn likes an almost neutral soil (pH6–7), rye can tolerate slightly acid soils better, but potatoes actually thrive in acid soils (pH5–6). Acid soil was an intractable problem, but we know it was not the primary brake on grain yields in the colonial era. We know this because grain yields *were* substantially improved in the decades following the Revolution, long before lime was added to the soil in any significant quantity. What was principally needed in Concord was simply more manure per acre. One remedy was to till fewer acres, the other was to apply more manure. Concord farmers could either reduce their grain cultivation, husband their existing manure better, or increase their supply of manure.

The first option, reducing the amount of land in tillage and looking elsewhere for a larger part of Concord's daily bread, was taken up after the Revolution, as we shall see.

But in the final decades of the colonial era, Concord farmers were pressing to supply the town's corn and rye from home ground, as they always had. Robert Gross calculated that an average farm required at least eight to ten acres in tillage simply to feed its own family, and this is about where the farmers along the Bay Road stood in 1771. Betty Hobbes Pruitt and Carolyn Merchant have argued that only about half that would have been adequate for subsistence, which suggests that these farmers, and even Concord as a whole, may have been marketing some surplus corn, rye, and potatoes to Boston.[37] But whether they were just feeding their own or responding to high grain prices, Concord farmers were struggling to maintain the level of grain production they had reached in 1749. To keep yields up, farmers needed more manure, which meant they needed to house more stock and feed more fodder. Could they?

Meadow and English Mowing

Concord farmers did keep more stock, it appears. Between 1749 and 1771, those along the Bay Road added about two cows to their herds (see tables 7.1 and 8.1). The same was true across Concord as a whole, where the total number of livestock "grazing units" stayed about even, in spite of the loss of a few dozen farms to Lincoln. Farmers added cows not primarily to supply more manure, of course, but because cows supplied more food in their own right. Cows were as essential as corn and rye to basic subsistence — they produced the cheese, butter, salt beef, and (indirectly) salt pork central to the colonial diet. Because demand was high, more cows were kept — but they must have been hungry cows. The Concord herd was up, but hay production was down.[38]

Farmers along the Bay Road mowed a few more acres in 1771 than they had in 1749. Most seem to have added an acre or two of English mowing — timothy, redtop, and clover. In spite of the extra acres, however, hay in the barn dropped by a few tons. Taking Concord as a whole, several hundred acres of meadow had been lost by the secession of Lincoln, but domesticated English hay was pushed upward by an almost equal amount. Thus on each farm, Concord yeomen were mowing more acres than they had been. Unfortunately, reported yields from the mowing had fallen significantly, from 0.82 to 0.71 tons per acre (see table 8.2). Some (or all) of this drop may have been simply the vagaries of the weather in 1771. Hay is a chancy crop to make even today, as every farmer knows — never mind the added risk of flooding that these meadowmen faced. But the downward trend would prove enduring: the yield from the meadows was slumping toward the soggy bottom where it would lodge throughout the early decades of the nineteenth century. Once again, these husbandmen were working harder to stay in the same place.

Three truths shine through the dusty old tax returns, truths that defined the world of Concord yeomen whose system of husbandry had been founded on their meadows. The first was that by the late colonial period almost every acre of meadowland that could be redeemed was already being mowed. The second was that the annual cut of hay from these meadows was declining. Hay, whether upland or meadow, is a demanding crop.

Continuous mowing reduces yield until it reaches a steady state determined by the rate of weathering of nutrients from soil parent materials, the action of legumes, deposition of atmospheric nitrogen, and, in the case of the meadows, annual flowage. Winter floods were beneficial but did not bring bumper crops—had the uplands surrounding Concord River been made of calcium-rich limestone, they might have, but these hardy folk had settled on granite. And meadows were never manured, except indirectly by flooding: the direction of nutrient flow was from the meadows, through the stock, onto the plowland. The meadows may have been inexhaustible but only at a modest output, and that floor was being reached.

The third truth was that the planting of domesticated English hay, the coming revolution that would transform this reality of constricted hay supply, was only beginning. It appears that many farmers were in fact simply replacing one kind of hay with another, seeding a few acres of their meadow to English grasses, at the upland fringes that were less subject to flooding. Others mowed some or all of their English hay within their orchards. Surely some were beginning to sow grass and clover to follow corn and a small grain (usually oats) on their plowlands, but it is not evident that this beneficial practice was widespread as of yet. More upland hay was being planted, but this was not yet leading to a substantial increase in hay production—and it would not do so for another half a century, at least in Concord and other valley towns.[39] Meanwhile, the limits of the meadows had been reached.

These limits were being encountered in many older New England towns, as Jared Eliot made clear in his second essay, published in 1751: "The scarcity and high price of hay and corn is so obvious, that there are few or none Ignorant of it. . . . This scarcity hath been gradually increasing upon us for sundry Years past. It is evident that the necessary stock of the Country hath out-grown the meadows, so that there is not hay for such stock as the present increased number of people really need."[40] That single phrase, "the necessary stock of the Country hath out-grown the meadows," says it all. It foretold the end of the yeoman era in New England.

Eliot went on to describe how this situation had come to pass. Many early towns (like Concord) had been planted by salt marshes or river meadows, where they "found so much mowing Ground more than they had Occasion for, that they Improved only such Parts as were best and nearest at hand, and let the Rest lie, and when by the increase of People they wanted more, they made use of what had been before Neglected, without any tho't or care to provide more; and Meadows not being easily or speedily bro't to, many are drove to great Straits."[41] The farming system in many older New England towns had rested on natural meadow and marsh grass from the outset and had continued to expand so long as there was meadow at hand to be improved for mowing. By the mid-eighteenth century, the outer limits of this original source of fodder were being reached throughout the country. The stock were outgrowing the meadows.

Eliot proposed two remedies for the scarcity of hay. The first was to go on draining. Cut a ditch into the most recalcitrant bogs and swamps, he advised, bringing yet more rich but previously inaccessible land into production; and drain more thoroughly, so that better crops than coarse meadow hay could be grown on that rich ground. The second was to plant cultivated hay: herdsgrass, fowl meadow grass, and, above all, red clover. "I believe it will not be well with New England, till every Farmer shall have a Bushel or two of Clover seed to sow every Year upon his own Land," Eliot wrote.[42] He had put his finger on two crucial improvements, draining of swamps and sowing of English hay, that would preoccupy Concord farmers for the next century.

Scarce meadow hay, far more than failing pasture, limited the number of livestock that could be kept in towns like Concord in the eighteenth century. More summer pasture could have been cleared, either in Concord or up-country, but without more hay it could not easily have been stocked. Limited winter fodder and limited herds in turn curtailed any further increase in the supply of dung. In Concord, this was far more important than scarcity or "exhaustion" of tillage land in capping the amount of bread corn that could be produced. By the third quarter of the eighteenth century, Concord was pressing hard against the limiting factor in its agroecological system: meadow hay. Planting of cultivated English hay increased, but not fast enough to keep pace with the continued expansion of plowland being pressed into grain production. By 1771, the amount of hay that was available to provide manure for each tilled acre was at an all-time low. As a result, grain yields fell (or at best remained at a very low level) as the manure supply was stretched over several hundred more acres of plowland than it could effectively fertilize.

This pattern can best be seen by viewing the late colonial period in a longer perspective of land use in Concord (see table 8.2). Grain yields tracked the available hay supply very well over the century from 1749 to 1850. The ratio of tons of hay being harvested to number of acres being tilled fell to its lowest point in 1771, and so did the yield of grain from that tillage land. Thereafter, things began to look up. Following the Revolution, Concord farmers cut back on the number of acres devoted to grain production, and yields accordingly rose—because more manure was available to fertilize each acre. Nineteenth-century farmers increased the supply of fodder by planting more English hay, kept more livestock, and handled their manure better; in consequence, grain yields rose even further.[43] The same positive relation between hay supply and grain yields appeared among farmers along the Bay Road: those with the most hay tended to reap the best corn crops (see table 8.1). The correlation between grain yield and the tons of hay each farmer could supply his cows to convert to dung for his tillage land is not perfect, but it is quite compelling.[44] It appears that in an agricultural system based on grass, cattle, and dung, a farmer needed to have access to about two tons of hay for every acre he tilled, in order to support enough livestock to manure his corn properly.[45]

Although they may not have calculated it to the decimal point, Concord farmers understood this formula very well. A century later it was expressed in an old Concord adage that came echoing down from the age of the yeoman: "No grass, no cattle; no cattle, no manure; no manure, no crops." A husbandman needed a good balance of tilled land, livestock, pasture, and mowing deployed across the landscape in order to achieve

a bare subsistence, let alone a reasonable chance at economic independence, a competency.[46] As the eighteenth century wore on, it became increasingly difficult for young
people coming of age in Concord to gain access to all the necessary elements of a working
farm. As a consequence, although not much more than half the forest was yet cleared—
and what remained was by no means all marginal land—the surplus children were leaving
Concord in droves. Before reaching any other limits, the town had outgrown the Great
Meadow.

Pasture

In contrast to meadow, pasture in Concord was plentiful enough. Just how much pasture
each farmer had and how its condition was changing are difficult to ascertain from tax
figures alone, however—in fact, in this case those figures prove deceptive. Farmers along
the Bay Road reported almost half again as much pasture in 1771 as they had in 1749 (see
table 8.1). Concord as a whole reported an even bigger jump in pasture. But this is wildly
misleading. Some land was undoubtedly added to the pasture base during this period,
but not 1,150 acres.[47] As we have already seen from a detailed examination of deeds and
probated estates, farmers owned far more pasture than had been reported to the assessor
in 1749 (see appendix 2). What they reported in 1771 appears to be a more accurate reflection of what they actually stocked (see table 8.2). Pastureland in Concord covered a
wide spectrum, from fallowed plowland, to stony upland "wood pasture" slowly emerging from forest, to scrubby brushland *returning to* forest. Stochastic jumps of 1,000 acres
up or down from decade to decade suggest that the lines between pasturage, woodland,
and unimproved land were difficult for the assessor to draw and not always drawn in the
same place.

Whatever their extent, Concord's pastures no doubt *were* declining in productivity—
slowly. The tax returns show a seemingly alarming drop between 1749 and 1771 from 0.7
cows per acre of pasture to 0.45 cows per acre, but this was mainly an artifact of the change
in the way pasture was reported.[48] Maintaining a cow on 1.4 acres of pasture (as the 1749
figures suggest) would have been doing impossibly well: keeping a cow through the grazing season on two acres was considered good performance for the day.[49] Maintaining a
cow (or her grazing equivalent) on 2.2 acres of pasture, as the 1771 figures suggest, is at
least plausible—Concord farmers, if those figures are accurate, were actually doing all
right. But if Concord pastures were not in a state of utter collapse by 1771, they probably
weren't getting any better, either.

As Concord farmers slowly enlarged their grazing lands by clearing forest, they ran
into long-term difficulties maintaining those pastures in a productive state. Managing
livestock so as to sustain a nutritious sward and fertile soil in New England is not easy,
and there is no indication that colonial yeomen were able to master it. Like woodland
management, it was a challenge that required more than an English mixed-husbandry
heritage and a few generations of experience with the New England environment to

fully work out. The problem involves a complex interaction of livestock grazing habits, vegetation, and soil chemistry. Concord soils are acidic by nature, which severely handicaps the pasture potential of even moist, mineral-rich glacial tills. Neutral soil is best for European pasture grasses and particularly for white clover, the most nutritious forage plant and the one with the leguminous capacity to improve soil fertility. A sweetening to merely slight acidity was perhaps initially achieved on some pastures, thanks to deposits of alkaline ash when the land was cleared and the slash burned. This lasted a few decades at best. Once subjected to continuous grazing, such pastures reverted slowly toward their native condition—that is, highly leached soil covered with oaks and pines. The gradual removal of nutrients by livestock brought the soil back down to a low, sour pH and to a low state of biological productivity. The prevailing acid condition, which was seldom (if ever) counteracted by liming, undercut the ability of white clover to improve matters by fixing nitrogen and translocating mineral nutrients from the subsoil.[50] Again, had Concord's uplands been formed of limestone there might have been a chance of developing and sustaining a reasonably productive legume-based grazing regime without any fertilizing, as had been done in some parts of Britain—but Providence was otherwise inclined.

Gradual soil depletion was bad enough, but continuous grazing also led directly to a slump in the quality of the vegetation. Grazing (unless very skillfully managed) encouraged unpalatable pasture weeds at the expense of more tender grasses. Livestock, left to themselves in the same pasture for weeks or months on end, nibbled down the regrowth of the best forage, exhausting it, and let the coarse stuff grow up unhindered. The farmer was forced to go out periodically and cut the invading brush with a stub scythe or bushwhack. We have little evidence concerning how Concord pastures were managed in colonial times. We only know what happened to those pastures in due course: they ran up to pine. Grazing was inimical to most hardwood tree seedlings, but it did favor woody pasture weeds such as blueberries and huckleberries (*Vaccinium* and *Gaylusacia* spp.), meadowsweet (*Spirea latifolia*) and steeplebush (*Spirea tomentosa*)—or, as it was also plaintively called, hardhack. In the long run, continuous grazing also encouraged the growth of juniper (*Juniperus communis*), red cedar (*Juniperus virginiana*), and white pine (*Pinus strobus*), seedlings of which the cows would not eat. This would become the common condition of much of Concord's pastureland during the nineteenth century and would lead in time to the return of the forest. Such brushy pastures probably began to appear here and there across Concord's landscape during the latter half of the eighteenth century.[51] The New England forest exerted relentless pressure to reassert itself, pressure that could be counteracted only by ceaseless labor or superb pasture management. In the long run, transforming the lowlands into verdant meadows proved much simpler than converting the forested granite uplands to productive permanent pasture.[52]

There was trouble to come, but during the late colonial period Concord pastures were still adequate to the summer grass exigencies of the basic subsistence herd of the town—the cows, working oxen, horses, some beef cattle, a few sheep. More pasture was steadily being cleared from the forest as older grazing land slowly deteriorated, but a severe decline in pasture productivity had not yet been encountered. The home herd probably could have been accommodated indefinitely without eating into the remaining forest any faster than the oldest pastures grew back up to pine. But that is not the whole story because, as William Jones observed in 1792, "The pasture land is not in proportion to the meadow land." That is, the "principal farmers" owned *additional* backcountry pastures where they grazed their beef herds. Judging by the better-off farmers along the Bay Road, some of these men ran as many cattle in hill towns as in Concord itself. These herds were no subsistence crop: the number of cattle being fattened by farmers such as Job and Asa Brooks and Ephraim Hartwell went far beyond their household needs, or even beyond what they could market to neighboring artisans—fall droves took these beasts to Boston, where some supplied the West Indies trade. Beef cattle were plainly the leading marketed commodity of most farmers, and they were partly raised on pastures beyond Concord. A system had emerged something like that linking Connecticut Valley farmers to surrounding hill towns, except that Concord did not have the surplus grain or high quality English hay to support a winter, stall-fattening business.[53] Concord stockmen merely maintained their steers through the winter on coarse meadow hay and brought them up to weight more slowly over several summers, on hill pastures they owned themselves.

Concord's home pastures did support some commercial production: the expanded cow herd, almost eight per farm along the Bay Road, not only reared the beef cattle but provided surplus butter and cheese and fattened pork for market. Almost all farmers raised some beef on their home pastures, and many smaller men also supplied the larger graziers with young stock and retired oxen to fatten. But a substantial part of the drive to clear more pasture for commercial beef production had been deflected from Concord to newly developing towns to the west and north. Acquiring those lands, clearing them for farmland, and marketing the beef they fattened was a synergistic solution to one of Concord's central problems: settling sons on farms of their own. As we have seen, backcountry pastures in Princeton, Pepperell, and beyond were often subsequently deeded or willed to one or more of the children. In the decades before these lots in new towns were permanently settled they generated income from wood products and beef, in the course of being cleared—often, no doubt, by the same boys who accompanied the cattle to the hills and later moved there themselves. In this way the new lands helped finance their own purchase and settlement.

Had pasture clearing for beef been confined to Concord, the town might have been driven to the limit of its pasture capacity by the end of the colonial era. But instead, this most commercial, expansive component of Concord's agricultural system was neatly bound up not so much with the problem of sustaining the inhabitants of Concord at a comfortable level, as with facilitating their demographic expansion beyond Concord and the development of new agrarian towns. This access to outside pasture, combined with the constraints to farm growth imposed by the shortage of meadow hay, kept grazing capacity from becoming a limiting factor in colonial Concord. Indeed, had there been a dire need for more pasture in Concord, more forest could have been cleared—as it was cleared in the nineteenth century, when an increased hay supply made it possible to keep more stock and a shift to commercial dairy farming made it necessary for more cows to spend the summer in Concord. One additional reason this land was not cleared in the eighteenth century, and one more reason backcountry pastures were preferred, may have been that Concord farmers were obliged to protect their woodlands.

Woodland

Given that Concord was feeling the pinch of farmland scarcity in the late colonial period, it may seem surprising that there was even a stick of timber still standing within the town. Yet there clearly was—woodland was not reported on the valuation of 1771, but valuations taken in 1784 and 1793 show some four thousand acres, a good third of the reported acreage, still growing trees. Farmers along the Bay Road generally reported ten to twenty acres of woodland each in the early 1790s. Probated estates indicate that these men usually had more acres in their woodlots than they reported to the assessor—at least twenty to twenty-five acres, as we have seen. Perhaps recently cut woods were allowed to repose for a time as unimproved land. Some men reported as much unimproved land as they did woodland, some less—but whether such land was cutover woods or overgrown pasture, it was on its way to becoming forest again. By expanding its farmland, Concord may have been cutting closer to the lower limit of a secure reserve of forest, but it was not yet facing an immediate shortage of timber and fuel.[54]

How much woodland was enough? Typical colonial wood consumption has been pegged as high as thirty cords per household per year in Hadley, in the western part of Massachusetts.[55] This seems high for longer-settled, milder Concord. When Ephraim Hartwell died in 1793 he allowed his widow, Elizabeth, "ten cords at her house cut fit for her use."[56] Hartwell was a wealthy man, and his allowance for his wife was generous across the board (it also specified white beans, malt, and winter apples); on the other hand, ten cords were probably not enough to supply all the fireplaces in the house, two-thirds of which Ephraim's son John had inherited. Were we to set average fuel consumption at twenty cords, Concord's two hundred or more houses would have required upward of four thousand cords of firewood every year during the latter half of the eighteenth century. By rule of thumb, the sustainable production of New England woodland is one cord per acre per year or perhaps a bit less. In the very broadest terms, it appears that by the late colonial era Concord's woodlands were just adequate to supply the inhabitants with fuel indefinitely—especially if brushy unimproved lands can be counted as part of a forest base of some five thousand acres. Concord's woodlands had reached a bare minimum and would remain there for the next half century.

Concord's forest was protected from more rapid depletion first of all because, as we have seen, the shortage of meadow hay put a brake on how much land could profitably be cleared for pasture or tillage. But beyond that negative check, woodland had positive value: it was a necessary part of the farm and an asset in and of itself. Woodland had been reduced to the point where it was in demand: from farm to farm, the value of woodland in the valuations of 1784 and 1793 ran sometimes just higher and sometimes just lower than pasture. By 1801 woodland would consistently exceed pasture in value. Had necessity or greed driven Concord farmers to clear the last of their forest to make cultivated land, it would have been at sharp cost to a diverse, intricately balanced agrarian economy. An individual yeoman might get by without sufficient woodland of his own, but the town as a whole could not exist without adequate forest, any more than it could without plowland or meadow. A few wood products might be carted in from other towns, but not very far—and not firewood, except at great expense. Farmers had a strong incentive to leave part of their land in trees.

That incentive went beyond fuel to fencing, timber, and other wood products, and it went beyond mere subsistence to the market. Concord was part of a larger forest economy, to a small degree in its source of supply but to a larger one in its outlets. Many yeomen from Concord and surrounding towns went to their woodlands not only for domestic requirements, but for exchange both within their circle of neighbors and with the wider world. By the late seventeenth century a market had developed for New England white oak barrels in the shipment of Madeira wine and Barbados rum, and farmers also supplied white oak timber for the shipping industry itself.[57] Farmers often paid for the goods they purchased at Ephraim Jones's store—buttons, gloves, raisins, nutmeg, thread, whips, knives, blank deeds, molasses, rum—with wood. Throughout the winter months they brought in loads of oak timber and boards, barrel hoops by the hundred and oak bark by the ton. One stranger picked up linen cloth and a pint of rum and departed without leaving his name, entering the ledger as simply "the man that brought ye claboards." Others—like Deacon Minot—settled their accounts by hauling these wood products to tidewater for the shipping trade. On April 19, 1743, Jones recorded, "Aaron Parker C[r] by Sam[l] Minott carting 2m 7c hoops to Medford." In the tangled manner of the times, it appears that Minot repaid a debt to Parker by carting twenty-seven hundred hoops against Parker's score with Jones.[58]

Concord farmers had long been producing timber. During the seventeenth century the first Joshua Brooks engaged in the ship timber business along with two men from Concord's South Quarter, Thomas Gobles and Nathaniel Billings.[59] For a time, there was a sawmill on the Brooks property.[60] The Brooks family continued cutting timber for market into the nineteenth century—when Asa Brooks died in 1816, he had forty tons of ship timber at the Moon lot in north Concord.[61] But at the same time many of their neighbors, including Ephraim Hartwell and the Meriams, were purchasing timber from Jones. Not all of that timber came from Concord. Some of the wood that passed through Ephraim Jones's yard came from younger towns to the north and west, such as Harvard and Acton—this may explain why the man with the clapboards wasn't recognized. Older stands of trees in Concord were still capable of yielding timber, but the local supply was being supplemented by production from hill towns on its way to the ports.

By the end of the eighteenth century, Massachusetts farmers were engaging in a debate that rings familiar to modern foresters. Was it better to cut woodlots clean, on short rotations, so as to encourage the most rapid, efficient growth of wood; or to selectively thin and let the most valuable trees grow for timber? We can assume that both methods were widespread in Concord, along with plenty of what is today called high-grading: cut the best and leave the rest. Cutting clean produced vigorous oak and chestnut sprout-lands that became commonplace by the nineteenth century (fig. 8.1). When these sprouts, which grew several to the stump, were only a few years old they could be thinned, thus yielding the slender poles that made hoops for hogsheads—sometimes several hundred per acre.[62] A New England version of coppicing was being invented. Hoops and, increasingly, firewood and even timber were products of managed woodlots, not of virgin forests. By the end of the colonial period Concord farmers were not simply cutting forest and clearing land willy-nilly, if they ever had been. They were working to maintain a complex regional forest economy that included a wide range of domestic uses for wood along with some marketed commodities. There were steady and growing demands upon the forest for both subsistence and sale, but those demands could no longer safely be met by cutting forest faster than it could grow. Concord's forest had arrived at a level where it would have to be sustainably managed if the inhabitants were to thrive, that is, so long as there were severe limits to what they could economically import.

The Concord forest was changing. By the end of the period, almost all of the primitive woods in Concord had either been cleared for farming or cut at least once for timber and fuel. Younger stands were growing up in their place, and the composition of the new growth may often have differed from the old. Slow-growing species like black ash and white cedar that were confined to a few wetlands had already grown scarce—references to lots in the Cedar Swamp had faded from probated estates and deeds of gift. In both woodlots and old pastures, the edge went to aggressive invaders, fast-growing species, those that could withstand grazing or those that sprouted most vigorously: red oak, chestnut, white pine, birch. There was still as much fire in the landscape as in Native times, helping pitch pine keep its hold on sandy land well into the nineteenth century. But since the first round of cutting was barely being completed and second growth was just getting started, these changes were only beginning to make themselves felt. They would become fully manifest in the great wave of reforestation a century later.[63]

Concord in 1771 had not arrived at a time of wood scarcity. But the inescapable requirement of keeping a sufficient part of the land in productive forest imposed yet another limit on the ability of the colonial system of husbandry to stretch and admit more yeoman farms. Tillage land, grassland, and woodland were closely balanced against one another. Had ways been found to expand the cultivated acreage there might have been more corn to eat and beef to sell, but the kitchen would have been colder. By the turn of the century, Massachusetts farmers were speaking of the need to "reserve" part of their acreage for woodland, and they were worrying about whether the annual growth of wood in their towns was equal to the consumption.[64] The Concord forest had been reduced

FIGURE 8.1.
Oak and chestnut sproutwood perhaps twenty years old, with sprouts from fresh-cut stumps
in the foreground. Within about five years all the chestnut stems would be killed back
by the blight. Herbert Gleason, 1918.

to its lowest comfortable level, and there it would remain for fifty years until the second quarter of the nineteenth century, when the advent of coal allowed it to be all but swept away.

Concord in the eighteenth century was a world closely bound by interlocking ecological limits. It was not a world of scarcity and want, much less one suffering from pronounced agricultural decline and environmental degradation. On the contrary, all the available evidence suggests that an environmentally well adapted system had been established. This system could provide a comfortable subsistence, which included a marketed surplus, for a certain population of farm and artisan families, and that population had been reached. All in all, Concord's land was not in bad shape. A substantial part of the landscape was still covered with forest, fish still mounted the rivers and brooks in the spring. The land continued to provide bread and cider, meat and milk, clothes and shoes, timber and fuel. There was little sign of serious soil erosion—indeed, the steady transfer to the tilled fields of nutrients from the meadows, via the ceaselessly carted dung of cattle and muck of ditches, may have increased the organic matter content and overall quality of many sandy plowlands, which typically rested on reasonably level ground and so did not wash much. Pasture management could have stood some improvement, but as long as limits in the supply of meadow hay checked the number of livestock that could be kept, and as long as Concord farmers had access to up-country pastures for part of the beef herd, the amount of land cleared for grazing at any given time would not go much higher than it had.[65] Given the system of husbandry that had been developed, a rough ceiling of sustainable production had been reached, one that could be held indefinitely.

In an integrated local economy such as Concord's, each part of husbandry was important to the whole. It became difficult for any one part to expand beyond a certain point if it could do so only at the expense of the others. These homeostatic, negative feedback loops worked to hold the system in balance. Furthermore, in a varied landscape such as New England, some of these elements—most notably, the meadows—were more or less restricted to certain suitable soils. In Concord, limited meadowland provided the first critical bar that held the spread of all improved acreage combined to about two-thirds of the town, preserving the rest in forest. This was in many ways a good thing, ecologically speaking, compared to what had happened to similar systems in England during earlier centuries and to what would happen in Concord itself under a new, more specialized commercial approach to farming during the nineteenth century. In colonial Concord, the meadow hay constraint in effect halted agricultural expansion before a set of more painful conflicts among plowland, pastureland, and woodland were encountered.

The meadow hay agroecological system in Concord appears to have been sound and stable, with two caveats. First, the inhabitants had not placed a corresponding limit on their own population growth; second, farmers were led to increase production for market beyond subsistence demands to meet rising material expectations and to finance the settlement of more children on farms beyond Concord. There was a sharpening conflict between the advantages of a large supply of family labor to run a New England yeoman farm and the available supply of land to pass on to those children. After the first four generations filled the town, fewer sons could inherit an adequate landholding on which to duplicate the economic independence and patriarchal success of their fathers. This became a social problem long before it reached a subsistence crisis in New England. Abundant land and the enormous task of making it into farms had encouraged large farm families, and a high degree of economic independence for each new generation had come to be an achievable social goal. Now it was becoming more difficult.

At root this demographic dilemma was simply the recurrence within a new social and environmental setting of an age-old problem in English village society. Such population pressure was nothing new. It had occurred before in England in the thirteenth century and again in the sixteenth: the demand for grain approached the limit of what the grass and livestock could support. Towns like Concord had reached it rapidly, with a comparatively high degree of demographic momentum and a high (and rising) level of material expectation. But the situation in eighteenth-century Concord was far less acute than it had been in more densely settled England, which had seen catastrophe in the fourteenth century and real suffering in the seventeenth. Concord was not yet pressing its natural resources to such painful limits. Furthermore, in populous English villages, increasing grain production had often meant not simply outrunning the manure supply but actually undermining it. Expanding the arable land ate up grazing land, which directly reduced the livestock herd and the amount of manure that could be delivered to the corn crops. Concord's version of mixed husbandry was more dependent on meadow hay to winter the stock and supply manure, and that snag was encountered while there was still some slack within other parts of the landscape and long before anybody was starving. Nevertheless, in Concord a well-established and steadily expanding ecological system had been brought up short, and an increasing part of the population was finding itself straining to meet social expectations. The yeoman world which had been comfortably settled in Concord had reached a crossroads.

Concord farmers increased their production for market as they worked to meet these expectations. Rising commercial production was not yet the dominant force in Concord's agrarian economy, and it was not degrading the environment, but it did give a firm indication of which road Concord would take. Concord appears to have marketed a small amount of surplus grain and probably some other produce such as onions and cider, but all these were relatively minor commodities that directly involved little more than 10 percent of the town's land area.[66] This had long been a pastoral and woodland economy, and cattle and wood remained its leading cash crops. But Concord's agriculture had not been reorganized around the commercial production of beef and wood, and the landscape had not been stripped of forest by burgeoning pastures or rampant logging. Instead, a goodly proportion—call it the cutting edge—of timber cutting and beef fattening had been redirected to clearing new farms in new towns, part of the process by which Concord's children were replicating Concord's agrarian society and system of husbandry across New

England. The old world of the yeoman was pushing its limits, and was just taking up the tools that would lead to its transformation.

Five generations earlier, seventeenth-century English agrarian society had responded to a similar set of ecological pressures with a set of adjustments that had far-reaching consequences: a reduction in births, the export of surplus population to the colonies, the adoption of more specialized, market-based farming as part of a more general capitalist economic expansion, and the gradual adoption of a set of new agricultural methods that made up the agricultural revolution. New England had been settled as part of this transformation of English society, but in an odd way had been able to recreate, in towns like Concord, an American freehold version of an older, yeoman way of life that England itself had overthrown. Now Concord faced, in its own way and in its own environment and as part of a much more thoroughly commercialized Atlantic world, some of the same fundamental constraints and conflicts inherent in a mixed-husbandry and woodland system. How would Concord respond? The answers to that question would help shape the American agrarian republic and in time would lend an undeserved taint to the way the husbandry of our colonial forefathers has come to be remembered.

CHAPTER 9

Epilogue: Beyond the Meadows

ON May 26, 1852, Henry Thoreau began surveying the old Noah and Joshua Brooks farm for its owner, the prominent Concord lawyer Nathan Brooks. One of fourteen children of Joshua Brooks, Nathan had grown up on the rocky farm by Elm Brook. Now he lived in the village. The land had come into his hands through the insolvency of his half-brother Isaac, who had remained on the farm until it broke him. Nathan Brooks's tenant Charles Bigelow assisted in the surveying and showed Thoreau the grubs that made the cattle jump in the spring sun by burrowing beneath the skin of their backs. The east wind was full of the scent of apple blossoms, and the meadows in the low places along the Bay Road smelled sweet to the surveyor as he walked home in the evening.[1]

Much of what Thoreau surveyed that day would have seemed familiar to Deacon Joshua Brooks a century earlier—or indeed to the first Joshua Brooks who made the farm, a century before that. Yet, upon closer inspection, all would have seemed deeply altered, too. The cows were of the same red breed, but their milk was headed (by rail) for Boston. The fruit trees blooming on the Lincoln hills were no longer Thoreau's astringent "wild apples" destined for the cider heap, the random vigorous seedlings planted by Deacon Brooks and his father, Noah, so long ago; but orderly ranks of grafted Baldwins and Porters bound for the city, with neat rows of peaches alongside them. The low meadows were still mowed, but many were now drained with clay tiles and planted in English hay. A great surge of upland hay had supplanted the meadows as the chief source of fodder, and pastures had driven most of the remaining forest from the surrounding hills. Concord, in 1850, had reached the low point of forest cover in its history—only about 11 percent—and Thoreau wondered how much longer he would be able to see woodlands on every horizon, encircling the town, from his attic study in the village.[2]

Neither were the farmers along the Bay Road the yeomen of old. Many of the sturdy old families—Minot and Fletcher, Jones and Stow—who had settled Concord, been granted land in the Second Division, and made it into farms had borne their own final crop and passed from the scene. Other families, like Meriam and Brooks, were in their last flowering at the places where they were first planted. Now the houses sheltered not only Yankee farmers, but also farmhands and domestics from New Hampshire, Nova Scotia, and Ireland. Several were occupied by wealthy businessmen such as William Leighton, a Cambridge glassmaker who had acquired what was once the homestead of Gershom Brooks, and "made the old place smile and blossom."[3] Some scions of the old Concord stock, men like the temperate lawyer Nathan Brooks, were flourishing in this new commercial landscape. Other shoots, like his brother Isaac, who had stayed with the family farm and tannery, had withered away.

The yeoman world of mixed husbandry had undergone an economic and ecological upheaval. This transformation had not come overnight, but by 1850 it was complete. It had begun slowly in the decades following the Revolution, as the inhabitants of Concord altered their husbandry and their economy to relieve stresses in the old system and at the same time changed the ways they thought and behaved. These changes came to fruition during the second quarter of the nineteenth century, when both the agrarian economy and the landscape of Concord were swiftly remade by a steadfastly progressive generation. Their recollections of the vanished world of their youth have left us with an indelible but subtly distorted image of the yeoman husbandry of their forefathers. To this day, we tend to view colonial agriculture largely through the sharply critical eyes of the self-confident nineteenth-century improving men who knew it in its dotage and who wrote its obituary.

The old Brooks farm that Thoreau surveyed in 1852 had fallen on hard times. The farm was remembered by Nathan Brooks, who was born there in 1785, as rough and rocky and at best a good place to learn habits of industry and perseverance—a family saying had it that the devil had dropped a sack full of stones there.[4] That memory was a far cry from the place which had, after all, produced a long string of prosperous yeomen who styled themselves gentlemen. But the Brooks farm was not floundering in the 1850s because its soil was worn out or even full of stones. Across the road, on very similar ground, Nathan's third cousin once removed, Asa Brooks Jr., continued to prosper on the farm of his ancestors until 1847, when he sold to a Danish dairyman named Emelius Leppleman.[5] The Noah and Joshua Brooks place faltered because its ancillary industries, a tannery and a tavern, failed to thrive in the new economic environment.

Nathan Brooks had left the rocky farm in 1804 and put himself through Harvard by teaching school. He then took up the practice of law in Concord, specialized in probate, and did well. In 1825, his father, the tanner Joshua Brooks, who half a century before had remarked, "I guess they are firing jackknives" when a musket ball sliced his coat at the North Bridge, died a gentleman deeply in debt.[6] The tannery had already passed to Nathan's brother Isaac, who (perhaps with Nathan's help) was able to redeem part of the home farm at the estate sale following their father's death. In 1828, Isaac purchased the neighboring Noah Brooks Tavern farm as well, mortgaging both properties heavily to his brother and others.[7] Thus, Isaac Brooks briefly reunited a farm that had been intricately subdivided among brothers and then cousins for three generations, since 1713. Such consolidations were under way throughout Concord as the agricultural economy commercialized.

But, for whatever reason, Isaac Brooks was not able to make a go of it as either a tanner or an innkeeper. Slaughterhouses and tanneries were flourishing in Concord village and near the great Brighton cattle market—perhaps the old Brooks tannery was too small and too far off the beaten track to compete. The tavern, like many others, may have fallen victim to the temperance movement of the 1830s—the innkeeper's brother and mortgage holder Nathan, ironically, was an early champion of sobriety. The Cambridge turnpike had been built in 1802 half a mile south of the tavern, taking some of the traffic from the old, winding road to the bay. This could not have helped business, either. Or perhaps, on top of these handicaps, Isaac Brooks simply lacked the acumen to succeed in the new world. He died in Boston in 1844, an insolvent debtor and lowly currier.[8]

Nathan Brooks looked after the old place for more than a decade, accommodating his stepmother, Sarah, in one of the family houses and leasing the remainder of the farm to various tenants. Finally he had it surveyed, split it up, and sold it off. The eastern part of the farm, where he had spent his youth, Nathan conveyed in 1859 to his nephew, Isaac's son Joshua—one last Joshua Brooks on the property. But Joshua sold up in 1862, and the land passed from the Brooks family at last.[9] The western part, the Noah Brooks Tavern farm, was sold in 1855. Two years later it was acquired by a young man from up the road named Samuel Hartwell: grandson of Samuel Hartwell the clockmaker, who was grandson of Samuel Hartwell the pioneer on Elm Brook Hill, who was grandson of William Hartwell from Bedfordshire. This Samuel Hartwell established a prosperous commercial orchard and market garden on the old Brooks farm. The land was not played out—it had simply been waiting for an owner who could march to the beat of the new economy.[10]

As Thoreau walked home to the village after surveying the Brooks farm he passed by Meriam's Corner, where much the same drama had been acted out over the previous half century and was now sliding toward its somber conclusion. The ancestral Meriam house, built by John Meriam about 1664, had vanished by 1805, its site lost beneath the plow. Across Bedford Road by the lane into the Great Field, the house built by John Meriam Jr. about 1691 had also disappeared. That property had been inherited in 1769 by another John Meriam, great-grandson of the first, who became an ambitious trader in land. In 1778, he acquired the neighboring farm, just west of the corner, from Daniel Taylor and bought and sold a great deal more land over the next three decades. When this John Meriam died in 1804 he owned some 182 acres, but his estate was heavily encumbered and had to be auctioned to pay the creditors. The family retained some interest in the property, and it eventually made its way back into the hands of one of John's sons, Tarrant Putnam Meriam. The place became steadily more deeply mired in mortgages and litigation (including rancorous suits within the family) and was finally auctioned for good in 1828. By the 1850s it was being farmed by a newcomer named Jabez Gowing. This land had strayed a long way from the close Meriam and Taylor family bonds of a century before.[11]

The third house built by the Meriam clan at the corner survived in 1852 and still stands today. Erected for Joseph Meriam about 1705, it was occupied by his son Nathan and then by his grandson Ephraim Meriam until Ephraim's early death in 1803. Ephraim's widow, Mary, lived on in the house until 1847, remarried, and became widowed again. With her lived two of her sons, Ephraim and Rufus Meriam, and her youngest daughter, Maria Swan. Ephraim Meriam Jr. became a successful businessman and a driving force in the commercial development of Concord. He ran a butchering and candlemaking enterprise in the village with his cousins Nathan and Cyrus Stow and was an investor in the Milldam Company, which in 1828 drained the millpond and turned it into Concord's central business district. Ephraim also had extensive real estate dealings—he took a mortgage of the Brooks Tavern Farm from Isaac Brooks, for example. Neighborly exchange now entailed not just recording and discharging petty credits and debits of goods and services, but holding mortgages which could break a man or dispossess his heirs. Ephraim Meriam made himself wealthy in this new world, but he was by all accounts civic-minded and a leading citizen, as such men were expected to be. He died unmarried in 1843. By then, ownership of the old place at Meriam's Corner had long since passed to his brother Rufus, who ran the home farm.[12]

Rufus Meriam operated the last Meriam place as a dairy—Thoreau marveled that the wealthy old man resolutely milked seventeen cows, by lantern light, on the coldest winter mornings.[13] Like his brother, Rufus never wed, but even after his youngest sister married and moved away and his mother died, he hardly lived alone. In 1850, he cohabited with a farmhand and a housekeeper. By 1860, *fourteen* people occupied the old Meriam property: Rufus, his housekeeper and farmhand, two older women, an Irish family of seven, and a younger couple from Ireland. But the living must have been plain in Rufus's part of the house: he was a legendary miser and died in 1870 seized of (it was said) less than one hundred dollars in household goods and farm implements appraised at sixty-nine dollars—a man lost between two worlds. Meanwhile he had accumulated well over thirty thousand dollars by investing his brother's legacy in railroad stocks. It was said that he had been swindled out of ten thousand dollars of those stocks by his own nephew Micky Ball, who ran away to Ohio with the money but was caught, convicted, and served ten years in prison. There is no simple accounting for human behavior, of course, but one has the feeling that an eighteenth-century Meriam would have sooner cut his throat than rob his uncle. Amidst these melancholy bankruptcies, lawsuits, and crimes within the family ended over two centuries of Meriams at the corner, fading away in a single generation.[14]

By 1850 Concord's agriculture, along with its agrarian culture, had been entirely remade. That complex transformation is not the subject of this book, but a brief look at the changes that took place in Concord's farming as it became thoroughly commercial will throw the preceding yeoman era into sharper relief. Even a glance at the town's land use (see table 8.2) reveals that dramatic ecological changes did not occur until the second quarter of the new century: a sharp rise in upland English hay and a collapse in woodland mirrored by an explosion of something called unimproved land. This marked the arrival of a new (manifestly unimproving) style of agriculture. But the roots of those changes lay in the decades following the Revolution, as the people of Concord adjusted to the ecological constraints of the old yeoman world and the increasing opportunities of the new world of commerce. By the 1820s, Concord farmers finally broke through the limits of locally based production and consumption and began extracting more profit from their land.

What forces led to this revolution in Concord's agriculture? By the late colonial period, Concord had felt the tightening of the same ancient binds that periodically gripped European village society, old frictions between large families and limited resources, between expanding arable and stagnant or shrinking grassland. As Concord encountered these stresses in the eighteenth century, the same set of responses were open to its people as to their English ancestors: control fertility, emigrate, intensify subsistence production, or specialize in commercial production. These were not mutually exclusive possibilities, and each was adopted to some degree. But the last would ultimately prove the most important in remaking the landscape of Concord.

After the Revolution, Concord became a demographically stable place, at least on the surface: its population grew hardly at all from 1775 through about 1820, and only slowly thereafter. This equilibrium was achieved both by lowering fertility and by leaving town. At first, the size of Concord families decreased because young people married later—that age-old response to straitened circumstances in English agrarian society. Between 1750 and 1790, the age at which women married rose from about twenty to twenty-three, and the average number of children fell from seven toward five. Fertility continued falling in the early nineteenth century to about four children, but by a new means: deliberate family limitation within marriage, particularly by spacing later births. By the 1820s Concord parents were planning their families and investing emotionally and economically in fewer children, allowing them to make good in a more individuated society. They were no longer counting so much on a large pool of family labor to run their farms. What had begun as a built-in yeomanly (and goodwifely) constraint on marrying was transformed into deliberate self-control within marriage.[15] Meanwhile, in spite of dampened fertility, a somewhat diminished but steady stream of offspring sought their fortunes elsewhere—

more children were still being born than the town's economy could support. Concord's young people went on moving to new towns in the New England hill country and beyond and to new opportunities in the growing cities as well. Providing a start for these sons drew fathers further into commercial activity to obtain the means to supply land and capital, so that they *could* make good. Thus, in replicating themselves, yeoman aspirations continued to lead many farmers into deeper involvement in the market.[16]

Concord's agriculture displayed a similar stability during the half century following independence. As we have seen, by the end of the colonial era the yeoman system of mixed husbandry was approaching maximum sustainable production of a diverse supply of food, fuel, and other goods for the town's inhabitants, along with a marketed surplus primarily of cattle and wood. It appears that this interlocking arrangement of necessary land uses crystallized, and there were few signs of either dramatic productivity increase or environmental decline over the next fifty years. But again, beneath the surface, changes were taking place—changes that both allowed the yeoman system to achieve the stability that had seemed so threatened and set the stage for a great transformation to come. Concord yeomen continued to improve their ability to provide the diverse elements of a comfortable subsistence but at the same time began importing basic commodities so that they could direct more of their own production to market.

Each generation had worked to adapt English husbandry more perfectly to Concord's environment, and late eighteenth-century farmers and their wives were no exception. The milking season was lengthened and the supply of butter and cheese extended through the winter, beans became a mainstay of the diet, and gardens produced a wider variety of foods. In particular, more root crops—turnips, onions, cabbages, and carrots—were grown for winter storage. Above all, far more potatoes were grown, not just in the garden but in the field. Potatoes yielded more calories and protein than the same acreage of corn, and by 1800 each Concord family was planting about an acre of them. Edward Jarvis tells us that at the turn of the century, either baked beans or baked potatoes would sometimes serve as the basis for a meal. Concord's locally grown subsistence production continued to become more ample, diverse, and reliable, although it could surely still be plain fare, particularly for poorer families.[17]

During the early Republic, Concord's agriculture saw modest gains in productivity and probably some increase in the marketed fraction of production—all within an ecological framework that had not radically altered. Grain yield went up a bit, but total grain production went down. By 1800, corn yield reached about thirty bushels per acre, as reported by Thomas Hubbard.[18] Yields rose after 1771 mainly because the area in tillage dropped by several hundred acres and remained at this lower level through 1850. This meant that the available dung could be concentrated on fewer acres of corn and potatoes: Concord's plowland was in better balance with its hay and manure supply. Hubbard also reported that oats were sown in succession to Indian corn at Concord. The normal course in Massachusetts by 1800 was to sow herdsgrass and clover along with the oats and then lay the land down for several years of hay mowing. The gradual spread of such beneficial rotations may have also helped boost the productivity of Concord's cropland, although it is difficult to be certain how widely such convertible husbandry was adopted.[19] But while yields were rising, overall grain production was declining because less land was being tilled and less tilled land was being planted to grain: several hundred acres of corn were replaced by potatoes. In a change that portended much, Concordians were beginning to buy wheat flour, imported from Maryland and Pennsylvania.[20] Through this combination of modified cropping and resource substitution, Concord farmers were able to achieve a modest increase in marketed production, on smaller cultivated acreage: they were growing a little less corn and rye but eating a lot less brown bread and more white bread and potatoes.

By the early decades of the nineteenth century Concord farmers may have been able to market more corn, rye, and oats, but it was clearly not a big jump or a big change in their system of husbandry. They probably continued to look to cattle and wood as their principal cash crops. New slaughterhouses and tanneries popped up in the village, and merchants searched for ways to send more salt beef, salt pork, tanned hides, and oak barrels to Boston. For a few years, the wood dealer Amos Wood tried boating firewood to Charlestown via the new Middlesex Canal—it was a long, expensive haul to Boston by either oxcart or canal boat, but as the urban market for firewood strengthened, some woodsmen kept at it.[21] But again, there was no sign yet of a dramatic increase in livestock on Concord farms or of a sharp decrease in Concord's forest. As commerce with Boston and milltowns picked up, Concord farmers expanded their production of familiar crops and undoubtedly managed to market a larger fraction of what they produced. It is hard to imagine that they were working much harder or smarter than their fathers, but they were directing their energies more systematically at the market. But until the 1820s, much as they strove to engage in this commercial economy (partly from desire and partly from necessity), they were still confined within the limits of their old mixed husbandry system. The modest improvements they made continued in a long tradition of steadily coaxing more from Concord's soil and did not represent a dramatic turnaround or a sudden conversion to more intensive farming. Their heads may have been increasingly in the world of enterprise, but by and large they still walked the furrows of yeoman husbandry plowed by their industrious forefathers.[22]

All this changed in the second quarter of the nineteenth century. A set of transformations of both consumption and production took place that allowed Concord farmers to break through these limits and concentrate overwhelmingly on farming for the market. Concordians all but abandoned their homegrown corn and rye johnnycake for store-bought wheat flour, and farmers largely shifted their tillage to oats for horses and corn for dairy cows. Oxen passed their peak and began to decline—they were steadily replaced by more expensive horses, geared for a faster, more businesslike pace of farming and transportation. Hard cider fell into disrepute and was replaced by imported tea and coffee—this was emblematic, as Edward Jarvis observed, of the new, temperate, rational approach to life in Concord. Farmers cut down their old cider orchards and planted fancy apples and peaches for urban consumers. Homespun linsey-woolsey was supplanted by cotton cloth from the mills springing up along rivers throughout the region, and farmers shifted from growing a little wool and flax for their wives to spin to growing Merino wool for the spinning mills. But sheep were never as big an item in Concord as in the hill country to the north and west. Concord farmers looked mainly to their cattle to find their market specialty: first beef, then butter and cheese, and, with the coming of the railroad in 1844, milk. The herd enlarged dramatically, and Concord became a town of dairy farmers.

The increase in commercial production by Concord's livestock was made possible partly by farmers' raising of fodder instead of bread on the tilled land, but mostly by a sharp rise in hay. Upland English hay production doubled in the second quarter of the century (see table 8.2). Some of this crop was marketed directly to urban stables, most went to feed the cows. With this expansion in grassland came an assault on Concord's remaining forest, which was reduced from one-quarter to one-tenth of the landscape. A strong urban market for firewood stimulated more rapid cutting of woodlots, but woodlots grow back: it was the explosion of hayfields and pastures that all but drove the forest from the hills. In effect, this was a time when cows ate trees. And this was enabled, once again, by a set of technological changes and resource substitutions, namely, iron, coal, nails, and dimensional lumber, that made local forests economically superfluous. Open wood fireplaces were largely replaced by coal grates, cook stoves, and parlor stoves, and, with the coming of the railroad, balloon frames of pine studs from Maine began to go up in place of local hand-hewn oak house timbers. To obtain what they ate, drank, wore, heated with, cooked on, slept in, inhabited, and generally used, Concord's residents increasingly turned from local to distant sources. In other words, they were turning into consumers in the modern sense of the word. To buy these things, Concord's farmers turned themselves into commercial producers—agricultural capitalism had arrived. In the process, Concord's system of husbandry and its landscape were remade.

This was the new world that Henry Thoreau surveyed in 1852. Concord agriculture was in the midst of a vigorous economic expansion but was also in ecological disarray. The production of corn and oats was up, the production of livestock and milk was up, the production of hay was up, but these increases meant the forest had to be sacrificed. Just as their embrace of the market economy required Concord farmers to specialize in a few cash crops, so it freed them from the ecological necessity of maintaining a diversified, balanced landscape. The woodlands were no longer needed to house and heat the inhabitants, so they could be stripped to supply the Boston firewood market and then converted to grass. Hayfields and pastures pushed out to the remotest hills, and the trees vanished from all but a few places—such as sandy Walden Woods, too coarse and droughty to be suitable for grass or any other crop, excepting Thoreau's dry beans. The loss of so much forest made life difficult for a wide range of woodland species and also contributed to increased flooding of the river and its meadows. Many of the upland hayfields and pastures themselves could not be long maintained and so began to grow up with brush, by 1850 encumbering the tax lists with "unimproved" land across a full quarter of the town. The new system was undoubtedly more productive (at least at first blush), but it also caused ecological trouble. It was, on the face of it, a far more extractive, *extensive* way of farming than what had gone before. A complete environmental evaluation of this commercial revolution is beyond the scope of this book, but it clearly carried Concord a long way from the tight local limits of the colonial era.[23]

Those who led this transformation—such enterprising men as Nathan Brooks, the lawyer, Ephraim Meriam, the businessman, and Edward Jarvis, the physician and public health crusader—had little doubt that the new, commercial world was an improved world. They had grown up at the turn of the century, as the old world of the yeoman pressed hard against its limits and groped for ways to satisfy an ambitious new breed of young Americans. It is no wonder these men saw the vanished world of their youth as quaint and interesting, even virtuous and admirable in its way, but also severely pinched and backward. They were unsentimental improvers dedicated to technological progress, economic growth, material prosperity, and rational management. To men like these, what stood in the way of improvement (in agriculture, as elsewhere) was precisely the stubborn survival of the clumsy, uninformed, intemperate ways of their forefathers. Because they were the first generation of Americans to write extensively about agriculture and country life and because our society continues, by and large, to share their values of prosperity and progress, their view of colonial agriculture has prevailed almost unchallenged for nearly two hundred years.[24]

But we need no longer depend on their jaundiced retrospective eye. In celebrating the increased prosperity of their own generation, the improvers sold the accomplishments of their ancestors short. When the yeomanry are allowed to speak for themselves through their own terse documents, their deeds and wills; and when these scattered records are drawn into a coherent picture of the way their husbandry was organized, a very different world is revealed. What Nathan Brooks remembered as a stingy pile of stones was, in his grandfather's day, a well-run, prosperous, stable farm. The same meadows and fields had been productively mowed and plowed, generation after generation. Colonial husbandry in Concord was not extensive farming, moving on constantly to fresh land as what lay behind was exhausted. On the contrary, it was intensive farming, in which a great deal of labor was concentrated on much the same lands, and a workable balance among these lands was established and carefully maintained. Plowlands, orchards, meadows, pastures, and woodlands were for the most part placed precisely on appropriate soils, and once placed, stayed put. Cornfields were diligently manured, meadows were laboriously ditched and diked, the flow of water was intricately regulated, nutrients were deliberately recycled. Family labor was distributed by seasons among a wide variety of agricultural and artisanal tasks in order to optimize production, and a healthy, diversified agrarian landscape was created and cared for. The persistence for half a century of one-quarter to one-third of the landscape in woodland, even after Concord became a crowded town, should catch the attention of any observer. Concord farmers, confronted upon their arrival with an unfamiliar abundance of land, had not adopted an extensive form of shifting cultivation, as far as any evidence can tell. They had instead systematically fashioned a remarkably sustainable version of a familiar English mixed husbandry system in a new land.

That yeoman world was bound by limits, and those limits could be painful—Jarvis speaks of them repeatedly as privations. Surely not many of us would relish returning to such uncomfortably close natural bounds while possessing no more than the simple tools with which the colonial yeomanry confronted them. But we needn't romanticize colonial life to respect its ecological fitness and stability: limits have their virtues, as well, and the achievement of that society in living within them at a very decent, widely shared level of comfort for its day can be admired on its own terms. Bypassing local limits may be essential to economic growth as we measure it, but this may have—and did have—and does have—disturbing ecological consequences, both at home and abroad. The remarkable thing about colonial Concord is that here, at the very moment when the English world was setting a capitalist course based on the denial of natural limits, long generations of new Americans put in place and steadily improved a workable version of an older mixed-husbandry village culture and economy, based on an ever-deepening understanding of their local environment. It took a century to fill the town, get fully acclimated, and reach those limits; whereupon Concord's yeoman husbandry adapted further and persisted for almost another century, supporting a stable population without undermining its environmental base well into the nineteenth century—well nigh unto the proverbial seventh generation.

It would be a distortion to see Concord's village culture as simply a throwback to a precapitalist state of society, deliberately holding out against the worldliness of the market—just as it is a distortion to see colonial New Englanders as simply profit-maximizing protocapitalists. The yeoman culture that appeared in New England is interesting precisely because it was something new in the world (and in truth, seldom seen since): a society of freeholders who lived within exacting social and ecological limits. Historians such as Stephen Innes and Daniel Vickers have argued that New Englanders were thoroughly committed from the beginning to mercantile capitalism, ownership of private property, and pious industriousness. Yet these potent market drives were held within bonds of family and community obligations and constrained by religious strictures against excessive greed and ostentation. New Englanders were not averse to bettering themselves economically, as long as such aspirations did not conflict too directly with the larger communitarian and spiritual goals upon which their towns rested. This proved a difficult balance to maintain, and tensions between individual economic advancement and communal and moral restraints increased as the generations passed, until they were resolved at the end of the colonial era in the form of a new, less inhibited Yankee acquisitiveness that gave shape to the nineteenth century. One need not pray for a return to Puritan patriarchal authority to see that there is something useful in a set of community obligations sufficient to keep us from being consumed by raw capitalism, the cruel marketplace fundamentalism that governed the slave plantation colonies, that dominated the industrial cities and westward boom of the nineteenth century, and that many feel is being suffered to run rampant again in our time. Guided by such restraints, by local control of their economies, and by a commitment to civic society, New Englanders achieved a modest prosperity that was remarkably broad for its time.[25]

These cultural reins on the runaway pursuit of profit combined with a similar set of ecological ones: Concord's mixed husbandry and woodland system also faced tight local limits. There were indeed social controls on how nakedly an individual could exploit the land for short-term gain—town regulations and covenants on the management of water and other common resources, expectations of reciprocity among neighbors and kin, and the strong desire to pass along an improved landholding to family heirs. While some children were always leaving for the frontier, others were fully expected to remain in Concord, on the same land, and prosper; and for almost seven generations they largely did. But the most powerful constraint of all lay in the land itself, which granted its rewards grudgingly: the New England environment did not present its inhabitants with the opportunity to produce a single staple export, such as sugar or tobacco, and run with it. Instead they had to cultivate a broad set of diverse, difficult, interlocking resources and thriftily conserve them, if they hoped to prosper at all. These ecological limits on the market drive to commodify natural resources and exploit them for short-term profit were in many ways parallel to social constraints upon taking undo advantage of other community members and acted to reinforce the social order. The farmers of Concord may have been willing to direct their production to distant markets, but for most of two centuries they were forced to look mainly within themselves for a comfortable subsistence, and the headlong rush of narrow market calculus was held in check.

The mixed husbandry system of Concord's yeomen was obviously very different from the Native economy of horticulture and foraging that had prevailed in Musketaquid before it, but it was arguably as appropriate and as sustainable—or at least, it had demonstrated the potential to be so. Obviously the English imposed greater changes upon the land, and one can hardly be certain that what had lasted for half a dozen generations could have endured as well as the Native system. Still, what emerged by the end of the colonial era had all the makings of a durable agrarian village economy on the ancient English model, a fundamentally sound agroecosystem. Yet for all of that, it inarguably was not sustained. It was bound up in and deeply committed to a broader world of European economic growth that ultimately led it to move in a new, arguably *less* sustainable direction as it encountered its limits. The Yankee version of yeoman husbandry carried within it the seed of something radically different: in time, the confining social and ecological shell was broken and a fully commercial agriculture emerged. But this should not blind us to what was achieved during the era when, partly by choice and partly by necessity, those constraints prevailed. As our society debates what ecological limits it might be wise to acknowledge and respect in our own time, the experience of Concord before independence is worthy of our attention.

On July 22, 1859, Henry Thoreau set off down the Concord River to the falls in Billerica. A southwest wind filled the little sail on his dory, so he made good speed—though he would have to row back. As he passed downstream he observed that "some have just begun to get the hay on our Great Meadows." Thoreau was no stranger to the farmers of Concord, and it is likely that as he sailed by many of the mowers gave him a friendly wave. Did they approve of the business he was transacting on the river? While men such as Edward Jarvis were praising the new enterprising spirit in Concord, a minority voice in American culture was speaking up in the same place. The Transcendentalists, especially Emerson and Thoreau, were uneasy about the changes sweeping over their town. These romantics were of the same generation as the improvers, but their feelings about the disappearing yeoman world were a good deal more ambivalent. Emerson was exuberant about technological progress and optimistic that the closeness of nature in the American countryside could redeem the encroaching squalor and barrenness of capitalism and industry; Thoreau was more skeptical about the prospects for that. Thoreau is remembered for his trenchant criticism of the spiritual and environmental poverty he believed the farmers of Concord, those improving men of quiet desperation, were bringing upon the landscape and themselves. But this day Thoreau had not gone boating to console himself with the bravery of muskrats or to wonder at the misspent industry of his farming neighbors: he was working for them. That summer, he had been hired by the meadow owners of Sudbury and Concord to survey the river, to help them make their case that worsened flooding of the meadows after summer rains was caused primarily by the Billerica milldam, not by natural bars and bridges holding back the water. Meticulous, reliable, knowing the river backwards (in which direction it occasionally ran), scrupulously fair yet sympathetic to the farmers' cause, and marching, in his own way, to as insistent a Yankee work ethic as any of his neighbors, Thoreau was the perfect man for the job.[26]

What did the surveyor see as he passed by in his "dead-river boat," which he had designed and built to a flat-bottomed pattern common along the quiet stream? The meadow looked about the same as it had for two hundred years. It was still divided into a myriad of narrow strips. It was still crowded with the farmers of Concord cutting their hay—although more of the stalwarts swinging scythes were hired men than formerly, replacing the sons, cousins, and neighbors with whom the yeomen once changed work. But the oxen still drew the hay to the hard ground, where it was made up into tall haystacks on wooden staddles or carted home to the barn, sometimes several miles away in the hills (fig. 9.1). The young men still swam the river, to fetch cool water from a well-known spring on the opposite hillside. But all that Thoreau saw that day was about to fade away, not long after his own death three years later, in 1862. That was the year the meadow owners finally lost their case, when an august scientific commission appointed

FIGURE 9.1.
Haystacks on Sudbury Meadow, above Concord. A few river meadows remained in production
into the early twentieth century. Now within the Great Meadows National Wildlife Refuge.
Herbert Gleason, 1925.

by the legislature to investigate the flowage controversy but dominated by the mill interests ruled against them. The river meadows, too often flooded by the dam and by deforestation to be worth much anymore, were steadily abandoned in the following decades.
In truth, the farmers were losing interest in their coarse native hay anyway—they were
mostly hoping that after removing the dam they would be able to drain the river sufficiently to reclaim their meadows as improved, cultivated ground. After two centuries at
the heart of Concord's husbandry, the river meadows had scant value in the new world
of commercial farming.[27]

But long before they were overgrown by sedges, buttonbush, and loosestrife, even before they were fatally overflowed with water, the meadows had undergone a small, barely
perceptible but decisive change. Thoreau and his contemporaries spoke of the Great

Meadows. To the yeomanry a century before, they were—it was—the Great Meadow. The singular had become plural. Before anything tangible about the place, or even how it was owned and managed, had much changed, something had changed in men's minds. Farmers still owned private mowing lots within the meadow and still took collective action in draining it, keeping its ways in repair, grazing its aftermath, burning it upon occasion, and struggling to protect it from the Billerica dam and other encroachments— just as they always had. But now, instead of a common entity of which each owned a part, the Great Meadows were thought of as a collection of individual pieces. In Thoreau's Concord men still paused to sharpen their scythes on the meadows, boys still turned the hay to the sun, oxen still pulled the heavily loaded carts, owners still dispatched their petitions to the Great and General Court. But the Great *Meadow*, and the yeoman world which had revolved around it, was already gone.

Genealogies

Appendix 1

Meriam Family Genealogy

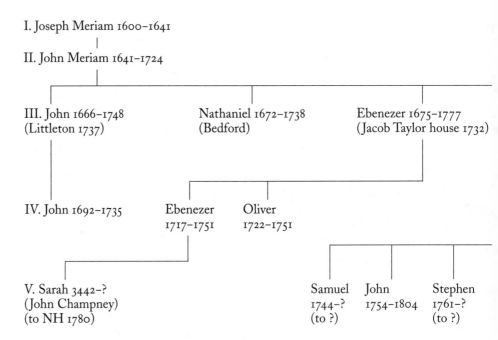

I. Joseph Meriam 1600–1641

II. John Meriam 1641–1724

III. John 1666–1748 (Littleton 1737)   Nathaniel 1672–1738 (Bedford)   Ebenezer 1675–1777 (Jacob Taylor house 1732)

IV. John 1692–1735   Ebenezer 1717–1751   Oliver 1722–1751

V. Sarah 3442–? (John Champney) (to NH 1780)   Samuel 1744–? (to ?)   John 1754–1804   Stephen 1761–? (to ?)

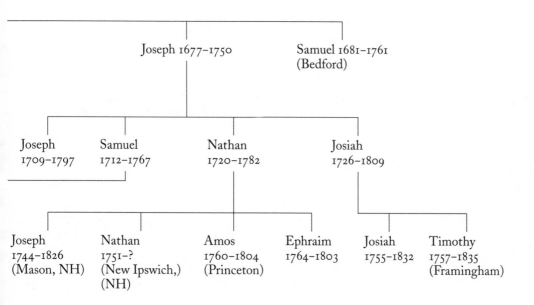

Joseph 1677–1750   Samuel 1681–1761
                         (Bedford)

Joseph   Samuel   Nathan   Josiah
1709–1797   1712–1767   1720–1782   1726–1809

Joseph   Nathan   Amos   Ephraim   Josiah   Timothy
1744–1826   1751–?   1760–1804   1764–1803   1755–1832   1757–1835
(Mason, NH)   (New Ipswich,) (Princeton)   (Framingham)
                (NH)

Brooks Family Genealogy

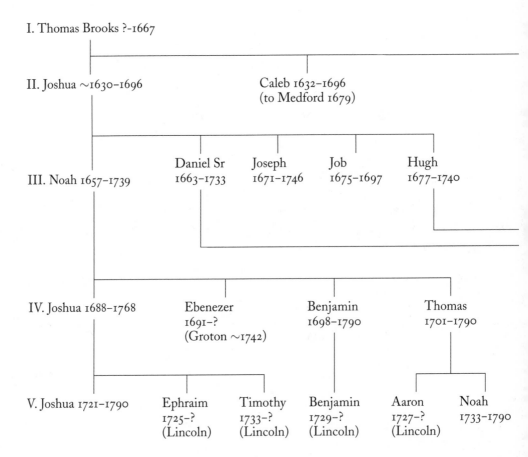

I. Thomas Brooks ?–1667

II. Joshua ~1630–1696   Caleb 1632–1696
(to Medford 1679)

III. Noah 1657–1739   Daniel Sr   Joseph   Job   Hugh
1663–1733   1671–1746   1675–1697   1677–1740

IV. Joshua 1688–1768   Ebenezer   Benjamin   Thomas
1691–?   1698–1790   1701–1790
(Groton ~1742)

V. Joshua 1721–1790   Ephraim   Timothy   Benjamin   Aaron   Noah
1725–?   1733–?   1729–?   1727–?   1733–1790
(Lincoln)   (Lincoln)   (Lincoln)   (Lincoln)

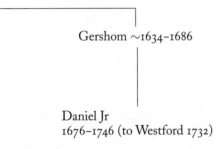

Gershom ~1634–1686

Daniel Jr
1676–1746 (to Westford 1732)

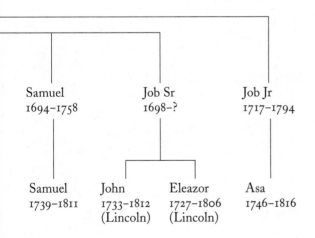

Samuel
1694–1758

Job Sr
1698–?

Job Jr
1717–1794

Samuel
1739–1811

John
1733–1812
(Lincoln)

Eleazor
1727–1806
(Lincoln)

Asa
1746–1816

Hartwell Family Genealogy

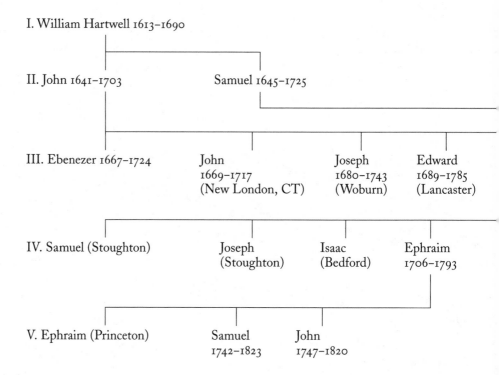

I. William Hartwell 1613–1690

II. John 1641–1703   Samuel 1645–1725

III. Ebenezer 1667–1724   John 1669–1717 (New London, CT)   Joseph 1680–1743 (Woburn)   Edward 1689–1785 (Lancaster)

IV. Samuel (Stoughton)   Joseph (Stoughton)   Isaac (Bedford)   Ephraim 1706–1793

V. Ephraim (Princeton)   Samuel 1742–1823   John 1747–1820

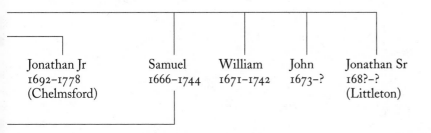

Jonathan Jr
1692–1778
(Chelmsford)

Samuel
1666–1744

William
1671–1742

John
1673–?

Jonathan Sr
168?–?
(Littleton)

East Quarter Land Use, by Deeds and Valuations—1749

TABLE A2.1   Ebenezer Meriam Sr & Ebenezer Meriam Jr—1749

Parcel	Tillage	Orchard	Pasture	Mowing	Wood	Improved	Total
2-acre houselot					2		
1/4 acre barn lot		0.25					
4 acre plowland in Great Field	4						
1 1/2 acre plowland below Sam. Meriam	1.5						
2 acre home meadow				2			
4 acre meadow N of road				4			
2 acre meadow in Virginia				2			
4 acre Potter's meadow				4			
5 acre river meadow in Bedfore				5			
8 acre pasture and meadow in Cranefield			4	4			
4 acre pasture in Ox Pasture			4				
1 1/2 acre Horse Pasture			1.5				
3 acre woodland Pine Hill					3		
4 acre woodland Flint's Pond					3		
5 acre woodland Chestnut field					5		
Total	5.5	0.25	9.5	21	14	36.25	50.25
EMSr-Val	2	0	4	6			
EMJr-Val	3	0	6	10			
Valuation Total	5	0	10	16		31	

Appendix 2

TABLE A2.2   Josiah Meriam—1749

Parcel	Tillage	Orchard	Pasture	Mowing	Wood	Improved	Total
1/2-acre houselot							
5 1/2 acres plowland	5.5						
3-acre meadow on Bay road				3			
5-acre meadow in Elm Brook Meadow				5			
2 1/2-acre river meadow in Bedford				2.5			
9 acres pasture on Billerica Road			9				
7 acres woodland at Chestnut Field					7		
Total	5.5	0	9	10.5	7	25	32
Valuation Total	3	0	7	7		17	

TABLE A2.3 Nathan Meriam—1749

Parcel	Tillage	Orchard	Pasture	Mowing	Wood	Improved	Total
4-acre home field	4						
3 acres Stow field	3						
3 acres orchard		3					
8 acres home meadow, w/ house, barn				7			
10 acres wood and meadow in Virginia meadow				4	6		
15 acres wood and meadow— Virginia woodlot				4	11		
3 acres Holt river meadow				3			
3 acres river meadow in Bedford				3			
3 acres Birch Island meadow in Bedford				3			
6 acres Stow pasture			6				
18 acres Hartwell pasture			13		5		
17 acres house swamp					17		
2 acres cedar swamp in Bedford					2		
4 acres woodland					4		
3 acres at Flint's Pond					3		
Total	7	3	19	24	48	53	101
Valuation Total	7	0	14	11		32	

Appendix 2

TABLE A2.4　Samuel Meriam—1749

Parcel	Tillage	Orchard	Pasture	Mowing	Wood	Improved	Total
6 acres with tenement	5						
3 acres river meadow				3			
10 acres Dam meadow (part of 41 3/4 acres)				10			
8 acres pasture on highway (part of 41 3/4 acres)			8				
16 acres Stow pasture			16				
part of 41 3/4 acres					24		
Total	5	0	24	13	24	42	66
Valuation Total	5	1	8	7		21	

Appendix 2

TABLE A2.5 Samuel Minot—1749

Parcel	Tillage	Orchard	Pasture	Mowing	Wood	Improved	Total
36-acre houselot N of road–upland, pasture, meadow	6	1	12	9	4		
2 acres plowland on Billerica Road	2						
8 acres pasture, meadow, upland, orchard S of road		1	3	4			
7 acres Prescott meadow				7			
3 acres in Great Meadow				3			
40 acres woodland on Elm Brook Hill					40		
Total	8	2	15	23	44	48	92
Valuation Total	8	2	15	23		48	

also 40 acres woodland and pasture in N Concord
also1/2 120 acres in Princeton (w/ Nathan Meriam)

Appendix 2

TABLE A2.6 Samuel Fletcher—1749

Parcel	Tillage	Orchard	Pasture	Mowing	Wood	Improved	Total
4-acre houselot	3						
Harwood lot adjoining house lot	2						
4 acres plowland (Great Field?)	4						
4 acres meadow adjoining				4			
1 1/2 acres orchard below road		1.5					
3 acres meadow below Hoar's dam				3			
5 acres river meadow				5			
5 acres river meadow in Bedford				5			
5 acres in Elm Brook Meadow				5			
5 acres Brickiln pasture—meadow and upland			5				
29 acres upland and meadow in Ox pasture (6 lots)			29				
3/4 acres woodland					0.75		
4 acres woodland					4		
17 acres woodland on Pine Hill					17		
Total	9	1.5	34	22	22	66.5	88.5
Valuation Total	7	1	14	14		36	

Appendix 2

TABLE A2.7   John Jones—1749

Parcel	Tillage	Orchard	Pasture	Mowing	Wood	Improved	Total
4-acre houselot	3						
2 1/4 acres Rigby lo t	2						
4 acre Stow lot	4						
4 acres Brickiln lot	4						
1 acre orchard S of road		1					
4 acres Dam meadow				4			
11 acres Island meadow				11			
4 acres swamp mowing				4			
6 acres Long meadow				6			
4 acres in Elm Brook meadow				4			
3 acres Brickiln pasture			3				
11 acres Island pasture			11				
1 acre Upper pasture S of road			1				
4 acres at head of ox pasture			4				
12 acres at Chestnut Field					12		
9 1/2 acres woodlot					9.5		
5 acres sawmill woodlot					5		
14 acres rough pine land on Pine Hill					14		
Total	13	1	19	29	40.5	62	102.5
Valuation Total	8	1	14	23		46	

TABLE A2.8   Joseph Stow—1749

Parcel	Tillage	Orchard	Pasture	Mowing	Wood	Improved	Total
10-acre homelot—meadow and upland		0.5	2	4	.		
9 acres plow and pasture land in Brickiln Field	9						
11 acres plow and woodland at Brickiln Island	3				8		
2 acres meadow				2			
6 acres Brickiln Island meadow				6			
John Stow home meadow				4			
Part of 32 acres Thomas Stow woodland at BKI				5	5		
5 acres Blosses Island pasture			5				
15 acres woodland at Chestnut Field					15		
Total	12	0.5	7	21	28	40.5	68.5
Valuation Total	4	0.5	4	10		18.5	

TABLE A2.11   Samuel Brooks—1749

Parcel	Tillage	Orchard	Pasture	Mowing	Wood	Improved	Total
9-acre houselot—upland N of road		1	6				
9 acres upland S of road	5	0.5	4				
2 3/4 acres upland on way to Island	2.75						
5 acres plowland and mowing at Brickiln Island	5						
10-acre meadow adjoining upland N of road				10			
3-acre meadow N on way to Island				8			
1 acre upland on way to Island				1			
12 acres Virginia meadow				12			
9 3/4 acres Fletcher meadow (of Ebenezer Brooks)				9			
4 3/4 acres meadow and swamp of Ephraim Hartwell				4			
4-acre meadow in Pine plain				4			
20-acre Suburb lot			10		10		
22 3/4 acres meadow and swamp at Suburbs				12	10		
20 acres upland and swamp near Chestnut Field					20		
Total	12.75	1.5	20	60	40	94.25	134.25
Valuation Total	7	1	14	40		62	

TABLE A2.12   Thomas Brooks—1749

Parcel	Tillage	Orchard	Pasture	Mowing	Wood	Improved	Total
13-acre houselot	10	1					
3 acres Further field	3						
Barren orchard		2					
10 acres Tanyard meadow				10			
19 acres meadow and upland			12	7			
7 1/2 acres upland and mowing—							
rocky pasture			3	4			
6 1/2 acres river meadow				6.5			
5 acres Garden pasture (part of 35 acres)			5				
2 acres Calf pasture (part of 35 acres)			2				
4 acres Little pasture			4				
2 acres Hog pasture			2				
35 acres mostly woodland "Oak Barron"					28		
3 1/2 acres at "14 Acres"					3.5		
right in Cedar swamp							
Total	13	3	28	27.5	31.5	71.5	103
Valuation Total	7	1	32	27		67	

TABLE A2.13   Ephraim Hartwell—1749

Parcel	Tillage	Orchard	Pasture	Mowing	Wood	Improved	Total
18-acre houselot—woodland and upland			7	2	11		
32 acres Samuel Hartwell home lot		1	30				
6 acres plowland	6						
15 acres plowland—Further field	15			2			
3 1/2 acres	3.5						
Orchard at 15-acre field		1					
4 acres Elm Brook Meadow in Bedford				4			
5 acres First Division Rocky meadow				8			
6 acres meadow and upland from Joshua Brooks				6			
Rocky meadow from Ebenezer Brooks				5			
Lower Rocky meadow				11			
5 acres from Stephen Davis				5			
Part of 6 acres meadow and swamp				5			
18 acres meadow and pasture at Suburbs			6	12			
Lot before house			15		9		
2 3/4 acres woodlot					2.75		
Total	24.5	2	58	60	23	144.5	167.5
Valuation Total	10	1	20	35		66	

TABLE A2.14   Nathaniel Whittemore—1749

Parcel	Tillage	Orchard	Pasture	Mowing	Wood	Improved	Total
30 acres adjoining old house	8		11	9			
3 1/2 acres chiefly plowland—part 25-acre home lot	3.5						
small corner plowland—part 25-acre home lot	1						
3 acres orchard and mowing		1		2			
3 acres meadow from Joshua Brooks				7			
3 acres meadow from Ebenezer Brooks				6			
8 acres Fletcher meadow				8			
17 acres upland and meadow			2	15			
rest of 25-acre home lot				11	8.5		
7.5 acres meadow lot below cross fence				7.5			
Fore swamp—pasture and orchard			8				
Great Pasture			16				
10 1/2 acres upland and swamp—Ebenezer Brooks			10		1		
Woodland and swamp at Silver Brook					14		
Total	12.5	1	47	65.5	23.5	126	149.5
Valuation Total	12	1	10	40		63	

Notes

Abbreviations

CFPL   Concord Free Public Library
CTR   Concord Town Records
MP   Middlesex County Court of Probate
MRD   Middlesex Registry of Deeds
SPNEA   Society for the Preservation of New England Antiquities

CHAPTER 1. Introduction

1. The translation is Thoreau's. Henry David Thoreau, *A Week on the Concord and Merrimack Rivers* (New York: Penguin Classics, 1998), 7–11.

2. How far the Cranefield extended is difficult to discern from the sketchy early records of the town. My impression is that it was the earliest field nearest the village and that the Great Field included later additions "beyond Cranefield," "in the Plain," "at the New Field." By the eighteenth century the Cranefield was seldom referred to and had apparently been subsumed by the Great Field.

3. Lincoln's fame for conserved farmland and forestland today flows from this simple fact. The story is told in John C. MacLean, *A Rich Harvest: The History, Buildings and People of Lincoln, Massachusetts* (Lincoln: Lincoln Historical Society, 1988).

4. This land held Job Brooks's "home pasture" by the early eighteenth century, although how completely the entire parcel was cleared of trees is impossible to tell.

5. Many of them rebuilt recently by the Park Service. The road here has also been restored to something like its original hard-packed earthen surface.

6. Timothy Dwight, *Travels in New England and New York,* vol. 1 (Cambridge: Harvard University Press, 1969), 76. But a few sentences later Dwight concluded, "Such, upon the whole, is our soil and such our culture that probably fewer persons suffer from the want either of the necessaries or the comforts of life than in any other country containing an equal population," 77. Overall, Dwight painted an attractive portrait of the well-kept New England countryside through which he traveled in the 1790s. It is telling that his criticism is so much more often repeated than his praise.

7. Harry J. Carman, ed., *American Husbandry* (New York: Columbia University Press, 1939); Percy W.

Bidwell, "The Agricultural Revolution in New England," *American Historical Review* 26 (1921); Percy Wells Bidwell and John I. Falconer, *History of Agriculture in the Northern United States, 1620–1860* (New York: Peter Smith, 1941); Clarence H. Danhoff, *Change in Agriculture: The Northern United States, 1820–1870* (Cambridge: Harvard University Press, 1969).

8. Kenneth Lockridge, *A New England Town: The First Hundred Years* (New York: W. W. Norton, 1970); Kenneth Lockridge, "Land, Population and the Evolution of New England Society, 1630-1790; and an Afterthought," in Stanley N. Katz, ed, *Colonial America: Essays in Politics and Social Development* (Boston: Little, Brown, 1971); Richard Bushman, *From Puritan to Yankee: Character and the Social Order in Connecticut, 1790–1765* (New York: W. W. Norton, 1967); James A. Henretta, *The Evolution of American Society, 1700-1815: An Interdisciplinary Analysis* (Lexington, Mass., 1973); Robert Gross, *The Minutemen and Their World* (New York: Hill and Wang, 1976); Christopher Clark, *The Roots of Rural Capitalism: Western Massachusetts, 1780–1860* (Ithaca: Cornell University Press, 1990).

9. Stephen Innes, *Creating the Commonwealth: The Economic Culture of Puritan New England* (New York: W. W. Norton, 1995), 7; Winifred Barr Rothenberg, *From Market-Places to a Market Economy: The Transformation of Rural Massachusetts, 1750–1850* (Chicago: University of Chicago Press, 1994); Gloria L. Main, *Peoples of a Spacious Land: Families and Cultures in Colonial New England* (Cambridge: Harvard University Press, 2001); John Frederick Martin, *Profits in the Wilderness: Entrepreneurship and the Founding of New England Towns in the Seventeenth Century* (Chapel Hill: University of North Carolina Press, 1991); Daniel Vickers, *Farmers and Fishermen: Two Centuries of Work in Essex County, Massachusetts, 1630–1850* (Chapel Hill: University of North Carolina Press, 1994); Virginia DeJohn Anderson, *New England's Generation: The Great Migration and the Formation of Society and Culture in the Seventeenth Century* (New York: Cambridge University Press, 1991). A recent synopsis of New England in this vein appears in Alan Taylor, *American Colonies* (New York: Viking, 2001), 158–203.

10. Innes, *Creating the Commonwealth*, 5.

11. William Cronon, *Changes in the Land: Indians, Colonists, and the Ecology of New England* (New York: Hill and Wang, 1983); Carolyn Merchant, *Ecological Revolutions: Nature, Gender and Science in New England* (Chapel Hill: University of North Carolina Press, 1989). Merchant's version aligns more precisely with the communitarian social historians in that she sees the colonial period as a transitional stage between the Native world and the full-blown capitalism of the nineteenth century. In her view, extensive colonial farming was less damaging than the intensive farming that followed.

12. Peter M. Vitousek et al., "Human Domination of Earth's Ecosystems," *Science* 277 (1997): 494–99.

13. Edward Johnson, *Johnson's Wonder-Working Providence* (1654; reprint, New York: Barnes and Noble, 1910).

CHAPTER 2. Musketaquid

1. Wallace S. Broecker and George H. Denton, "What Drives Glacial Cycles?" *Scientific American* 262 (1990); Steven M. Stanley and William F. Ruddiman, "Neogene Ice Age in the North Atlantic Region: Climatic Changes, Biotic Effects, and Forcing Factors," in National Research Council, Board of Earth Sciences and Resources, *Effects of Past Global Change on Life* (Washington: National Academy Press, 1995); Thomas J. Crowley and Gerald R. North, *Paleoclimatology* (New York: Oxford University Press, 1991), 110–47; Raymond S. Bradley, *Paleoclimatology: Reconstructing Climates of the Quaternary*, 2d ed. (San Diego: Academic Press, 1999), 213–14, 281–83. Glacial cycles are keyed to small orbital changes (called Milankovitch cycles after their discoverer), but the geophysical processes by which such small fluctuations in the amount of sunlight reaching different parts of the earth are amplified into dramatic climate swings are as of yet uncertain. They could involve ocean circulation patterns, atmospheric carbon dioxide levels, and feedback from the ice sheet itself.

2. William Jones, "A Topographical Description of the Town of Concord, August 20th, 1792," *Collections of the Massachusetts Historical Society*, (1792), 1:237.

3. Tage Nilsson, *The Pleistocene: Geology and Life in the Quaternary Ice Age* (Stuttgart, West Germany: Ferdinand Enke Verlag, 1983); Richard Foster Flint, *Glacial and Quaternary Geology* (New York: John Wiley and Sons, 1971). An excellent recent account is in Robert M. Thorson, *Stone by Stone: The Magnificent History in New England's Stone Walls* (New York: Walker, 2002). The frame for the Surficial Geology map (and for many of the maps to follow) is the original town of Concord as granted in 1635. Present-day Concord is smaller, having lost corners to Carlisle, Bedford, and Lincoln. The Geology Map is compiled from Carl Koteff, *Surficial Geology of the Concord Quadrangle, Massachusetts* (Washington, D.C.: USGS Map GQ-331, 1964); Arthur E. Nelson, *Surficial Geologic Map of the Natick Quadrangle, Middlesex and Norfolk Counties, Massachusetts* (Washington, D.C.: USGS Map GQ-1151, 1974); Wallace R. Hansen, *Geology and Mineral Resources of the Hudson and Maynard Quadrangles, Massachusetts* (Washington, D.C.: USGS Geological Survey Bulletin 1038, 1956). The map lacks data for the northern corner of Concord because no USGS Surficial Geology maps have been completed for the Billerica and Westford quadrangles.

4. Koteff, *Surficial Geology of Concord;* Nilsson, *The Pleistocene.* There are some low-lying areas where only a thin layer of lake bottom sediment overlies till.

5. "Soils and Their Interpretation for Various Land Uses, Town of Concord, Massachusetts," (Soil Conservation Service, USDA, 1966). Many till slopes are a complex mixture of lush swales and sparse knolls.

6. Koteff, *Surficial Geology of Concord;* James W. Skehan, S.J., "Walden Pond: Its Geological Setting and the Africa Connection," in Edmund A. Schofield and Robert C. Baron, eds., *Thoreau's World and Ours: A Natural Legacy* (Golden, Colo.: North American Press, 1993), 222–29.

7. Henry D. Thoreau, *Walden* (Princeton: Princeton University Press, 1973), 175.

8. This overview of southern New England paleoecology has been drawn from Marjorie Green Winkler, "Changes at Walden Pond During the Last 600 Years: Microfossil Analysis of Walden Pond Sediments," in Schofield and Baron, *Thoreau's World and Ours,* 199–211; Marjorie Winkler, "A 12,000-Year History of Vegetation and Climate for Cape Cod, Massachusetts," *Quaternary Research* 23 (1985); Margaret B. Davis, "Quaternary History of Deciduous Forests of Eastern North America and Europe," *Annals of the Missouri Botanical Garden* 70 (1983); Denise C. Gaudreau, "The Distribution of Late Quaternary Forest Regions in the Northeast," in George P. Nicholas, ed., *Holocene Human Ecology in Northeastern North America* (New York: Plenum Press, 1988). For a similar treatment, see John F. O'Keefe and David R. Foster, "An Ecological History of Massachusetts Forests," in Charles H. W. Foster, ed., *Stepping Back to Look Forward: A History of the Massachusetts Forest* (Petersham, Mass.: Harvard University Forest, 1998), 19–32.

9. Sara Webb, "Potential Role of Passenger Pigeons and Other Vertebrates in the Rapid Holocene Migration of Nut Trees," *Quaternary Research* 26 (1986).

10. Looking at ecological processes at larger spatial and temporal scales that include the role of natural and human disturbances follows the approach offered by *landscape ecology,* which treats individual ecosystems as elements in a broader system, the landscape. A landscape is defined as a "heterogeneous land area composed of a cluster of interacting ecosystems that is repeated in similar form throughout." Richard T. T. Forman and Michel Godron, *Landscape Ecology* (New York: John Wiley and Sons, 1986), 11.

11. Anonymous quotation in Harry J. Carman and Rexford G. Tugwell, "Editors' Foreword," in Harold Fisher Wilson, *The Hill Country of Northern New England: Its Social and Economic History 1790–1930* (New York: Columbia University Press, 1936).

12. Shirley Blancke, "Survey of Pre-Contact Sites and Collections in Concord," Report for Massachusetts Historical Commission Survey and Planning Project, 1980.

13. A recent discussion of the arrival of Paleoindians in New England is Mary Lou Curran, "Exploration, Colonization, and Settling In: The Bull Brook Phase, Antecedents, and Descendants," in Mary Ann Levine, Kenneth E. Sassaman, and Michael S. Nassaney, eds., *The Archaeological Northeast* (Westport, Conn.: Bergin and Garvey, 1999), 6–9. Concord sites are outlined in Shirley Blancke, "The Archaeology of Walden Woods," in Schofield and Baron, *Thoreau's World and Ours,* 244.

14. Thoreau, *Walden,* 182. The possibility that early inhabitants saw Walden mountain was pointed out

by the geologist James W. Skehan, S.J., *Puddingstone, Drumlins, and Ancient Volcanoes: A Geologic Field Guide Along Historic Trails of Greater Boston* (Dedham, Mass.: WeStone Press, 1979), 47. Given the gap of about two thousand years between the currently accepted dates for the departure of ice and the arrival of people in Concord it seems a bit of a stretch, but it makes a nice story—and surely, Paleoamericans were broadly familiar with postglacial geomorphology.

15. Dena F. Dincauze, "A Capsule Prehistory of Southern New England," in Laurence M. Hauptman and James D. Wherry, eds., *The Pequots of Southern New England: The Fall and Rise of an American Indian Nation* (University of Oklahoma Press: Norman, 1990), 19–32; Dean R. Snow, *The Archaeology of New England* (New York: Academic Press, 1980), 101–56.

16. Winkler, "History of Vegetation and Climate for Cape Cod," 301–12.

17. On the adaptation of species to the climate cycles of the Pleistocene, see Thompson Webb III, "Pollen Records of Late Quaternary Change: Plant Community Rearrangement and Evolutionary Implications," in *Effects of Past Global Change on Life,* 221–27. For two recent discussions of the Pleistocene extinctions that reach opposite conclusions, see Shephard Krech III, *The Ecological Indian: Myth and History* (New York: W. W. Norton, 1999), 29–42; and Jared Diamond, *Guns, Germs, and Steel: The Fate of Human Societies* (New York: W. W. Norton, 1997), 46–47. As Krech insists, this debate needs to steer clear of the gruesome ideological tar pit of labeling Native Americans as either ecological saints or hit men.

18. Dincauze, "Capsule Prehistory of Southern New England"; Snow, *Archaeology of New England,* 159–72; Blancke, "Archaeology of Walden Woods," 244. It is possible that the fringes of the former glacial lakes in Concord provided richer foraging mosaics, and that Early Archaic sites are simply harder to find. See George P. Nicholas, "Ecological Leveling: The Archaeology and Environmental Dynamics of Early Postglacial Land Use," in George P. Nicholas, ed., *Holocene Human Ecology in Northeastern North America* (New York: Plenum Press, 1988).

19. Pollen profiles for eastern Massachusetts include Lesley Sneddon and Lawrence Kaplan, "Pollen Analysis from Cedar Swamp Pond, Westborough, Massachusetts," *Archaeological Quarterly* 9 (1987); and Winkler, "History of Vegetation and Climate for Cape Cod."

20. This discussion is compiled from Snow, *Archaeology of New England,* 172–233; Dincauze, "Capsule Prehistory of Southern New England"; Blancke, "The Archaeology of Walden Woods," 244–46; and Mitchell T. Mulholland, "Territoriality and Horticulture: A Perspective for Prehistoric Southern New England," in Nicholas, *Holocene Human Ecology.*

21. Winkler, "Changes at Walden Pond During the Last 600 Years," 199–211; Winkler, "History of Vegetation and Climate for Cape Cod"; William A. Patterson and Kenneth E. Sassaman, "Indian Fires in the Prehistory of New England," in Nicholas, *Holocene Human Ecology.* The alleged Indian use of fire has been a controversial issue since it was proposed by Stanley Bromley in the 1930s. Patterson and Sassaman's essay is a good overview of the issues involved in the debate. Important articles include Stanley W. Bromley, "The Original Forest Types of Southern New England," *Ecological Monographs* 5 (1935); Gordon M. Day, "The Indian as an Ecological Factor in the Northeastern Forest," *Ecology* 34 (1953); Hugh Raup, "Recent Changes of Climate and Vegetation in Southern New England and Adjacent New York," *Journal of the Arnold Arboretum* 18 (1937); Emily W. B. Russell, "Indian-set Fires in the Forests of the Northeastern United States," *Ecology* 64 (1983); Richard T. T. Forman and Emily W. B. Russell, "Commentary: Evaluation of Historical Data in Ecology," *Bulletin of the Ecological Society of America* 64 (1983); Ronald L. Meyers and Patricia A. Peroni, "Commentary: Approaches to Determining Aboriginal Fire Use and Its Impact on Vegetation," *Bulletin of the Ecological Society of America* 64 (1983); Stephen J. Pyne, *Fire in America: A Cultural History of Wildland and Rural Fire* (Princeton: Princeton University Press, 1982); Gordon G. Whitney, *From Coastal Wilderness to Fruited Plain* (New York: Cambridge University Press, 1994), 107–20.

22. Winkler, "History of Vegetation and Climate for Cape Cod"; Arthur A. Joyce, "Early/Middle Holocene Environments in the Middle Atlantic Region: A Revised Reconstruction," in Nicholas, *Holocene Human Ecology.*

23. Dincauze, "Capsule Prehistory of Southern New England"; Blancke, "The Archaeology of Walden Woods," 246; and Mulholland, "Territoriality and Horticulture." Not all archaeologists working in New England are convinced that Indian populations declined during this period or that the shift toward lowland and coastal sites and the development of new cultural systems that took place over the next two thousand years were necessarily driven by environmental change. See Snow, *Archaeology of New England*, 253–57.

24. Snow, *Archaeology of New England*, 278–90; Dincauze, "Capsule Prehistory of Southern New England"; Mulholland, "Territoriality and Horticulture"; Kevin A. McBride and Robert E. Dewar, "Agriculture and Cultural Evolution: Cause and Effects in the Lower Connecticut River Valley," in William F. Keegan, ed., *Emergent Horticultural Economies of the Eastern Woodlands*, Center for Archaeological Investigations, Occasional Paper No. 7 (Carbondale: Southern Illinois University at Carbondale, 1987). On the development of native cultivars in eastern North America, see Richard A. Yarnell, "The Importance of Native Crops during the Late Archaic and Woodland Periods," and other essays in C. Margaret Scarry, ed., *Foraging and Farming in the Eastern Woodlands* (Gainesville: University Press of Florida, 1993).

25. Southern New England densities were in the neighborhood of two hundred people per one hundred square kilometers. Presumably densities had been somewhat higher in the Mast Forest than in the harsher northern forest all along. Snow, *Archaeology of New England*, 75.

26. Mulholland, "Territoriality and Horticulture." Some anthropologists argue that New England Indians did not turn decisively to corn cultivation until about 1300, or even until European contact. See Kathleen J. Bragdon, *Native People of Southern New England, 1500–1650* (Norman: University of Oklahoma Press, 1996), 38, 88–97.

27. Blancke, "Survey of Pre-Contact Sites and Collections in Concord."

28. Shirley Boothe Blancke, "Analysis of Variation in Point Morphology as a Strategy in the Reconstruction of the Culture History of an Archaeologically Disturbed Area" (Ph.D. diss., Boston University, 1978); S. F. Cook, *The Indian Population of New England in the Seventeenth Century* (Berkeley: University of California Press, 1976). Cook places the Concord River Indians with the Wamesit group, the same group also called Pawtucket. He places the Mystick village with the Massachusetts. Snow, *Archaeology of New England*, considers the Lower Merrimack Indians essentially a branch of the Massachusetts.

29. Cook, *Indian Population of New England*. The evidence here is very thin. Estimates of Pawtucket population run as high as three thousand. One archaeological survey cited by Cook suggests there were fourteen village sites along the Concord River between Sudbury and the Merrimack, with perhaps one-third of them in use at any given time.

30. Snow, *Archaeology of New England*, 75–76. A family's corn supply of forty bushels would weigh about a ton, so it is unlikely that the underground pits that stored the corn would have been too far from the fields, or that the longhouses were too far from the pits during the late winter months.

31. Peter A. Thomas, "Contrasting Subsistence Strategies and Land Use as Factors for Understanding Indian-White Relations in New England," *Ethnohistory* 23 (1976): 1–18. Also see discussion in Merchant, *Ecological Revolutions*, 74–87. Cultivating an acre or two by hand would not have been particularly taxing for a skilled gardener. See Howard Russell, *Indian New England before the Mayflower* (Hanover, N.H.: University Press of New England, 1980), 99–100, 169–70, on the great care, but also the reported pleasantness and sociability of women's work in their cornfields.

32. Cronon, *Changes in the Land*, 45, 48; Russell felt some fields were kept in cultivation much longer, *Indian New England*, 140–41. The cultivation period of eight to ten years seems to be derived primarily, if not exclusively, from a single source, William Wood, *New England's Prospect* (1634; reprint, Boston: Prince Society, 1865).

33. The idea that New England Indians regularly put fish under their corn hills was questioned by Lynn Ceci, "Fish Fertilizer: A Native North American Practice?" *Science* 188 (1975): 26–30. But Thomas Morton observed that the Indians did dung ground with fish, Russell, *Indian New England*, 166–67. Carolyn Merchant calculated that each woman would have needed to carry about one thousand pounds of fish from the

weir to her one-acre field. But that was light work, much less than the two thousand pounds of corn (forty bushels) she carried back out. If Concord's sandy, nutrient-poor soils were kept in cultivation for anything like eight years, it seems highly likely that they were fertilized with fish. Merchant, *Ecological Revolutions*, 78; Snow, *Archaeology of New England*, 75; Thomas, "Contrasting Subsistence Strategies"; William Wood, *New England's Prospect*. For a discussion of how farming and fishing may have worked together at Algonquian villages, see Robert J. Hasenstab, "Fishing, Farming, and Finding the Village Sites: Centering Late Woodland New England Algonquians," in Levine, Sassaman, and Nassaney, *The Archaeological Northeast*, 150.

34. Getting a handle on the length of the period of forest fallow is even more problematic than the length of the cultivation period—observations by Europeans simply do not exist. The twenty-five- to fifty-year range is from Peter Thomas, "Contrasting Subsistence Strategies."

35. Russell, *Indian New England*, 126–28, 140–41.

36. Blancke, "Survey of Pre-Contact Sites and Collections in Concord"; Concord Town Records, vols. 1 and 2, *passim*. These areas are approximations. Short of detailed parcel-by-parcel research covering the whole of Concord, it is not possible to show the precise boundaries of these Indian planting fields and pine plains as they were when they passed to the English in 1635.

37. The Native population may have been smaller—although there was certainly ample room for a group of the size suggested. Shirley Blancke believes that the scant archaeological evidence from the late Woodland period argues against a dense Native horticultural population in Concord. See Blancke, "Survey of Pre-Contact Sites and Collections in Concord." But horticultural people moving their winter villages every decade or so and dispersing for the summer to single-family homesteads may have left behind fewer recoverable traces of themselves over a few centuries than their predecessors, who gathered at the same key foraging sites over thousands of years. See Hasenstab, "Fishing, Farming, and Finding the Village Sites," on the obstacles to discovering Late Woodland sites in New England.

38. The original grant of Concord was officially a six-mile square, but it was really closer to seven miles, totaling about 29,000 acres. Of this, about half, or 14,500 acres, was sandy outwash. Compare this to the 100–200 acres a Native band of five hundred would have cultivated in a given year. Even doubling the Indian population wouldn't have changed things by much more than 1 percent.

39. Gordon G. Whitney and William C. Davis, "Thoreau and the Forest History of Concord, Massachusetts," *Forest History* 30 (1986); Winkler, "Changes at Walden Pond During the Last 600 Years."

40. Robert Tarule, "The Joined Furniture of William Searle and Thomas Dennis: A Shop-based Inquiry into the Woodworking Technology of the Seventeenth Century" (Ph.D. diss., Graduate School of the Union Institute, 1992), 34–42, has a good discussion of early descriptions of the New England landscape. See also Cronon, *Changes in the Land*, 49. Pollen evidence on grasslands from David R. Foster, personal communication.

41. Indian firewood use is discussed by Cronon, *Changes in the Land*, 48–49. My aim is simply to get a rough estimate of the outer limit of wood demand and its ecological impact. Rob Tarule suggests that early travelers' accounts of low thickets and coppices may indicate that Indians deliberately coppiced red maple in swampy areas, Tarule, "Joined Furniture of William Searle," 39–40. Coppicing of tens of thousands of hardwood poles one to four inches in diameter has been associated with the Bolyston Street Fish Weir in Boston during the Archaic period—see George P. Nicholas, "A Light but Lasting Footprint: Human Influences on the Northeastern Landscape," in Levine, Sassaman, and Nassaney, *Archaeological Northeast*, 32–33. Coppicing of mangroves for fuel by Florida Indians is discussed in Lee A. Newsom, "Plants and People: Cultural, Biological, and Ecological Responses to Wood Exploitation," in Scarry, *Foraging and Farming in the Eastern Woodlands*, 130–33. An idealized circle containing twelve hundred acres (two hundred acres of planting ground surrounded by one thousand acres of fuelwood) has a radius of 4,079 feet, or about 3/4 of a mile. So a circle a mile or two in diameter could theoretically have supplied a village of five hundred people with crops and wood for about a decade. It is interesting that archaeological surveys put village sites in the

Concord valley 1.4 miles apart on average and that the horticultural tracts on my map lie between one and two miles from one another. Cook, *Indian Population of New England.*

42. Combining meat (over 10 percent of diet), nuts and seeds (over 8 percent of diet), and part of fruits (4 percent of diet), Snow, *Archaeology of New England,* 334. See also Cronon, *Changes in the Land,* 47–48.

43. Russell, *Indian New England,* 103, 125.

44. Richard J. Eaton, *A Flora of Concord* (Cambridge: Museum of Comparative Zoology, Harvard University, 1974); John W. Brainerd, "The Vegetation of the Town of Dover, Massachusetts: An Ecological Survey" (Ph.D. diss., Harvard University, 1949); and personal observations in Concord, Lincoln, Weston, Wayland, Sudbury, and other Middlesex County towns.

45. In some cases they may have attempted to keep these burned areas open as long as possible through repeated burning, creating special hunting grounds remote from the croplands. Russell, *Indian New England,* 125.

46. That, at least, is the impression left by early town records. The hills surrounding Flint's Pond were known for their chestnuts down to Henry Thoreau's time. The way to the Chestnut Field, which passed from the edge of the sand plain through the till uplands to Flint's Pond, ran by the following trees when laid out in 1736: one small pine, ten walnuts (i.e., hickories), ten chestnuts, ten black oaks, three white oaks, two gray oaks, one red oak, one *blue* oak, and two poplars. The poplars were probably aspen; I'm not sure about gray oak and blue oak. CTR 2:112. For an overview of the aboriginal upland forests of Massachusetts, see O'Keefe and Foster, "An Ecological History of Massachusetts Forests," 26–31. The widespread oak forest of the entire Atlantic coastal plain and Piedmont has been attributed by some to periodic fire discouraging northern species, particularly red maple. See Marc D. Abrams, "Fire and the Development of Oak Forests," *Bioscience* 42 (1992): 346–53.

47. Daniel Gookin, *An Historical Account of the Doings and Sufferings of the Christian Indians in New England, in the Years 1675, 1676, 1677* (New York: Arno, 1972). Hirtleberry is an archaic English term for blueberry.

48. Some researchers doubt whether salmon in particular was ever an important element in New England fish runs. Catherine C. Carlson, "'Where's the Salmon?'—A Reevaluation of the Role of Anadromous Fisheries in Aboriginal New England," in Nicholas, *Holocene Human Ecology.* This discussion and the following paragraphs are based largely on Shirley Blancke, "The Archaeology of Walden Woods," 245; Shirley Blancke and Barbara Robinson, *From Musketaquid to Concord: The Native and European Experience* (Concord: Concord Antiquarian Museum, 1985); Snow, *Archaeology of New England,* 334; and Russell, *Indian New England,* 123–25.

49. Russell, *Indian New England,* 126, 131, 178–79; Blancke and Robinson, *From Musketaquid to Concord.*

50. Francis C. Golet and Joseph S. Larson, *Classification of Freshwater Wetlands in the Glaciated Northeast,* Resource Publication 116, Washington Bureau of Sport Fisheries and Wildlife, Fish and Wildlife Service (Washington, D.C.: U.S. Department of the Interior, 1974).

51. On beaver, see Nicholas, "A Light but Lasting Footprint," 31; Krech, *Ecological Indian,* 173–74. In Minnesota, ponds occupy about one-quarter of the cycle, wetlands (primarily meadows) the remainder. C. A. Johnson, "Ecological Engineering of Wetlands by Beavers," in W. J. Mitsch, ed., *Global Wetlands: Old World and New* (Amsterdam: Elsevier, 1994), 379–83.

52. I have seen no references in the literature to New England Indians firing meadows. Wetlands in other parts of the country, including canebreaks in the South and marsh grasses in prairie sloughs, were certainly burned, whether deliberately or not. See Ann Vileisis, *Discovering the Unknown Landscape: A History of America's Wetlands* (Washington, D.C.: Island Press, 1997), 22, 24. Meadow burning by Concord farmers is mentioned by Thoreau, *Journal* 9:128.

53. This is a hypothetical model. It is consistent with what is known about this landscape and its inhabitants, but what is known is so very little that the model is still largely conjectural. Its purpose is to bring some

ecological structure to what has been written and debated about the Indians' relation to their environment by many other scholars.

54. Peter Thomas, at the conclusion of his analysis of native subsistence systems, states that "such strategies characterize the interaction of man with integrated floral and faunal associations within *balanced* communities." Similarly, Carolyn Merchant writes that "the Indian production system of horticulture, gathering, and hunting had evolved in *symbiosis* with the local ecology," while William Cronon has it that "for New England Indians, ecological diversity, whether natural or artificial, meant abundance, *stability,* and a regular supply of the things that kept them alive" (emphasis added). Thomas, "Contrasting Subsistence Strategies," 10; Merchant, *Ecological Revolutions,* 85; Cronon, *Changes in the Land,* 53.

55. Mulholland, "Territoriality and Horticulture." A Malthusian interpretation of Native New England is elaborated in Diana Muir, *Reflections in Bullough's Pond: Economy and Ecosystem in New England* (Hanover, N.H.: University Press of New England, 2000), 4–23.

56. Five hundred people inside the forty-five-square-mile box called Concord (as purchased by English settlers in 1637) works out to some 427 people per one hundred square kilometers—twice the density suggested by Snow for the horticultural people of southern New England, *Archaeology of New England.* The abundance of game probably reflected an ecological system in which people were able to maintain and enjoy food resources that were at most seasons beyond what they had any reason to harvest, given their relatively light population density, economic system, and technology. For a similar discussion, see Krech, *Ecological Indian,* 73–78, 95–99.

57. See Dean R. Snow and Kim M. Lanphear, "European Contact and Indian Depopulation in the Northeast: The Timing of the First Epidemics," *Ethnohistory* 35 (1988), for a discussion of New England population depletion issues. But see also Russell Thornton, Tim Miller, and Jonathan Warren, "American Indian Population Recovery following Smallpox Epidemics," *American Anthropologist* 93 (1991), for a critique of population crash assumptions. Krech, *Ecological Indian,* 83–93, reviews this debate.

58. This marriage indicates that the Musketaquid Indians had an alliance with a coastal group of the kind suggested by Mulholland, "Territoriality and Horticulture." See Blancke and Robinson, *From Musketaquid to Concord;* Blancke, "Analysis of Variation in Point Morphology"; and Lemuel Shattuck, *A History of the Town of Concord* (Boston: Russell, Odiorne, 1835).

59. Shattuck, *History of Concord;* Gookin, *An Historical Account.*

CHAPTER 3. Mixed Husbandry

1. A recent overview by a group of leading English agricultural historians is Joan Thirsk, ed., *The English Rural Landscape* (New York: Oxford University Press, 2000).

2. In many regions herding and tilling complement one another in the way they utilize resources and in the products they provide. But there is often conflict over use of overlapping lands by both plowmen and pastoralists, never mind frequent cultural and ethnic antagonisms. See, for example, Cynthia Nelson, ed., *The Desert and the Sown: Nomads in the Wider Society* (Berkeley: Institute of International Studies, University of California, 1973).

3. If population rises so that fallow periods are shortened, crop yields decline. If cultivation is pressed onto less stable soils, and fields are worked so hard that the ability of desirable vegetation to regenerate is destroyed, ecological havoc can result. For a discussion of cropping systems, see Miguel A. Altieri, *Agroecology: The Scientific Basis of Alternative Agriculture* (Berkeley: University of California Press, 1983), 47–50.

4. On nutrient flow in Chinese agriculture, see Wen Dazhough and David Pimental, "Energy Flow in Agroecosystems of Northeast China," in Stephen R. Gliessman, ed., *Agroecology: Researching the Ecological Basis for Sustainable Agriculture* (New York: Springer-Verlag, 1990). The dense populations working irrigated fields commonly suffer from parasitic diseases, especially when food, wastes, water, and people are

thoroughly mixed and in frequent contact. See Eric L. Jones, *The European Miracle: Environments, Economics and Geopolitics in the History of Europe and Asia* (Cambridge: Cambridge University Press), 1981.

5. The agricultural historian Eric Kerridge states that English farmers eschewed the use of human waste, but that seems too fastidious. Eric Kerridge, *The Agricultural Revolution* (London: George Allen and Unwin, 1967), 241. There was apparently some carting of urban waste from larger towns back to the fields but much wasteful dumping as well. Granville Astill and Annie Grant, "The Medieval Countryside: Efficiency, Progress and Change," in Grenville Astill and Annie Grant, eds., *The Countryside of Medieval England* (Oxford: Basil Blackwell, 1988), 214. It appears that some human waste was recycled in English farming, but that this was not a hallmark of the system.

6. Examples range from Irish tenants putting seaweed with their potatoes in "lazy beds" and New England Indians putting fish in their corn hills, to the modern application of tons of mined and chemically synthesized nutrients to each acre annually.

7. On average, grazing livestock return about 75 or 80 percent of the nutrients they ingest directly to the pasture in their waste, presuming they stay on it day and night. If they are passing evenings in the fold or barnyard, they take more nutrients away with them.

8. Modern agronomy has instead succeeded in turning manure into a major pollutant.

9. That is, fields close to home generally received the most manure. Robert A. Dodgshon, *The Origin of British Field Systems: An Interpretation* (New York: Academic Press, 1980); also see essays in Alan R. H. Baker and Robin A. Butlin, eds., *Studies of Field Systems in the British Isles* (Cambridge: Cambridge University Press, 1973).

10. Maximizing the efficiency of grazing within a largely arable landscape was not the only factor that gave rise to open fields, but it was a central ecological feature. See Joan Thirsk, "The Common Fields" (1964), in R. H. Hilton, ed., *Peasants, Knights and Heretics: Studies in Medieval Social History* (Cambridge: Cambridge University Press, 1976); Dodgshon, *Origins of British Field Systems;* Richard C. Hoffman, "Medieval Origins of the Common Fields," in William N. Parker and Eric L. Jones, eds., *European Peasants and Their Markets: Essays in Agrarian Economic History* (Princeton: Princeton University Press, 1975); Carl J. Dahlman, *The Open Field System and Beyond: A Property Rights Analysis of an Economic Institution* (New York: Cambridge University Press, 1980); Oliver Rackham, *The History of the Countryside* (London: J. M. Dent, 1986), 178, 332; and Grenville Astill, "Fields," in Astill and Grant, *Countryside of Medieval England,* 64–68. Some regions combined both commons and private pastures with smaller open fields and enclosed fields held in severalty. On Kent, the origin of many Concord settlers, see Alan R. H. Baker, "Field Systems of Southeast England," in Baker and Butlin, *Field Systems.*

11. Heath is dominated by heather and furze rather than grass. "Wold" derives from *wald,* meaning wood, which the walds presumably once were. "Downs" means "uplands" mainly to charm tourists but ultimately comes from the Celtic word "dun," meaning a small fort or walled homestead, which tended to sit atop hills.

12. J. R. Wordie, "The South: Oxfordshire, Buckinghamshire, Berkshire, Wiltshire, and Hampshire," in Joan Thirsk, ed., *The Agrarian History of England and Wales,* vol. 5, part 1, *Regional Farming Systems* (Cambridge: Cambridge University Press, 1984), 333. See also other essays in this volume and in its predecessor, Joan Thirsk, ed., *The Agrarian History of England and Wales,* vol. 4, *1500–1640* (Cambridge: Cambridge University Press, 1967); and in Baker and Butlin, *Field Systems;* and chap. 2, "The Farming Countries of England," in Kerridge, *Agricultural Revolution.*

13. For discussions of grass within the arable fields, see Rackham, *History of the Countryside,* 165; Kerridge, *Agricultural Revolution,* chap. 2; David Roden, "Field Systems of the Chiltern Hills and Their Environs," in Baker and Butlin, *Field Systems,* 334–35, 349; and Dahlman, *Open Field System and Beyond.*

14. Rackham, *History of the Countryside,* 332.

15. James Greig, "Plant Resources," in Astill and Grant, *Countryside of Medieval England,* 119. This

removed nutrients from the arable just as surely as food for humans did, although animal manure was most certainly returned to the fields.

16. Rackham, *History of the Countryside,* 332.

17. Roden, "Field Systems of the Chiltern Hills," 327.

18. Baker, "Field Systems of Southeast England," 387. The Weald was the upland, once heavily wooded central region of Sussex and Kent ("weald," like "wold," means "woods").

19. Eric Kerridge, *The Farmers of Old England* (Totawa, N.J.: Rowman and Littlefield, 1973), 22.

20. Rackham, *History of the Countryside,* 333–37; Christopher Dyer, "Documentary Evidence: Problems and Enquiries," in Astill and Grant, *Countryside of Medieval England,* 18. Meadow acreage increased through drainage of fens and marshes. Unlike many other points of English agrarian history, the distinctive role of meadowland seems undisputed. See also essays in Baker and Butlin, *Field Systems;* and Thirsk, *Agrarian History of England and Wales,* both volumes.

21. Joan Thirsk, chap. 3, "Farming Techniques," in Thirsk, ed., *Agrarian History of England and Wales,* 4:167–68.

22. Rackham, *History of the Countryside,* 295. English archaeologists complain that they can't find enough telltale middens to dig up in medieval villages because the rubbish all got shifted with the manure. Astill and Grant, "The Medieval Countryside," 2.

23. Astill, "Fields," 79. Mucking and folding had to be integrated into the complex field rotations of fallow, winter corn, and spring corn. In the Midlands winter corn usually followed the sheepfold while spring corn was mucked with barnyard manure. See Eric Kerridge, *The Common Fields of England* (New York: Manchester University Press, 1992), 74–86.

24. Until the advent of modern forestry in recent centuries, "forest" had a meaning in England that had little to do with trees. It meant land on which deer were protected by special laws and which might or might not have been wooded. See Rackham, *History of the Countryside,* 129–39.

25. Ibid., 72–88, 106. William Hoskins wrote that what the Celts and Romans cleared reverted to woodland during the Dark Ages of the fifth and sixth centuries, that the pioneering Anglo-Saxons had the woods about half cleared by 1086, and that the great wave of deforestation occurred during the "land hunger" of the twelfth and thirteenth centuries. W. G. Hoskins, *The Making of the English Landscape* (London: Hodden and Stoughton, 1955). Rackham's revision benefited from several decades of research inspired by Hoskins. There is evidence lending renewed support for Hoskins's assumption that cultivation declined after the Roman collapse, and that the Saxons did have to do a good bit of reclearing. Paul Stamper, "Woodlands and Parks," in Astill and Grant, *Countryside of Medieval England,* 128–29.

26. Rackham, *History of the Countryside,* 86.

27. Ibid., 65–67, 85–88. Very little English timber was sawn into boards. By the Middle Ages most pine boards were imported from Scandinavia and oak boards from commercial sawmills in the great woodlands of central Europe. See also Stamper, "Woods and Parks," 130–33.

28. Rackham, *History of the Countryside,* 65–66, 119–22; Stamper, "Woodlands and Parks," 132–33.

29. Rackham, *History of the Countryside,* 78, 80, 84, 88.

30. Ibid., 185–90; Stamper, "Woodlands and Parks," 130; Greig, "Plant Resources," 124. Parts of England that had more enclosed fields and hedges than the open field country of the Midlands were sometimes referred to as "woodland" regions.

31. Greig, "Plant Resources," 121, 124–25; Stamper, "Woodlands and Parks," 138–40, 143–45; Annie Grant, "Animal Resources," in Astill and Grant, *Countryside of Medieval England,* 165–66. Rabbits and deer belonged legally to the wealthy or to the king, but their bones turn up in peasants' privies.

32. Rackham, *History of the Countryside,* 167–69. Starting along the center line of the strip the plow went around and around, always turning the furrow slice inward. Because it was difficult to turn the heavy plow

and the six- or eight-head plow team, and the turns got longer each time around, strips tended to be long and narrow: the classic dimensions were 11 yards by 220 yards (an eighth of a mile, or "furlong"), making exactly half an acre. Few strips in the real world had exactly the classic dimensions.

33. The great centralized drainage projects that took place in the Fens and other large coastal marshes after 1600 overrode centuries of smaller but already elaborate drainage systems designed to augment the natural richness of this environment. H. C. Darby, *The Changing Fenland* (New York: Cambridge University Press, 1983); Joan Thirsk, *English Peasant Farming: The Agrarian History of Lincolnshire from Tudor Times* (London: Routledge and Kegan Paul, 1957); Rackham, *History of the Countryside*, 374–93; Kerridge, *Agricultural Revolution*.

34. John Fitzherbert, *The Boke of Surveying and Improvements* (London, 1523), as quoted in Rackham, *History of the Countryside*, 338–39. Any homeowner with an overflowing leaching field is familiar with the fertilizing effect of warm, nutrient-rich water, even in winter, although perhaps insufficiently grateful for the blessing of this organic method of lawn feeding and moldywarpe control.

35. Kerridge, *Agricultural Revolution*, 250–55; Rackham, *History of the Countryside*, 338–39; J. H. Bettey, "The Development of Water Meadows in Dorset during the Seventeenth Century," *Agricultural History Review* 25 (1977); John Sheail, "The Formation and Maintenance of Water-meadows in Hampshire, England," *Biological Conservation* 3 (1971). Floated watermeadows were especially productive in regions where streams laden with calcium ran down from chalky uplands.

36. Quote from Grant, "Animal Resources," 170. In some regions, most notably the Fens, wild fish and fowl were very important. See Darby, *Changing Fenland*, and Thirsk, *English Peasant Farming*. Foraged foods were always more prominent at the tables of common people than is visible to historians because such fare was commonly foraged under cover of darkness and made it into the record only in the small percentage of cases in which some poor villain was clumsy enough to get caught. Rackham, *History of the Countryside*, 365–68.

37. James Frederick Edwards and Brian Paul Hindle, "The Transportation Systems of Medieval England and Wales," *Journal of Historical Geography* 17 (1991); John Langdon, "Inland Water Transport in Medieval England," *Journal of Historical Geography* 19 (1993).

38. Dyer, "Documentary Evidence," 29–30.

39. Jules N. Pretty, "Sustainable Agriculture in the Middle Ages: The English Manor," *Agricultural History Review* 38 (1990): 1–19; Donald N. McCloskey, "The Persistence of the English Common Fields," and Richard C. Hoffman, "Medieval Origins of the Common Fields," in Parker and Jones, *European Peasants and Their Markets*.

40. Garrett Hardin, "The Tragedy of the Commons," *Science* 162 (1968): 1243–48. Tragedies of the commons generally result from a conflict between common resources and unbridled free enterprise, not common resources and human nature. See Arthur McEvoy, "Toward an Interactive Theory of Nature and Culture," in Donald Worster, ed., *The Ends of the Earth: Perspectives on Modern Environmental History* (New York: Cambridge University Press, 1990); and B. McCay and J. M. Acheson, eds., *The Question of the Commons: The Culture and Ecology of Communal Resources* (Tucson: University of Arizona Press, 1987).

41. There is evidence that at least in some areas population was no longer growing by the beginning of the fourteenth century, so perhaps fertility was damping down. Richard Smith, "Human Resources," in Astill and Grant, *Countryside of Medieval England*, 188–211; B. M. S. Campbell, "People and Land in the Middle Ages, 1066–1500," in R. A. Dodgshon and R. A. Butlin, eds., *An Historical Geography of England and Wales*, 2d ed. (London: Academic Press, 1990), 70–71.

42. Charles R. Bowles, "Ecological Crisis in Fourteenth-Century Europe," in Lester J. Bilsky, ed., *Historical Ecology: Essays on Environment and Social Change* (Port Washington, N.Y.: Kennikot Press, 1980); Rackham, *History of the Countryside*, 86, 88, 359; Dyer, "Documentary Evidence," 33; Greig, "Plant Resources," 124; and Astill, "Rural Settlement," 54, in Astill and Grant, *Countryside of Medieval England*.

43. The Middle Ages saw a decline in the size of stock, probably because of poor pasture and inadequate winter feed. Overgrazing in the thirteenth and fourteenth centuries is indicated by a high incidence of gum disease in sheep. Grant, "Animal Resources," 154, 176–77.

44. The issue of how well the medieval economy was able to respond to population growth is far from settled. In East Anglia and the Home Counties, peasants were able to raise grain yields by more careful manuring and cultivation and in particular by replacing bare fallows with legume fodder crops such as peas and vetch. See Dyer, "Documentary Evidence," 25–33; Greig, "Plant Resources," 125; Astill and Grant, "The Medieval Countryside: Efficiency, Progress, and Change," 213–33; Campbell, "People and Land in the Middle Ages, 1066–1500," 92–102; and Bruce M. S. Campbell, "Ecology Versus Economics in Late Thirteenth- and Early Fourteenth-Century English Agriculture," in Del Sweeny, ed., *Agriculture in the Middle Ages: Technology, Practice, and Representation* (Philadelphia: University of Pennsylvania Press, 1995), 76–94. An older overview of the crisis is Edward Miller and John Hatcher, *Medieval England: Rural Society and Economic Change, 1086–1348* (New York: Longman, 1978). The growth of the market allowed some regions to specialize in either corn or livestock and thus partly break free of local ecological constraints on subsistence production. See Bruce M. S. Campbell, "Economic Rent and the Intensification of English Agriculture, 1086–1350," in Grenville Astill and John Langdon, eds., *Medieval Farming and Technology: The Impact of Agricultural Change in Northwest Europe* (Leiden: Brill, 1997), 225–49.

45. Grant, "Animal Resources," 154; Smith, "Human Resources," 192–93; Campbell, "People and Land in the Middle Ages," 100–01. While some historians see this decline as the result of a decisive long-term shift to a cooler, wetter climate, reversing a "warm epoch" that had allowed medieval expansion, more see the bad weather as a triggering event that fell upon a society that had outgrown its ecological limits. Astill and Grant, "Efficiency, Progress and Change," 232–33.

46. John Hatcher, *Plague, Population and the English Economy, 1348–1530* (London: Macmillan, 1977); Campbell, "People and Land in the Middle Ages," 102–13; Smith, "Human Resources," 192–93; David Herlihy, "Ecological Conditions and Demographic Change," in Richard L. DeMolen, *One Thousand Years: Western Europe in the Middle Ages* (Boston: Houghton Mifflin, 1974).

47. For an overview of population trends, see R. M. Smith, "Geographical Aspects of Population Change in England, 1500–1730," in Dodgshon and Butlin, *Historical Geography of England and Wales*, 151–73. Smith's essay incorporates the findings of E. A. Wrigley and R. S. Schofield, *The Population History of England, 1541–1871*, 2d ed. (Cambridge: Cambridge University Press, 1988).

48. The decline in fertility closely followed the decline in real wages that lasted to the mid–seventeenth century and then continued for a generation after wages began to recover. Wrigley and Schofield, *Population History of England;* Smith "Geographical Aspects of Population Change in England," 153, 165–67.

49. Smith, "Human Resources," 155–62, 172–75; Joan Thirsk, "Patterns of Agriculture in Seventeenth-Century England," in David D. Hall and David Grayson Allen, eds., *Seventeenth-Century New England* (Boston: Colonial Society of Massachusetts, 1984); Victor Skipp, *Crisis and Development: An Ecological Case Study of the Forest of Arden, 1570–1674* (Cambridge: Cambridge University Press, 1978). Migration may have played a role in keeping fertility low because differing male and female migration patterns created unbalanced sex ratios in many regions, depressing the rate of marriage.

50. Joan Thirsk, "Patterns of Agriculture," 39–53. Thirsk expanded on her picture of English agriculture as a long series of swings between periods of high demand for grain and periods of low demand that stimulated specialty crops and other innovations in *Alternative Agriculture: A History from the Black Death to the Present Day* (New York: Oxford University Press, 1997).

51. For a review of the enclosure literature, see Mark Overton, "Agricultural Revolution? Development of the Agrarian Economy in Early Modern England," in Alan R. H. Baker and Derek Gregory, eds., *Explorations in Historical Geography* (New York: Cambridge University Press, 1984). See also J. R. Walton, "Agriculture and Rural Society, 1730–1914," 334; and J. Yelling, "Agriculture 1500–1730," 182–83, in Dodgshon and Butlin, *Historical Geography of England and Wales;* J. M. Neeson, *Commoners: Common Right, Enclosures*

and Social Change in England, 1700–1820 (Cambridge: Cambridge University Press, 1993); John Broad, "The Verneys as Enclosing Landlords, 1600–1800," in John Chartres and David Hey, eds., *English Rural Society, 1500–1800: Essays in Honour of Joan Thirsk* (Cambridge: Cambridge University Press, 1990); and R. A. Butlin, "The Enclosure of Open Fields and Extinction of Common Rights in England, circa 1600–1750: A Review," in H. S. A. Fox and R. A. Butlin, eds., *Change in the Countryside: Essays on Rural England, 1500–1900* (London: Institute of British Geographers, 1979).

52. Joan Thirsk, "Farming Regions," in Thirsk, *Agrarian History of England and Wales,* vol. 4; David Hey, "The North-West Midlands," Brian M. Short, "The South-East," and R. C. Richardson, "The Metropolitan Counties," in Thirsk, *Agrarian History of England and Wales,* vol. 5; Alan Baker, "Field Systems of Southeast England," and David Roden, "Field Systems of the Chiltern Hills and Their Environs," in Baker and Butlin, *Field Systems in the British Isles.*

53. Yelling, "Agriculture 1500–1730," 181–96; Thirsk, "Patterns of Agriculture," 39–53.

54. Some communities were able to buy their replacement stock from the upland regions, but they still had to grow something to feed the stock once they arrived. A few arable parishes were able to import manure supplies by joining in a symbiotic relation with large-scale private graziers and folding large sheep flocks on their fields. Kerridge, *Agricultural Revolution,* chap. 2.

55. M. A. Havinden, "Agricultural Progress in Open-field Oxfordshire" (1961) in E. L. Jones, ed., *Agriculture and Economic Growth in England, 1650–1815* (New York: Barnes and Noble, 1967); Skipp, *Crisis and Development;* Yelling, "Agriculture 1500–1730," 191, 195. There is little evidence of *dramatic* improvement in the overall productivity of English farmland before the middle of the seventeenth century, as some, for example, Eric Kerridge, have argued. Kerridge, *Agricultural Revolution,* chap. 3, "Up and Down Husbandry." For an overview of the often acrimonious "agricultural revolution" debate, see Mark Overton, "Agricultural Revolution? Development of the Agrarian Economy in Early Modern England," in Baker and Gregory, *Explorations in Historical Geography,* 118–39.

56. Joan Thirsk, "Agricultural Innovations and Their Diffusion," in Thirsk, *Agrarian History of England and Wales,* 5:553.

57. John Broad, "Alternate Husbandry and Permanent Pasture in the Midlands, 1650–1800," *Agricultural History Review* 28 (1980); Yelling, "Agriculture 1500–1730," 194–95; E. L. Jones, "Introduction," and "Agriculture and Economic Growth in England, 1660–1750: Agricultural Change" (1965), in Jones, *Agriculture and Economic Growth in England;* Joan Thirsk, "Industries in the Countryside," in F. J. Fisher, ed., *Essays in the Economic and Social History of Tudor and Stuart England, in Honour of R. H. Tawney* (Cambridge: Cambridge University Press, 1961).

58. Broad, "Alternate Husbandry and Permanent Pasture."

59. It is certainly possible to employ legume rotations without livestock. This occurs on many organic market gardens today. In this case "green manures" and cover crops are simply grown and then plowed in, and no manure handling is involved.

60. Claims that the charcoal industry destroyed woodlands to supply iron forges appear to be exaggerated because these fuel supplies were largely maintained by coppicing. But the growing English iron industry outran its available fuel supply even if it didn't destroy its base. In the process it caused additional scarcity for people trying to heat their homes. Rackham, *History of the Countryside,* 89–91; H. C. Darby, "The Age of the Improver: 1600–1800," in H. C. Darby, ed., *A New Historical Geography of England* (Cambridge: Cambridge University Press, 1973), 363.

61. Richard G. Wilkinson, *Poverty and Progress: An Ecological Perspective on Economic Development* (New York: Praeger, 1973), chap. 6, "The English Industrial Revolution."

62. Claims of scarcity in supplies of ship timber appear to be exaggerated. Scarcity in building materials was real enough. Rackham, *History of the Countryside,* 89–90.

63. Thirsk, "Patterns of Agriculture," 53.

CHAPTER 4. The First Division and the Common Field System

1. Concord shares this distinction with Lexington, of course.

2. Henry Thoreau shares this distinction, let us say, with George Perkins Marsh.

3. Edward Johnson, *Johnson's Wonder-Working Providence* (1654; reprint, New York: Barnes and Noble, 1910), 110–13. For a nice placement of Johnson's own intricate work within the history of the founding of New England towns, see David Jaffee, *People of the Wachusett: Greater New England in History and Memory, 1630–1860* (Ithaca: Cornell University Press, 1999), 16–22.

4. Charles H. Walcott, *Concord in the Colonial Period* (Boston: Estes and Lauriat, 1884), 3. Johnson's breathless style, which compresses several years into two reeling paragraphs, does have a melodramatic effect. But Johnson is only employing the same literary device later used by Thoreau when he distilled several decades of exploration within Concord into an account of a single round of seasons at Walden Pond. Johnson just does it a shade less skillfully than Thoreau, which is why there was never a chain of *Wonder-Working* bookstores in malls across America. The poor settlers of Concord seem to pass all their trials on a single long trek down one narrow sentence winding through Johnson's bewildering syntax.

5. Karen Ordahl Kupperman, "Climate and Mastery of the Wilderness in Seventeenth-Century New England," in David D. Hall and David Grayson Allen, eds., *Seventeenth-Century New England* (Boston: Colonial Society of Massachusetts, 1984), 5–6.

6. Cecelia Techi, *New World, New Earth: Environmental Reform in American Literature from the Puritans through Whitman* (New Haven: Yale University Press, 1979), 37–60, articulates the Puritan drive to reform the New World. Thanks to Richard Judd for the thought that follows.

7. The details of Concord's settlement are sketchy because the town's first book of records has been lost. In January 1664, the town voted to transcribe "all that is useful from the old booke into the new booke," but not much was deemed useful. The skeletal outline given here is derived from secondary accounts, including Shattuck, *History of Concord*; Walcott, *Concord in the Colonial Period*; Ruth R. Wheeler, *Concord: Climate for Freedom*, (Concord: Concord Antiquarian Society, 1967); and Blancke and Robinson, *From Musketaquid to Concord*.

8. Anderson, *New England's Generation*, 92; CTR 2:1, CFPL; Wheeler, *Climate for Freedom*.

9. Not that having a deed from the Indians would have done much good had Andros gotten his way—he held that the only valid grants of land were those that had come directly from the Massachusetts Bay colony and bore the corporate seal, and that towns were not legal entities that could hold or grant land. Andros was overthrown in 1689. See the discussion in John Frederick Martin, *Profits in the Wilderness: Entrepreneurship and the Founding of New England Towns in the Seventeenth Century* (Chapel Hill: University of North Carolina Press, 1991), 260–67.

10. Shattuck, *History of Concord*, 22 ff.

11. Ibid., 15.

12. Walcott, *Concord in the Colonial Period*, 14. In 1641, Willard was named superintendent of trade with the Indians by the General Court. He was later instrumental in the founding of Chelmsford, Groton, and Lancaster to the west of Concord and spent his life dealing with Indians and acquiring land—the very model of an early frontiersman. See Jaffee, *People of the Wachusett*, 55–57; Martin, *Profits in the Wilderness*, 19–20.

13. In general, the Massachusetts Bay fur trade was in decline by the 1660s, and Concord must have been largely trapped out long before then. See Peter A. Thomas, "The Fur Trade, Indian Land and the Need to Define Adequate Environmental Parameters," *Ethnohistory* 28 (1981). Judging by the names the English gave to a handful of places in their new hometown, it appears that beaver were plentiful in the early days. Several Concord settlers received First Division mowing lots in Pond Meadow, at the foot of which the road to Sudbury (now Route 126) crossed the Beaver Dam Bridge, First and Second Division grants of John Billings and Nathaniel Billings Jr., CTR 2:36. Today this extensive wetland has become a red maple swamp. Beaver have recently returned to Concord, although not to universal welcome.

14. Howard S. Russell, *A Long, Deep Furrow: Three Centuries of Farming in New England* (Hanover, N.H.: University Press of New England, 1976) 128–29. See also Joseph S. Wood, *The New England Village* (Baltimore: Johns Hopkins Press, 1997), 22–24, 34–36.

15. Johnson, *Wonder-Working*, 110. Many of the houselots lay within the general field fences.

16. Studies of why many (though not all) early New England towns installed commons systems, of the differences among these systems, and of the conflicts that arose within them over time include Chilton Powell, *Puritan Village: The Formation of a New England Town* (Middletown, Conn.: Wesleyan University Press, 1963); David Grayson Allen, *In English Ways: The Movement of Societies and the Transferral of English Local Law and Custom to Massachusetts Bay in the Seventeenth Century* (Chapel Hill: University of North Carolina Press, 1981); Lockridge, *A New England Town;* Anderson, *New England's Generation.*

17. The First Division map is a composite of two decades of land grants in Concord before the Second Division began in 1653 and of exchanges among these parcels lasting into the 1660s. Since not all landholdings were recorded at the same time and some were never recorded at all, it is impossible to present a snapshot of common field Concord at a single date. No plat of land divisions in Concord is known. This map was drawn from metes and bounds given in the town book. Boundaries are approximate.

18. That order required that dwelling houses be within one-half mile of the meetinghouse, for defensive purposes as well as to maintain spiritual and communal values. The order was repealed in 1640. The order and its context are discussed in Wood, *New England Village*, 44–46.

19. Johnson, *Wonder-Working*, 113–14.

20. Peter Bulkeley to William Hunt, MRD 3-103, 1654.

21. The 1654 deed of Samuel Adams of Charlestown to Richard Temple, MRD 1-129, mentions only "land in Concord that was William Spencer's," not buildings or other improvements. It appears that Temple built the sawmill (later Barrett's mill) there, although he may not have been the only man interested in it, for he conveyed only a single one-eighth share of the mill to his son Isaac when dividing his estate among his children in 1688, MRD 10-116. Several of the largest early investors in towns such as Concord held property but never resided there.

22. MRD 2-222, 1658, Peter Bulkeley to Thomas Dane; MRD 3-128, 1663, Grace Bulkeley to Timothy and George Wheeler; MP12305, 1667, William Hunt. There are to my knowledge no surviving town records, deeds, or probated estates for Concord before the 1650s.

23. Concord was divided into three quarters—meaning neighborhoods—for the purpose of making the Second Division in 1652. I am using this quartering of the town in describing the First Division lots for convenience, although it may not yet have been in use at that time.

24. Roger Draper to Dolar Davis, MRD 3-188, 1655; Isaac Hunt to Samuel Hunt, MRD 7-289, 1671.

25. Great lots were often recorded as being located on pine plains. This may indicate that at the time of the Second Division they had only been recently granted and were not yet cleared.

26. Buckinghamshire Record Society, vol. 21, 1983. As recorded in Potter family material collected by John Threlfall, CFPL, 1992.

27. The first Humphrey Barrett died in 1662. The record of land grants made in 1666 was actually that of his son, Humphrey Barrett Jr. His other son, Thomas, had died in 1652, drowned in the North River. Thomas's son Oliver Barrett received "16 acres belonging to his house on the plain beyond the millpond" in his grandfather Humphrey's will, MP1182, 1662. It is possible that additional First Division grants of Humphrey Barrett Sr. had earlier passed to this side of the family. I can discern no correlation between the English regions of origin of Concord settlers and those who received more consolidated or more dispersed lots—the North Quarter does not appear to be an enclave of settlers from Kent, for example.

28. One can also imagine that from the sandy plains of the Cranefield the farmers could sometimes see great blue herons flying ponderously over the Great Meadows to the north and were inspired by these "cranes."

29. The first inventory of Nathaniel Ball done in 1725 included "10 acres of pasture in the great field, with 5 acres of plowland adjoining," which a second inventory done in 1726 to set off the widow's third rendered as "15 acres at New Field bounded west on the way leading to the plain," MP 915. In 1746, his son Nathaniel Ball sold part of a larger piece of land across the lane described as "4 acres plowland in the great fields, bounded east on the way leading to a field called the far plain," MRD 46–338. It is difficult to tell whether these fields were originally considered part of the Cranefield or adjuncts to it. See also the records of the Proprietors of the Great Field, discussed below.

30. Ruth Wheeler places the kiln on Brickiln Island, which lies across a stretch of meadow to the east of the Brickiln field, but I have seen no documentary evidence of this. Wheeler, *Concord: Climate for Freedom.* There was a gravel pit on the island (CTR 2:96), not a clay pit, which makes sense given the coarse soils there. Clay pits were mentioned in a resurvey of the Bay Road done in 1716, north of the road on the Minot farm—that is, just west of the Brickiln Field. Probably Brickiln Island was named after Brickiln Field, and the kiln was near (or within) the field.

31. Mentoo is an odd name with no discernible English antecedents. It may have been a corruption of the Indian word "manitou," meaning spiritual power. This raises interesting questions about this place, which was by the river opposite a major Indian site of great antiquity, the Clamshell Bank. For more on the subject of Native people's relation to wetlands, see the provocative treatment in James W. Mavor Jr. and Byron E. Dix, *Manitou: The Sacred Landscape of New England's Native Civilization* (Rochester, Vt.: Inner Traditions International, 1989).

32. The complements and conflicts among the individual, family, and community motivations of the New England yeomanry are addressed, among other places, in Anderson, *New England's Generation;* Daniel Vickers, "Competency and Competition: Economic Culture in Early America," *William and Mary Quarterly* 47 (1990); and Wood, *New England Village.*

33. Johnson, *Wonder-Working,* 114. Note the wonderful assurance, so natural to Johnson, that the Lord created maize and delivered it to New England (employing in this enterprise a few hundred generations of Native Americans spread across half a continent) with a special eye to the coming needs of his chosen people upon their arrival in 1636.

34. MRD 3–103, 1654. Bulkeley had bought this land of William Judson, who had apparently left Concord. Indian corn was "planted," rye and other small grains "sown."

35. For discussions of the merits of various grains in New England, see Sarah F. McMahon, "A Comfortable Subsistence: The Changing Composition of Diet in Rural New England, 1620-1840," *William and Mary Quarterly* 42 (1985); and Russell, *Long, Deep Furrow,* 40–42.

36. I have encountered only one reference to barley in seventeenth-century Concord, and only two to wheat. Middlesex Probate 17185, Samuel Potter, 1676, two bushels of mault; MP 1182, Humphrey Barrett, 1662, includes wheate and pease.

37. CTR 2:1–2. The rate was cut to one shilling in 1672, but there was no indication of what it was to be paid in. Caleb Brooks, who moved from Concord to Medford in 1679, had "barley and oats in the straw" when he died there in 1696. MP 2798. Grains besides maize and rye were not unknown in Concord, just comparatively uncommon.

38. For example, MRD 64–100. Chestnut trees were common (though not ubiquitous) as witness trees on colonial deeds in this area, and chestnuts remained common in this area into the twentieth century.

39. MRD 9–78 (emphasis added). Dressing the corn in this context may refer simply to cultivating it but probably implies manuring it. I shall return to manure in more depth in chapter 6.

40. Jared Eliot, *Essays Upon Field Husbandry in New England, and Other Papers, 1748–1762,* Harry J. Carman and Rexford G. Tugwell, eds. (New York: Columbia University Press, 1934), 29; Cronon, *Changes in the Land,* 151.

41. Howard Russell mentions common sheep flocks and folding on common arable fields in Newbury, Ipswich, and Deerfield, lasting into the eighteenth century. Russell, *Long, Deep Furrow,* 155-56. The cow

herd was typically returned to the farmsteads for evening milking. Thus they could dung corn land only if it lay close to the barn, as I will describe later. Samuel Deane recommended cattle as well as sheep for folding in New England but states that folding was little practiced. Samuel Deane, *The New England Farmer,* 3d ed. (Boston: Wells and Lilly, 1822), 153–54.

42. CTR 3:19.

43. Hoar to Wright, MRD 4-409, 1672. "Mitten" is a misspelling of midden.

44. Rackham, *History of the Countryside,* 337. This is obviously a very rough figure and varied greatly from region to region.

45. At the end of the colonial period in Concord, when the agroecological system was fully developed and hay was in fact in very short supply, meadow still occupied about 25 percent of the improved land (excluding forest), and at least 15 percent of the total land in town. A typical eighteenth-century Concord farmer mowed from ten to twenty acres of hay and cut two or three acres of meadow for every acre of arable he tilled. I will return to the matter of the meadows and their maintenance at greater length in chapter 6.

46. See discussions in Eaton, *Flora of Concord;* 22–27; and Ann Zwinger and Edwin Way Teale, *A Conscious Stillness: Two Naturalists on Thoreau's Rivers,* (New York: Harper and Row, 1982), 130–35.

47. "Proprietors of the Great Field," March 1, 1691. Collection in CFPL.

48. *Oxford English Dictionary,* vol. 5 (Oxford: Clarendon Press, 1933). There was also a "willow swamp" in the midst of Fair Haven Meadow, where black willows are still to be seen today.

49. As quoted in Shattuck, *History of Concord,* 15. "Abate the Falls" implies that the Concordians intended to cut away downstream bedrock at (or above) the falls in Billerica to improve the drainage of the river. Had they been sufficiently foresighted they might have sought to gain legal control of the falls themselves, because this turned out to be a choke point at which their entire system of husbandry could be effectively throttled.

50. Ibid., 15–16.

51. Some towns still have sewer committees, but now they deal with waste pipes and treatment plants. They still have a considerable impact on the rivers, though.

52. Eaton, *Flora of Concord.* See also Daniel A. Romani Jr., "'Our *English Clover-grass* sowen thrives very well': The Importation of English Grasses and Forages into Seventeenth-Century New England," in Peter Benes, ed., *Plants and People: Annual Proceedings of the Dublin Seminar for New England Folklife* (Boston: Boston University, 1996), 25–37. Romani's discussion is mostly cogent, but he fails to adequately distinguish between native *upland* grasses, which were ill-adapted for grazing and were soon replaced by European grasses, and native *meadow* grasses, which were coarse but serviceable and remained New England's leading hay source (as well as providing fall grazing) throughout the colonial period.

53. William K. Kruesi, *The Sheep Raiser's Manual* (Charlotte, Vt.: Williamson Press, 1985), 56–58, is one handy source on the seasonal growth curves of pasture grasses.

54. As quoted in Walcott, *Concord in the Colonial Period,* 50. This petition for more land was endorsed by the General Court in 1651.

55. CTR 1:83.

56. *Watertown Records: First and Second Books of Town Proceedings* (Watertown: Press of Fred G. Barker, 1894), 1:8, 21–24, 52, 94–99, 104–05, 111, 146–47; 2:3–5; Russell, *Long, Deep Furrow,* 34–37.

57. MRD 4-409, 1672.

58. "Proprietors of the Great Field," March 1, 1691. Papers in CFPL.

59. "Articles of Agreement of the Proprietors of ye Great Fields," March 28, 1757, CFPL.

60. Allen, *In English Ways,* 50. Deane, *New England Farmer,* 367, mentions fall grazing of winter grains. I have pastured rye into early December—but we've probably gained a week or two on the seventeenth century as the climate has warmed.

61. See Kupperman, "Climate and Mastery," 20–25, for a discussion of English reaction to seventeenth-century New England climate. The climate of northwest Europe is moderated by massive ocean currents that transport warm water from the tropics. New England is warmed, too, but not as much.

62. Wood, *New England's Prospect*, 4.

63. Russell, *Indian New England*, 121.

64. Tarule, "Joined Furniture of William Searle," 43–82. Tarule's picture of Ipswich contrasts with that of a highly privatized East Anglian town given by Allen, *In English Ways*.

65. Bernard Bailyn, *The New England Merchant in the Seventeenth Century* (Cambridge: Harvard University Press, 1955).

66. Kupperman, "Climate and Mastery," 6.

67. John Winthrop, *The History of New England from 1630 to 1649* (Boston: Little, Brown, 1853), 88.

68. As quoted in Shattuck, *History of Concord*, 15.

69. Peter Bulkeley to John Cotton, December 17, 1640, as quoted in Walcott, *Concord in the Colonial Period*, 43.

70. Johnson, *Wonder-Working*, 110.

71. As quoted in Shattuck, *History of Concord*, 16.

72. See, for example, Powell's classic work on neighboring Sudbury, *Puritan Village*.

CHAPTER 5. The Second Division

1. CTR 2:4.

2. CTR, *passim*. Concord had only three quarters because the term was used in the sense of a group of residences, as we might say "the Latin Quarter." In time the term was applied to the entire section of the town each quarter had assigned to it, yielding the odd result of a town divided into three equal quarters.

3. A five-acre parcel of woodland (mostly pine) along the road just across the river was reserved to supply wood for the North River bridge. Maintaining these bridges was a heavy expense, and many of the surviving early records of the South Quarter relate to the periodic rebuilding of the South Bridge with oak and pine drawn from a nearby hill. In 1665 and 1666, they were getting timber, framing the arch, and raising a new bridge. In January 1673 they agreed that the bridge was to be covered with pine stringers and railed with white oak. CTR 1:32, 88.

4. CTR 2:5. It thus appears that the hog walk had been a joint endeavor of a group of Concord men during the time of the First Division.

5. It is possible that the "enlargements" referred to were to be considered First Division, since the enlargements were to "have Second Division as others have had." In that case "paying 12s per acre as others have done" might indicate that this was what had been paid for *First* Division holdings, at least by some. Either way, it appears that men *paid* for at least some of their "grants." It is also possible that "others" refers to newcomers to Concord who were not proprietors.

6. In 1654, William Hunt bought seventy acres from Peter Bulkeley for £7 10s, or a bit more than 2 shillings an acre. In the 1667 inventory of William Hunt's estate, Second Division land was valued at 3 shillings. Both sources indicate meadow and improved upland was worth £1 an acre or more. By comparison, a common laborer earned about 2½ shillings a day. MRD 3-103, MP 12305.

7. This was for the North and East quarters—the South Quarter had to pay 4p/£. This strange discrepancy may have been a temporary measure as part of the complex agreement worked out to equalize the quarters.

8. This obviously benefited large landholders most. This rate was equivalent to 2p/£ if the land were assessed at 3s/acre. The word "only" was used in the town records, indicating either that by 1667 the going

price for wasteland was beginning to surpass 3s or that the new rate was a correction for the South Quarter paying that higher 4p/£ rate. CTR 2:3.

9. CTR 1:1, 2. Until 1753 the old year lumbered on until March 25, sometimes being written 1655/6 during January and February, for example. For the first generation or so the Puritans declined to use the pagan names for months as well. So January 1, 1656, was written as 1:11:55. I have converted all dates to their modern equivalent in the text. I have not corrected dates for the ten- to eleven-day discrepancy of the Julian calendar, except to mention it when discussing seasonal matters such as field closings.

10. CTR 1:3.

11. Ibid.

12. CTR 2:1.

13. CTR 1:14.

14. See Shattuck, *History of Concord*, 13–14; and MacLean, *Rich Harvest*, 29.

15. Helen Fitch Emery, *The Puritan Village Evolves: A History of Wayland, Massachusetts* (Canaan, N.H.: Wayland Historical Society, Phoenix Publishing, 1981).

16. Virginia was the name conferred on a swamp in the East Quarter of Concord.

17. See discussion in Innes, *Creating the Commonwealth*, 214–19, of the novel and potent balance stuck in New England between private property ownership and the common good embodied in town government. The "town" as a public institution that could own and distribute land was a new departure in English common law and was later challenged by Governor Andros. As a result, much of the remaining common land in New England was transferred from towns to ownership by groups of proprietors. This occurred in Concord in the 1680s and 1690s. See Martin, *Profits in the Wilderness*, 260–80.

18. Walcott, *Concord in the Colonial Period*, 63. The great struggle in Concord was with the Blood family. The town also quarreled with the "pretended successors of Major Willard" over some land in another "Virginia Meadow" in the New Grant. Part of Concord's cost in pressing their case was five shillings and seven pence for "drink for ye men that went to mow" to enforce the town's claim. CTR, vol. 2, July 5, 1703. Note the early date of mowing (July 15, New Style—meadow hay was normally cut about August 1), probably intended to steal a march on the contesting party.

19. Edward M. Flint Jr., S.M., and Gwendolyn S. Flint, *Flint Family History of the Adventuresome Seven*, vol. 1 (Baltimore: Gateway Press, 1984); and CTR 1:55. Part of this farm is still occupied and farmed by the Flint family, a rare example of continuity in ownership from the Second Division until today. Some of their land is now conservation land, owned by the Town of Lincoln once again. The honor of the Flint family name in conservation circles has long since been fully redeemed. All boundaries on the map are approximate.

20. The subsequent history of Bulkeley's farm is followed in Alan Emmet and Christine Fernandez, *The Codman Estate "The Grange": A Landscape Chronicle* (Boston: Society for the Protection of New England Antiquities, 1980).

21. An investor named Biggs who lived in Cambridge received 666 acres lying west of the Bulkeley farm, bounding south on the Sudbury line and west on the river. This became the farm of Thomas Goble and Daniel Dane. CTR 2:50, 51. Maj. Simon Willard collected several grants of land at Nashawtuc Hill adjoining his First Division holdings near the meeting of the rivers. These were consolidated into a 350-acre farm by Willard himself and then by Henry Woodis, a Derbyshire man who bought Willard out in 1661 after the fur trader moved on to other frontier towns. This estate eventually passed to Woodis's son-in-law John Lee and became the Lee Farm. CTR 1:55; 2:40. In the North Quarter, Richard Temple received 298½ acres of Second Division in a sprawling, irregular tract along Spencer's Brook, adjoining 86 acres of First Division meadow and upland that had belonged to Richard Spencer. Temple acquired the "land in Concord that was William Spencer's" from Samuel Adams of Charlestown in 1654, CTR 2:15; MRD 1-129. James Blood received 500 acres of Second Division bounding on the Blood family's farm that lay beyond Con-

cord's northern border, CTR 2:11. William Hunt put together a long strip of First and Second Division lands nearby, totaling some 426 contiguous acres, MP 12305, 1667.

22. It appears that Hartwell should have had more land coming to him, but his sons John and Samuel also received land in the Second Division, which may account for the shortfall. Just how much of a man's First Division land qualified for Second Division in the eyes of his quarter seems to have been a matter of intense negotiation. Few men entered exactly three times as much Second as First Division in the town record book. By the time the grants were recorded a great deal of land had been bought and sold, further confusing the picture.

23. CTR 2:18-19, 21-22.

24. CTR 2:31-32. It is, of course, possible that not all these meadows were yet being mown, or that Potter leased some of them to others. Dunge Hole was also spelled Dung Hole, Dungie Hole, and other variants in various documents. Certain Concord historians have carefully explained that this name derives from "dungeon," and they are free to believe that if they wish.

25. CTR 2:14.

26. Powell, *Puritan Village*, is an account of these disputes in the town of Sudbury.

27. The month is obscured in the records, so we cannot be sure. Similar woodland divisions, for which the records have been lost, may have been undertaken in the other quarters.

28. CTR 1:30. The bounds on the map are approximate.

29. This observation about the likely origin of "Walden" has already been made by Thomas Blanding, "Historic Walden Woods," in *Walden Woods* (Concord: Thoreau Country Conservation Alliance, 1989), 9. Blanding credits Walden historian Richard O'Connor with noticing it earlier and cites Adams Tolman earlier still, in Adams Tolman, *Indian Relics in Concord* (Concord, 1902), 11.

30. CTR 1:64. Since many men recorded their divisions more than once, there is ample opportunity to check whether a parcel labeled "pine plain" in one list later turned up as "woodland." It never happened.

31. Thirty-six parcels of woodland in the South Quarter, along with seventeen in the East Quarter are outlined on the Second Division Woodlands map. These parcels were either explicitly called woodland in one of several records of land divisions or can be confidently included because of their context among abutting parcels or their description in subsequent deeds.

32. A few deeds that date from the same era do identify woodlands, particularly on Elm Brook Hill—another glacial till upland where we would expect to find a heavy, closed-canopy forest. CTR 2:57; Timothy Wheeler to Gershom Brooks, MRD 4-459, 1665; Nathaniel Stow, MP 21771, 1684.

33. MRD 10-609, 1681. Wheeler and Ball reconfirmed the bounds of the lot that had been granted to Thomas Fox, apparently as part of his estate settlement. These lots connect with a similar string of small woodlands granted by the South Quarter.

34. Several north Concord lots were described as woodland in seventeenth-century deeds and inventories. With its preponderance of glacial till uplands, the North Quarter may have been better endowed with mature woodlands than the other quarters, and so less punctilious about subdividing small parcls. CTR 2:5.

35. More common white ash (*Fraxinus americana*) likes rich, moist upland soils, but it can grow in swamps. Black ash is strictly limited to wetlands. It is now extremely rare in Concord.

36. MP 10598.

37. MRD 10-494, 10-438, 10-533, 12-71. These rights had presumably passed to Joshua from his father, Thomas, and his brother Caleb.

38. Eaton, *Flora of Concord*, 10, 62-63.

39. CTR 1:34. In 1709, the owners of the ironworks in Concord Village petitioned for permission to "improve the pine trees yt were blooded to make cole for their forge." That is, they wanted to make charcoal from trees that had been killed in tar making. Permission was granted, but they were enjoined to reserve

any trees suitable for timber and were prohibited from felling living trees. CTR 2:163. See also Charles F. Carroll, *The Timber Economy of Puritan New England* (Providence: Brown University Press, 1973).

40. White pine is vulnerable to frequent fires, although it can thrive in a regime of large fires every century or so. For discussions of pine in Concord, see Winkler, "12,000-Year History of Vegetation and Climate," 206–07; and Whitney and Davis, "Thoreau and Forest History," 70–81.

41. MRD 14-166, 1702; Benjamin Barrett, MP 1147, 1728.

42. CTR 2:39, 43, 48, 54, 58, 99–100.

43. MRD 10-519, 1695. Judah was Luke's only surviving son—an older boy, Samuel, was killed by the Indians in the ambush at Sudbury in 1676.

44. Luke Potter, MP 17812, 1697. Judah Potter, MP 17811, 1731.

45. MP 10598, 1690; MRD 22-269, MP 10586, 1725.

46. He also received the more easterly of his grandfather's two one-hundred-acre blocks of Second Division. MRD 13-681, 1702.

47. MRD 14-166, 1702. Col. James Barrett of the 1775 Concord militia was Benjamin Barrett's son. To follow the story of this family into the next generations, see Gross, *Minutemen and Their World*, 74–83.

48. Joseph Wood, *New England Village*, 53–70. Beyond three miles the journey to the center, particularly for Sunday meeting, became a burden. Accordingly, the corners of Concord were sliced off to join new communities during the eighteenth century. It is easy to see why larger towns like Dedham or those like Watertown and Cambridge that were originally settled at one end began to hive off daughter communities earlier.

49. CTR 2:2.

50. Records of the Proprietors of the Great and Commons Fields, March 1, 1691, CFPL. The Great Field apparently encompassed the old Cranefield, a name which gradually faded from land records. This shift from management of the Great Field by the selectmen to a group of proprietors may have been partly in response to the threat posed by Governor Andros; however, the land itself within the field was already privately owned. It is more likely that the change reflected the realization that only a diminishing fraction of Concord's inhabitants had an interest in the field and that management ought to be devolved upon that group.

51. Ibid., March 28, 1757. This stricture could not have applied to cowyards and paddocks within the included houselots, although that exception is not stated.

52. CTR 2:2, 1672.

53. MRD 4-409, 1672. No specifications are given for fences in the Concord records. By 1653 Watertown required a four-rail fence three and a half feet high, but many towns later insisted on higher. Russell, *Long, Deep Furrow*, 105–06.

54. Thanks to Rob Tarule for a discussion of colonial fencing. Fencing the two-square-mile Great Field required about two thousand rods of fence. Fencing the same area into rectangular eight-acre lots (about the average size) would have required over ten thousand rods of fence, or five times as much. Keeping the field fence in sufficient repair to turn away swine must have been serious business.

55. Ebenezer Meriam Jr., MP15051, 1751. The inventory includes eight acres pastureland in the Cranefield, and one and a half acres called the Horse Pasture. Jonathan Stow to Nathan Meriam, MRD 45-345, 1745, includes twenty acres pasture and plowland in the Great Fields.

56. Records of the Proprietors of the Great and Common Fields, June 10, 1731; November 8, 1731; October 27, 1738; no date (about 1740). As the surviving records are incomplete and sporadic, it is impossible to tell exactly on what date these changes in boundaries were agreed to.

57. Ibid., October 27, 1738.

58. Ibid., March 24, 1774. The assessors returned 237 acres. According to my rough map of the field boundary, closer to 400 acres were actually still enclosed.

59. Ibid., February 24, 1774; October 26, 1778.

60. Ibid., no date, about 1740.

61. CTR 2:39.

62. CTR 2:55. A bridle way along the upland edge of the meadow was laid out and accepted in 1684. Getting a way "laid out" often meant having an existing way surveyed and accepted as a town way, to be maintained by the highway tax. There was some difference between a bridle way, a cartway, and a driftway, but it does not appear that the terms were used in a consistent manner in Concord. In general, bridle ways were narrower, usually only one rod wide (16.5 feet). Driftways (or open ways) were for cattle, cartways for carts, but most such ways were used for both purposes. These ways seemed to run about two rods wide, although one running to Brickiln Island for carting and driving creatures to pasture began at twenty-two feet wide and then "grows wider and wider" (this is official language) until it reached four rods wide. CTR 2:94.

63. See Brian Donahue, "'Dammed at Both Ends and Cursed in the Middle:' The 'Flowage' of the Concord River Meadows, 1798–1862," *Environmental Review* 13 (1989).

64. Shattuck, *History of Concord*, 15.

65. Walcott, *Concord in the Colonial Period*, 52.

66. Shattuck, *History of Concord*, 22 ff.

67. CTR 2:3. See also Walcott, *Concord in the Colonial Period*, 55 ff. It thus appears that Concord acquired 5,000 acres for £15, to raise which sum they sold 610 acres, a tidy profit of 4,390 acres free and clear. Obviously the natives were not considered to fully own the land in a legal sense, but only to have some residual interest in it. Wheeler got his tract for a bit less than half a shilling per acre.

68. Harold R. Phalen, *History of the Town of Acton* (Cambridge: Middlesex Printing, 1954), 3–7; Shattuck, *History of Concord*, 274–75. Stephen Hosmer's survey in 1730 measured 12,965 acres; the modern town of Acton is reckoned to be 13,085 acres.

69. See Wood, "Village and Community," 333–46, for an overview of the establishment of villages. In early records the village at the center of Concord itself was referred to as the town, in what would remain the normal sense of a town outside New England—i.e., a small urban place. In New England, "town" came to mean both the people of the community and the place where they lived. When the peoples' dwelling places dispersed, the all-encompassing New England sense of a town as the entire area of land politically controlled as a unit by a single community emerged. "Village" came to mean the cluster of settlement often found near the center of the town.

70. CTR 2:3.

71. The lease had to be amended in 1672 because the townsmen were not yet supplying Wheeler with the agreed upon number of cattle (they could come up with only about fifty instead of the expected eighty). They weren't paying their bills, either. The rate per head was cut to one shilling presumably to induce more Concord stockowners to participate, and Thomas's end of the bargain was sweetened by reducing his rent to nothing. It is hard to guess whether Thomas's right to take timber was a valuable concession, a mechanism to clear land, or simply a convenience for constructing buildings and fences.

72. Shattuck, *History of Concord*, 274–80; Phalen, *History of Acton*, 8–21.

73. This was not a stocking density of one cow per twenty acres of commons. It was a stocking allowance of one cow on the commons for every twenty acres of *private* land a man already held. The discussion that follows is based on the records of meetings held January 2, 1653 (2-11-52); March 7 and March 9, 1654. CTR 2:4–5.

74. It is possible that the number of "cow commons" allowed was to be based on First and Second Division holdings combined. But since on the order of twenty thousand acres were distributed, if that were the case one wonders where the theoretical common herd of one thousand head was to have been pastured.

75. Johnson, *Wonder-Working*, 110.

76. CTR 2:3.

77. Subsequent divisions of the Twenty Score indicate that these rights were closer to 17 acres each, or a total of 340 acres. The very rough, preliminary map of the Twenty Score suggests it might have actually amounted to 460 acres, on the other hand.

78. CTR 2:85. The sheep common was divided into 3¼-acre lots sometime before 1719. Nehimiah Hunt, MP 12274, 1719.

79. CTR 2:104, 105, 108, 109; MRD 42–476. Common land is hard to pin down because it was not recorded as such in the land division records, but rather appears obliquely when it abutted land that had been divided.

80. CTR 2:2.

81. CTR 1:83. Today this driftway is called Williams Road after a nineteenth-century Irish immigrant.

82. *Watertown Records* 1:94, 99, 111; 2:3.

83. The history of the retention, acquisition, and management of common forests in New England is detailed in Robert McCullough, *The Landscape of Community: A History of Communal Forests in New England* (Hanover, N.H.: University Press of New England, 1995). This tradition was extremely attenuated until the twentieth century. See also Brian Donahue, *Reclaiming the Commons: Community Farms and Forests in a New England Town* (New Haven: Yale University Press, 1999).

CHAPTER 6. Settling the East Quarter

1. The details of this day's hay drawing are of course invented, although many trips very much like the one depicted here surely occurred in reality.

2. The exact date of Joshua Brooks's birth is not recorded, but it was about 1630—his younger brother Caleb was born in 1632. In actual fact, Joshua may not even have been living at his father's house in Concord in 1653—he may, for example, have been living at Capt. Hugh Mason's in Watertown, where it is likely he learned the tanning trade.

3. The difference in elevation between these two major landforms is only fifty-odd feet. The uplands, composed mostly of glacial till, continue east for several miles through the eastern part of Concord (now Lincoln) and the western part of Cambridge (now Lexington and Arlington).

4. Some of these lots may have already been granted in the First Division—for example, Nathaniel Ball's four-acre meadow closest to the Bay Road. Concord's land records are inconclusive on this point.

5. It is not clear whether Wheeler received this block of land in the First or Second Division because his landholdings were not recorded in the town book.

6. The actual use of the Ox Pasture before it was divided is not specified in Concord's Town Records, but the name speaks for itself.

7. It is not entirely clear whether this land of Hartwell's lay within the field fence or immediately adjacent to the field. It is also possible that for convenience' sake the fence enclosed part of the meadow as well.

8. The survey of the Bay Road in 1716 mentions white oak and maple as bound markers in this stretch. CTR 2:95.

9. The order accepting the way to "Oburnd" as a town highway was passed in 1665. CTR 2:53.

10. There is no sure evidence that the husbandmen of Concord had yet made improvements in Rocky Meadow at this date. Perhaps it was not being mown at all, but only grazed. It is also possible that the meadow was being mowed, at least in part, but that it was not entirely fenced. Perhaps only the haystacks themselves were fenced. For example, when Isaac and Jacob Wood bought four acres in still more remote

Fairhaven Meadow from Nathaniel Jones in 1704, they also secured the "liberty to set hay cut from the meadow on the upland to the north, to fence the hay in, and to cart it away at their pleasure." MRD 17-599.

11. Today it lies under Hanscom Field, an airport.

12. Joshua Brooks was almost certainly a tanner and was later involved in the ship timber business in Medford, although he lived in Concord.

13. It is not entirely clear from vital records how many of these Hartwell children survived, or indeed if all were living in 1653. There is no way of knowing that Priscilla Wright was ever a servant in the Hartwell household, although such a thing would have been normal for a girl beginning adolescence.

14. Information on diet from MacMahon, "Comfortable Subsistence," amply confirmed in Concord probate inventories. William Hartwell's inventory included hops and barley, MP 10598, 1690.

15. This process has been described previously by such authors as Philip J. Greven, *Four Generations: Population, Land, and Family in Colonial Andover, Massachusetts* (Ithaca: Cornell University Press, 1970), and Bushman, *Puritan to Yankee*. The precise mapping here delves deeper into its social and ecological texture.

16. Vickers, "Competency and Competition," has a good discussion of these tensions.

17. If any of Hartwell's other sons were still living, they had left Concord and were not mentioned in his will.

18. It was also called Poldin or Polding and was later corrupted to Poland. That name persisted into the nineteenth century. This part of the Great Field lies beyond the map and cannot be shown in detail. Today it is a pond again. CTR 1:195-99.

19. "Proprietors of the Great and Common Fields, March 1, 1690/91," CFPL.

20. MP 10598. From the wording in several deeds and probated estates, it appears that these small pieces of the Cedar Swamp may have been undivided rights in a larger common.

21. Ibid.

22. John Hartwell's eldest son, Ebenezer, married Sarah Smedley, received a 150-acre Second Division tract in the North Quarter from his father-in-law in 1698, and moved three miles from the meetinghouse. John Hartwell Jr. moved to New London, Connecticut, Joseph to Woburn, and Edward to Lancaster. Youngest son Jonathan was still in Concord when he married upon reaching his majority in 1713, but he then disappeared, apparently moving to Chelmsford. John F. Hartwell, *The Hartwells of America* (Saginaw, Mich.: Privately published, 1956).

23. Ibid. Third son John was close enough by to share with his brothers in the division of his father's apparel but apparently not living in Concord, MP 10586. Jonathan was called Senior to distinguish him from his cousin Jonathan Hartwell Jr., born next door a few years later.

24. MRD 22-269. This may indicate that John Hartwell's piece of the Cedar Swamp had been acquired by his brother. Both the deed and tax records indicate that Samuel Hartwell made a very even split of his property with his son.

25. MRD 22-268, 22-270.

26. MP 10586. Some of this land may have been distributed to Hartwell's other unmarried daughters, Elizabeth and Sarah, who are named in the will written in 1722. But neither of these women appears subsequently in Samuel's place in the tax rolls, as Ruth does.

27. The deed for this sale is not recorded. My date for it is deduced from tax valuations. This land was later referred to as the "Hartwell pasture" in Meriam family deeds and probate inventories.

28. This homestead was later owned by Joseph Wheeler, who married Joseph Meriam's widow. John Meriam was born after his father died. Twenty-six years later, in 1667, he and his older brothers William and Joseph quitclaimed all their rights in their father's estate to their "father-in-law" Joseph Wheeler, presumably after they had been established themselves, MRD 9-259. William later moved to Lynn, and Joseph to Cambridge Farms. For further details on the Meriams and the house at Meriam's Corner, readers may

consult Barbara A. Yocum, *The Meriam House, Minute Man National Historical Park, Concord, Massachusetts, Historic Structure Report* (Lowell, Mass.: National Park Service, North Atlantic Region, 1994); and Brian Donahue and Heidi Hohmann, *Cultural Landscape Report for Meriam's Corner, Minute Man National Historical Park* (Boston: National Park Service, North Atlantic Region, 1994).

29. By contrast John's uncle, the gentleman Mr. Robert Meriam, recorded over 600 acres for himself in the Town Book (he was the town clerk); John's uncle George had 249 acres recorded.

30. Most of John Meriam's deeds were not recorded, so the exact date of many of his acquisitions is uncertain. The land to the east of the house became Meriam's home meadow, which abutting deeds show was in his hands by no later than 1698, MRD 25-373. The small piece across Billerica Way, on the hill south of the lane into the Great Field, was planted with an orchard by 1685, CTR 1:268–69.

31. The Cambridge lot was contributed by his father-in-law, John Cooper. MRD 27-409.

32. MRD 13-39, 39-59, 49-260, 55-36, 64-100.

33. MRD 14-106, 27-409. Nathaniel and Samuel each received half of their father's ninety-one-acre Second Division lot as a starting point. Samuel was also given half of the forty-acre lot in Cambridge Farms, so perhaps Nathaniel was given the other half by an unrecorded deed.

34. Wheeler, *Concord: Climate for Freedom.*

35. Unfortunately, the deeds by which they were provided for were not recorded, so the details are uncertain. By the time of the earliest tax records in 1717 the three sons owned property at the corner and the father no longer did.

36. Joseph married Noah Brook's oldest daughter, Dorothy. Ebenezer married Dorothy's first cousin once-removed Elizabeth, youngest daughter of Gershom Brooks.

37. Investigation of the surviving house at Meriam's Corner confirms that it was built about 1705. The other two houses have since disappeared. See Yocum, *Meriam House.* Gloria Main observes that a young married couple moving in with parents was rare in New England, but this appears to have been quite common among families I have studied in Concord. The domestic arrangements by which the household was shared are not revealed in my sources. Main, *Peoples of a Spacious Land,* 172.

38. MRD 19-403, 25-567, 26-47.

39. For several years the two brothers had exactly the same assessment on the tax rolls, except for Joseph's additional "faculty" as a blacksmith. Joint ownership appeared occasionally in Concord deeds and probate inventories. A well-documented example occurred in the Wheeler family at Nine Acre Corner, later in the eighteenth century, MRD 77-411, 77-412, 80-3. There was manifestly a large degree of close functional integration of landholding, husbandry, and artisan trades within families in Concord. For several years Ebenezer Meriam owned a team of oxen, while his brother Joseph didn't, for example. This is of great importance in the way one interprets tax valuation data and thinks about the matter of economic independence.

40. This purchase was not recorded. The date, from tax records, was 1732, two years after the death of Meriam's mother. Taylor moved up the road to Gershom Brooks's old quarters near Elm Brook Meadow. The original Meriam house at the corner, which Ebenezer had vacated, was subsequently occupied by a variety of tenants, including newly married Ebenezer Meriam Jr. (1717–51) from 1741 to 1743, before being passed along to Joseph's son Josiah in 1747. The deeds of the Taylor family suggest a pattern of land fragmentation and intermingling even more intense than that of the Meriams.

41. The New Grant rights are not shown. In 1735, Acton was just being formed, and these rights were being turned into actual acreage. By this time the Cedar Swamp lay almost in the center of the new town of Bedford.

42. Seventeen thirty-four is the last year John Meriam appears on the tax roll. His move may have been somehow linked to the death of his only son, John, in 1735. In 1737, John Meriam of Littleton, "late of Concord," sold his Concord lands, MRD 39-59. He survived until 1748, styling himself a gentleman.

43. MRD 38-106, 38-115.

44. "Proprietors of the Great and Common Fields," CFPL. The records are fragmentary, so it is a fair guess that the Meriams served in other capacities in other years.

45. MRD 50-73.

46. Although these women both came from Bedford, they were not closely related. Elizabeth was born Elizabeth Fletcher and was the widow of Stephen Davis. Sarah may have been the daughter of Stephen's first cousin Ebenezer Davis. It would have made a nice story if the two women had been sisters or mother and daughter, but the facts are not compliant.

47. MP 15051; Concord Tax Rolls.

48. Ebenezer Jr.'s younger brother Oliver, a recent graduate of Harvard College, died the same year as his brother, 1751.

49. He was either that or a serial killer.

50. Champney signed a covenant to maintain Ebenezer in exchange for his land in 1758. The story grows even more complex because in 1767 the Champneys in turn signed over Meriam's land to Phineas Blood, who had married Ebenezer Meriam Jr.'s widow, Sarah: that is to say, she was Sarah Champney's stepmother and Ebenezer Sr.'s daughter-in-law. By 1780, three years after Ebenezer Meriam Sr.'s death, the Champneys had moved to New Hampshire. Sarah Blood, by then a widow for the second time, was once again living in the house where she began married life nearly forty years before. MRD 73-208, 69-275, 147-165.

51. There is no complete inventory of Joseph's land, but it appears that by the 1740s he owned at least ninety-two acres, having acquired some forty acres to add to what his father had passed on to him at the turn of the century. By then, he had also probably helped his son Samuel acquire his brother John's farm across the road when John moved to Littleton in 1737. Joseph's was a more robust version of a dispersed eighteenth-century farm than his brother Ebenezer's. Joseph Meriam's holdings were worked out from MRD 19-403, 38-106, 49-259, 49-260, 56-508, 55-332, 74-177, 158-163, 122-441.

52. MRD 38-115.

53. The house where he had grown up came into Joseph's possession through a series of complex trans-fers of land between 1743 and 1747, as Josiah and his older brother Nathan came of age. By 1744 the house and its seven-acre lot had apparently been sold to a neighbor named Jonathan Stow, who in turn sold house and land to young Nathan Meriam in 1745, along with a twenty-acre block of tillage and pastureland in the New Field (it had been new back in about the 1640s) adjoining the Meriam tillage lands across Bedford Road. Then, through a bit of unrecorded intrafamily property shuffling, Josiah moved into the old house on a reduced lot with the adjoining meadow split off, while his brother Nathan inherited the newer house and the attached home meadow from his father. Part of the other newly acquired land in the New Field stayed with Nathan, and the rest was parceled off to his brother Samuel. Jonathan Stow moved to Townsend. There were no apparent family ties between Stow and the Meriams, although the whole thing seems highly orchestrated. MRD 45-345, 49-259, 49-260.

54. Deeds in 1737 and 1756 describe Samuel as being a blacksmith. MRD 38-115, 53-528. Unfortunately, we have no account books to shed further light on the Meriams' smithing business. The Concord Museum, according to Curator David Wood, owns a fowling piece apparently assembled by Joseph Meriam in 1709 and fitted with a new lock by Josiah Meriam about 1779.

55. A 1779 deed denominates Josiah as a locksmith, MRD 91-155. In the 1749 Valuation, Josiah Meriam's farm acreages and stock are only about half those of his yeoman brother Nathan, though not far behind his blacksmith brother Samuel.

56. MRD 49-260. Josiah was given the use of only half the shop, but it is unclear whether the other half remained with Joseph until he died or had already been passed to Samuel. The location of the shop is unknown.

57. Minot married Timothy Wheeler's daughter Rebecca—a child of Wheeler's second marriage, late in life, to Mary Brooks, MRD 26-61; Joseph Grafton Minot, *A Genealogical Record of the Minot Family in*

America and England (Boston: Privately published, 1897), 14. The mill was purchased by Timothy Wheeler from Grace Bulkeley, the Reverend Peter Bulkeley's widow.

58. CTR 2:56; MRD 26–61; Concord Tax Rolls. The tenant at the earlier period was named John Mack.

59. MRD 28–301. James Minot, who died "happy in a virtuous posterity" (so reads his gravestone on the Hill, Minot, *Minot Family*, 15), left his son James his homestead in the center and his son Timothy the mill, MRD 30–234. He also saw his seven daughters well married—including the twins Love and Mercy, born April 15, 1702, and both married December 13, 1722. Timothy Minot was a minister who occasionally assisted Reverend Whiting (Shattuck, *History of Concord*, 164). Col. James Minot Jr. took his father's place as justice of the peace and a leading citizen of his generation. Samuel Minot moved to a new house at the farm upon his marriage in 1732.

60. MRD 28–302, 35–424, 35–425, 38–302, 39–453, 51–230, 51–231, 51–232, 51–233, 53–507, 54–123, 54–169, 54–170, 54–230, 55–186, 63–222, 67–189.

61. The Brickiln Field contained about a dozen tillage lots, most of which seem to have been multiples of four acres in size. The entire field may have originally been laid out in four-acre lots, although few things in Concord were that regular. Robert Fletcher and Timothy Wheeler may have bought up several lots in the field when other owners left Concord during the 1640s, helping set the stage for later consolidation.

62. CTR 2:2.

63. Samuel Fletcher appears to have been the first husbandman to settle at the Brickiln Field sometime between 1682, when he married, and 1694, when his father confirmed to him (at age thirty-seven) the land upon which he already dwelt. He recorded few of his deeds, so the dates of acquisition of most of his property beyond what his father gave him are unknown. The full extent of his holdings was documented when he passed the homestead to his son Samuel Fletcher Jr. in 1729. MRD 35–619, 65–351.

64. The next to settle at the Brickiln Field appears to have been Nathaniel Stow, sometime before his marriage to Ruth Meriam in 1690. Stow set about buying many more small lots at the Brickiln Field, Brickiln Island, the Ox Pasture, and elsewhere in the vicinity. MRD 25–444, 25–346, 25–445, 25–380, 26–46, 25–381, 13–127, 24–269, 14–292, 26–47, 26–48, 26–49, 26–50, 26–135.

65. The final settler at the Brickiln Field, dwelling between Samuel Fletcher and Nathaniel Stow, was John Jones—presumably upon his marriage to Anna Brooks in 1716. John Jones appeared at the Brickiln Field location on the earliest surviving Concord tax list in 1717. Much of Jones's homestead came from First and Second Division grants of his grandfather John Farwell. The deeds by which Jones received most of his land were not recorded. A small residue came in his father's will probated in 1726, MP 12859.

66. Stow began in 1717 by providing his eldest son, John Stow, with a houselot adjoining his own, along with several other small lots. These twenty-five scattered acres were deemed a "good perfect estate of inheritance," MRD 23–439. John Stow died in 1724, the same year as his father. Nathaniel Stow's estate was then divided into three "livings" among his sons Thomas, Benjamin, and Joseph, MP 21772. Thomas and Benjamin sold up and left Concord, MRD 35–501, 25–567, 25–531, 32–550, 39–711. Joseph Stow acquired some of this land, and neighbors like John Jones the rest.

67. Joseph Stow had no surviving children from his first marriage and thus finally began a family in 1759, at age sixty-seven. When he died in the early 1770s his children were minors, and it is likely that the widow, Olive, managed the Stow farm with the help of her brother and neighbor, Farwell Jones.

68. The local name for buckthorn, *Rhamnus frangula*, an invasive European shrub with very astringent berries.

69. Shattuck, *History of Concord*, 265.

70. MRD 4–433, 4–459, 5–3. This counts land of Timothy Wheeler that passed to Gershom Brooks about the same time the land grants were recorded in 1666. In 1654, Brooks was granted land in the South Quarter, but after that he faded from the records on that side of town. Early in 1655 he made a "proposion" for land adjoining his houselot by the millpond, but that was "defaced" in the book and not granted. CTR

1:1, 2, 14, 40. In 1664, after his wife, Grace, died, Thomas Brooks sold his village houselot to John Wheeler, son of George Wheeler who lived a couple houses down, MRD 3-169. Thomas presumably accompanied his youngest son, Gershom, to Elm Brook to end his days. He died in 1667.

71. The exact dates Joshua and Caleb arrived are unknown and could have been as early as the 1650s. Joshua Brooks married in 1653, Caleb in 1660. Their brother-in-law Timothy Wheeler owned a homestead just to the west. In 1665, Wheeler sold his "house and barn and other housing on the hill this side of Elm Brook, with all the pasture ground and plow ground it stands upon, and the orchard on the end of it" to his "brother" Gershom Brooks, MRD 4-459. Wheeler had purchased Peter Bulkeley's village homestead from the minister's widow. Gershom Brooks married Hannah Eckles in 1667.

72. MRD 5-34. This settlement was recorded by deed.

73. MRD 7-199. Joshua had inherited the lion's share of Thomas Brooks's land in Medford, Henry Bond, *Genealogies of Watertown, Massachusetts* (Boston: Little, Brown, 1855), 719; MRD 10-128. It is likely that Caleb received at least part of this land from Joshua in exchange for his Concord estate.

74. He was called Daniel Brooks *Jr.* because he was several years younger than his cousin Daniel Brooks *Sr.,* Joshua's son, who lived next door. Gershom also had several daughters, one of whom, Elizabeth, married Ebenezer Meriam. Gershom's youngest son, Joseph (born only four years before his father's death), inherited his father's land in Medford, MP 2828. He apparently later came into possession of his father's land in Concord's South Quarter, sold it, and bought a farm in Weston in 1725, MRD 17-382, 26-216. He subsequently bought land in Lexington and was an important figure in the founding of Lincoln, giving "liberally" to both town and church, Bond, *Genealogies of Watertown,* 722 (Bond credits this to the wrong Joseph Brooks—Joshua's son Joseph had already died by the time of Lincoln's founding).

75. MRD 17-3, 10-494, 10-438, 10-533, 12-71. Joshua had already bestowed some land upon these sons, who ranged in age from thirty-eight to twenty. On this occasion he sold them the remainder of their inheritances. Although the portions were quite similar, they ranged in price from £59 to £32—Job, the youngest, who inherited the home place and its duties, paid the least. These were clearly not full market sales. Job Brooks died less than two years later in 1697, at which time the appraisers (including John Meriam Jr.) put the actual value of his housing and lands at £80, MP 2841.

76. This Pine Plain lay to the north of the Suburbs and had been land of Caleb Brooks. The 1695 deed from Joshua to Daniel didn't mention Caleb's house (which had been standing in 1679), but it is likely that Daniel was already occupying it or had built another because he had married Anna Meriam in 1692, and their children were already being born. MRD 10-494, 25-173.

77. MRD 10-533, 17-3.

78. Again, no dwelling was mentioned in the deed, but Noah had doubtless built a house there (or on adjoining land which Joshua had given him earlier) because by 1695 he and his wife Dorothy already had three children. MRD 12-71.

79. MP 2841. It isn't clear what member of the weasel family that "sabel" was. The Brooks's tanning business must have handled furs as well as domestic hides.

80. MRD 14-353.

81. Noah Brooks was called a tanner on the deed from his father in 1695 and on several subsequent deeds, MRD 12-71, 18-82, 18-79, 18-275, 21-221.

82. Joshua Brooks's father-in-law, Hugh Mason, was a tanner, Bond, *Genealogies of Watertown,* 720. I have no deeds or other records that firmly establish that Joshua Brooks was a tanner. However, seventeenth-century deeds did not often specify the grantor or grantee's occupation.

83. Since the map cannot show all Brooks landholdings, the prosperity of an individual cannot be judged from this map alone.

84. His farm went to Jacob Taylor Jr. (the same man who had sold his place near Meriam's Corner to Ebenezer Meriam), MRD 34-273. When Jacob Taylor died in 1753, the bulk of the farm was inherited by

his daughter Mary, who was married to Eleazer Brooks, a Lincoln physician who was to become a distinguished general in the Revolutionary War. Thus much of this land was gathered back in by the long arm of the Brooks family, MRD 49-66, MP 22191. Eleazer Brooks was son of Job Brooks Sr. and grandson of Daniel Brooks Sr.

85. MRD 32-402, 26-310, 38-239, 39-362, 64-100, 38-241, 39-575. Daniel's daughter Anna married John Jones in 1716 and took up residence at the Brickiln Field just to the west. In 1725, Daniel purchased a part of the old Iron Works farm in Acton and passed this along to his youngest son, John, MRD 25-178.

86. No deed or probated estate—1729 by Concord Tax Roll. Ensign Brooks died in 1733, while his wife, Anna, lived on to 1759—long enough to see her grandson Samuel Brooks inherit the home place from her son Samuel, MP 2891. Samuel Brooks Sr. died young, in 1758.

87. Hugh split his large lot at the Suburbs with his brother Daniel but then acquired upland and woodland throughout the remainder of his life, MRD 25-173, 45-667, 35-501, 35-210, 37-574, 39-52. Hugh's elder son Jonathan acquired a farm in Concord's North Quarter. Job Brooks Jr. was some twenty years younger than his cousin Job Brooks Sr.

88. Hugh, who called himself a gentleman, gave Job all his land for love at age twenty-three, reserving, however, the "power and liberty to dispose of the abovesaid premises or any part thereof as I shall think best," MRD 45-577.

89. MRD 57-262, 55-95.

90. MRD 18-82, 18-79, 18-80, 18-81, 18-275, 21-221.

91. MRD 39-512. Lt. Joshua Brooks subsequently purchased additional parcels from his father, swapped one for another, and was finally given the tanyard itself in 1725. MRD 39-513, 39-515, 39-516, 39-517.

92. MRD 39-514, 39-518, 39-519, 39-520, 39-521, 39-522, 39-523. By 1749, Lt. Brooks himself had in fact already passed the tanyard and home place along to *his* eldest son, Joshua Brooks Jr., had moved to his brother Benjamin's farm (lying off the map to the south), and was styling himself a gentleman, MRD 45-622, 45-623. Concord Tax Rolls suggest that Lt. Joshua Brooks moved about 1743; the homeplace and tanyard passed to Joshua Jr. in 1745.

93. The gift amounted to some eighty-seven acres, including ten acres of good tillage land by the house, thirty-five acres of woodland on Elm Brook Hill, a little three-acre field lying south of that, the ten-acre Tanyard Meadow across the road, nineteen acres of upland and meadow lying east of the Bay Road on the north flank of the hill, three and a half acres at "Rocky Pasture" near the Suburbs, six and a half acres of river meadow, and the obligatory piece of the Cedar Swamp. MRD 35-193.

94. MRD 35-193.

95. No deed was recorded—from 1716 survey of the Bay Road, CTR 2:95; and 1717 Concord Tax Rolls. In 1714, Ebenezer married Sarah Fletcher, the daughter of Samuel Fletcher who had settled at the Brickiln Field.

96. MRD 32-550. Benjamin's seventy-one acres lay beyond the map. He was married in 1719 and had a child in 1720, so he may have built the dwelling before the formal grant of the land from his father. Benjamin sold this farm to his brother Joshua in 1744 and moved to another part of what would soon become Lincoln.

97. Eldest daughter Dorothy settled with Joseph Meriam in his new house at the corner in 1705, Mary wed the wealthy miller (and sometime minister) Timothy Minot and lived in the village, and youngest daughter Elizabeth married young John Miles in 1728, moving across Concord to his new farm at Nine Acre Corner.

98. MRD 36-60. Ebenezer apparently also owned a piece of mowing in a more distant part of Elm Brook Meadow in Bedford, adjacent to Samuel Hartwell, MRD 35-187. He recorded few of his deeds. The boundaries of his parcels at the southern end of the map are approximate in size, shape, and location. If anything, this sketch may exaggerate the size of these lots.

99. Joshua and Ebenezer had very similar assessments in 1719 of £21 and £20, respectively. By 1732 (the

last year that includes assessments in which Ebenezer appeared) Joshua was assessed at £35, Ebenezer at only £26. Ebenezer also owned slightly fewer livestock. Meanwhile, younger brother Benjamin was assessed at £30, while youngest brother Thomas was assessed for half of the homeplace, worth £48 in total.

100. Ebenezer sold to his brothers and cousins and to his neighbors Ephraim Hartwell and Nathaniel Whittemore. MRD 37-164, 39-520, 44-80, 44-152, SPNEA 1741 (Minute Man National Historical Park collection); Concord Tax Rolls.

101. A mere ten days after receiving his inheritance in 1695, Joseph sold several parcels at the Suburbs to his brother-in-law Benjamin Whittemore—along with his acre of the family lot in the Cedar Swamp, MRD 13-82. (Whittemore married Esther Brooks in 1692 and built up a substantial farm on the uplands a mile to the east.) Other sales followed, whittling the property away, MRD 13-81, 45-667, 19-410, 39-362. It is possible that Joseph was strategically building up land in other towns—but if so, he did not record the deeds.

102. Two acres to eldest son Joseph in 1734, four and a half acres of meadow to son Jonas in 1746, two acres of pasture to son Isaac that same year, MRD 46-195, 46-195b, 46-5. There appear to have been other unrecorded gifts of land, and Joseph may have been helping to provide for his children in other ways. Still, the property was being dribbled out slowly and not in sufficient quantity to allow these young men to establish themselves in Concord.

103. MRD 45-155, 47-251. James paid for the second transfer.

104. A forty-acre houselot, ten acres across the way, eight acres in Elm Brook Meadow, and three acres of woodland at Sandy Pond, MRD 57-385. James, as required, reserved the right of their mother, Rebecca, to the use and improvement of the property. Two pieces are shown on the 1749 map.

105. CTR 2:9. The parcel is identified as an orchard in records of Lt. Joseph Wheeler, CTR 2:26. Presumably Rice decided to hold onto this orchard until one planted at his new home could come into bearing. Another trace of Rice's original holdings was "two acres bought of St. Rice" by Nathaniel Stow in the Brickiln Field, MRD 25-346, MP 21771.

106. Son Peter Rice at the west end of the property, and daughter Abigail across the Bay Road with her husband Dr. Philip Reed. These farms have not been mapped for this study.

107. MRD 10-371. The property included fifteen acres of upland along with "4½ acres of meadow in the Suburbs belonging to it" nearly half a mile to the north. It had passed into the hands of Moses Whitney, a third-generation native of Watertown who moved to Concord and then to Stow. Whitney may have pioneered the property because it already had a house on it when he sold it to Whittemore in 1692.

108. MRD 10-370, 10-369, 13-82, 13-81, 13-80, 16-521, 37-158, 37-159, 37-160, 52-456, 37-161, 21-377, 47-708, 44-709, 44-711, 44-78, 24-418, 44-712, 37-163, 44-74.

109. Samuel Hartwell Jr.'s brother William took up another part of his father's outlands in a similar manner, north of Samuel in what would become Bedford.

110. MRD 10-358. The land had been occupied by Rice's son-in-law Dr. Philip Reed. Hartwell may have lived here for a time. Whether there was a house on the parcel is unclear, however. Reed's house may have stood on land farther west, conveyed in 1710 to Benjamin Whittemore, MRD 44-708.

111. MRD 37-161, 37-163, 39-161, 52-456, 55-509. Samuel Hartwell also acquired several parcels for which the deeds were never recorded.

112. MRD 36-198, MP 24776. Tax rolls indicate that Benjamin Jr. received land across the Bay Road between 1724 and 1728, and he bought a right of way into the property from Samuel Hartwell in 1731, MRD 37-163. Nathaniel Whittemore received forty acres at the west side of his father's property in 1728, and he had a house already on it, MRD 44-713. The remainder of the land passed first by sale, then in Lieutenant Whittemore's estate, and finally in the division of the widow Esther's third upon her death in 1743, MRD 44-76, 36-198; MP 24776, 24786. Benjamin Whittemore Sr. had many more children, but they are not mentioned among his heirs.

113. The division was complicated by the death of Benjamin Whittemore Jr. in the same year as his father. His estate was held together for fifteen years until the heirs came of age, whereupon it was sold intact to cousin Aaron Brooks, MRD 55-169. Aaron Brooks was Thomas Brooks's son and Noah Brooks's grandson.

114. MRD 21-164, 39-161, 44-80, 44-80b, 55-37, 56-193, 71-38. Nathaniel Whittemore had also acquired one of the Lamson farms to the east by this time.

115. Samuel and Joseph settled in Stoughton, Isaac in Oxford. Daughters Abigail, Mary, and Lydia found husbands in Stoughton, Lexington, and Concord. MP 10587; Hartwell, *Hartwells of America*.

116. The gift included eighteen acres of woodland and upland to the west of Samuel's homelot with a new dwelling on it, three acres in Rocky Meadow, half of eighteen acres of meadow and upland at the Suburbs, six acres of plowland in the field on the Lexington line, and half of eight acres in Elm Brook Meadow in Bedford, MRD 35-87.

117. MP 10587, MRD 39-160, 39-161, 48-605.

118. MRD 10-519. Like his father's, Judah's will contained a lengthy and precise list of duties for Samuel to perform, "the deacons of the church to decide any controversies or differences," MP 17811. Samuel Potter lived on to 1800, an extraordinary instance of the grandson of one of Concord's original English settlers surviving into the nineteenth century. Given the tendency to leave the homestead to the youngest son, by the end of the colonial period many of those yeomen living in the center of Concord were literally a generation closer to their forefathers than those living on the outskirts of the town.

119. Settlement in the South Quarter is detailed in Donahue, "Plowland, Pastureland, Woodland and Meadow: Husbandry in Concord, Massachusetts, 1635–1771" (Ph.D. diss., Brandeis University, 1995). Some fragmentation was beginning to appear by the 1740s. Virtually all farms had outlying meadow lots.

120. Main, *Peoples of a Spacious Land*, 62.

121. Wheeler, *Concord: Climate for Freedom*, 72–73.

CHAPTER 7. The Ecological Structure of Colonial Farming

1. Historians have employed tax valuation figures to gain insight into farming in Concord and elsewhere. See Merchant, *Ecological Revolutions;* Bettye Hobbs Pruitt, "Self-Sufficiency and the Agricultural Economy of Eighteenth-Century Massachusetts," *William and Mary Quarterly* 41 (1984); Gross, *Minutemen and Their World;* James Kimenker, "The Concord Farmer: An Economic History, 1750-1850," and Brian Donahue, "The Forests and Fields of Concord: An Ecological History, 1750-1850," both in David Hackett Fischer, ed., *Concord: The Social History of a New England Town, 1750–1850* (Waltham: Brandeis University, 1983).

2. These elements were universal in Massachusetts towns. The 1749 valuation listed all mowing as a single category but also recorded tonnage of fresh, salt, and English hay, which allows a rough partitioning of mowing acreage by assuming a (perhaps high) one-ton-per-acre English hay yield. No one called the meadows fresh in everyday speech in Concord—they were listed that way in the valuation to distinguish them from salt marsh.

3. This map is about two and one half miles long and a mile wide. It was compiled by taking the property maps prepared for the previous chapter and adding finer detail of land use collated from deeds, probated estates, and the 1749 tax valuation. While many of the parcels were precisely described by these sources at dates very near 1749, others had to be interpolated. Many boundaries are approximate and not verified in the field. Farms for which detailed surveys or estate divisions are available invariably turn out to be more intricately subdivided than imagined, so this map should be regarded as an oversimplification. Sources for characterizing each parcel are annotated on the database underlying the map, which I will make available to interested scholars.

4. This was not a process that could ever arrive at a completely rational conclusion. Farmers' needs and

capabilities, like their economic situations and their ideas about what was the best land, were always in flux as their families passed through their life cycles.

5. For example, Ebenezer Meriam Jr. had two acres of plowland in Polden but also eight acres of pasture in the Cranefield, MP15051. Nathan Meriam owned the three-acre Stow field just west of the home field, and along with it the six-acre Stow pasture, which adjoined his brother Samuel's sixteen-acre Stow pasture, MRD 45-345, 158-163; MP15056, 15076.

6. The Meriams tilled Wareham and Deerfield loamy sands, whose only limitation is a seasonally high water table. This was (and still is) mitigated by drainage ditches reaching out from the adjoining meadows. The Brickiln Field consists of prime Paxton and Woodbridge sandy loams—among the most workable soils that form in glacial till. These fine soils overlie a hardpan that impedes drainage, but a ditch reached from the north nearly to the top of the rise to carry away seepage. Much of the land tilled by the Meriams and the Brickiln neighbors remains in cultivation to this day.

7. The Brookses plowed Windsor loamy sands, grading into a Canton sandy loam upslope onto glacial till at the south end.

8. Brooks left in 1741, and most of his land was bought up by his neighbors. Tax lists indicate the home place was occupied for several years by a cooper named Timothy Cook, who scarcely farmed, but he vanished in 1747. The place was purchased by Joseph Mason from Benjamin Farley of Boston in 1754, MRD 52-431. The soil is an extremely stony Montauk till with hardpan, grading into an absurdly stony Hollis with bedrock outcrops.

9. MRD 10-358, 35-187, MP10587, 10560, MRD 3020-128. Deeds and probated estates identify this as plowland throughout the colonial period. The soil is a Haven silt loam, as good an arable soil as can be found in the region. It is still cultivated today. The houselot, by contrast, was described as "orchard, pasture, and mowing," MP10587.

10. A man named John White appears to have been occupying the Benjamin Whittemore house and farm in 1749. White also owned four acres of plowland east of the house, MP 24776, MRD 39-161, 48-605. The location of the home field was indicated in a deed of sale of the adjoining woodland to the west, MRD 36-198.

11. This field had been split for a time between Nathaniel and his father, Benjamin Whittemore Sr., MRD 44-713, 1728; MP24776, 1735. The east side was enclosed by a rail fence and called Nathaniel Whittemore's cornfield in 1735. It is a Merrimac fine sandy loam, somewhat droughtier than Haven silt loam but still a prime arable soil.

12. See, for example, Matthew Patten, *The Diary of Matthew Patten of Bedford, N.H., from 1754 to 1788* (Camden, Maine: Picton Press, 1993), 127. Patten was forever fixing his plow.

13. By 1749, Concord had 1,471 acres in tillage. Under the Native forest-fallow system this would have required a base of 7,355 to 14,710 acres of tillable land—more than all the improved land in Concord.

14. Johnson, *Wonder-Working*. There is no subsequent mention of the use of fish for fertilizer in Concord.

15. MRD 4-409, 1672. Given the complexity and vagueness of this deal—which also involved Wright's supposed inheritance back in England—it is no surprise that another agreement had to be signed thirteen years later in which Wright was to pay Hoar £30, half in corn and half in cattle, to "settle controversies." MRD 13-73, 1685.

16. MRD 79-509, 1774.

17. Russell, *Long, Deep Furrow*, 126-28, 193-94. Dung forks in Noah Brooks, MP2886, 1790; Asa Brooks, MP2790, 1809; Samuel Hartwell, MP10586, 1725; John Meriam, MP15076, 1804.

18. Patten, *Diary*, May 23, 1759, "began to Plant and had my Bros oxen to Cart Dung," 64; May 2, 1760, "Jonas and I Carted 12 Load of Dung," 79; May 19, 1761, "I had my brors oldest oxen and I carted Dung," 95; May 18, 1767, "Andrew Walker and the boys planted the piece back of the Barn with Corn and Potatoes

and Dunged it they carried the Dung from the Barn," 194; May 14, 1771, "Thos Newman Mended the Tang of my Dung fork and I began to furrow for planting corn," 266; April 26, 1781, "I got Wm Newman to put a peice of iron across the shoulders of my Dung fork and a new tang and two Rings for our two dung forks and two nails for nailing them in the handles," 430; May 19, 1785, "we plowed some for potatoes and got out some Dung for planting them," 505; May 27, 1788, "we hald dung and got all the Dung out from the end of the barn and the most of it laid and planted some potatoes," 542. Patten does not mention carting dung every spring, but this may have been because he was often away on other business (especially at the fish run) and left the planting to his sons. More important, the dunging appears so routine that it was surely often subsumed in the simple entry "planted our corn."

19. Ibid. Patten planted peas and a little wheat in the cowyard on April 13, 1765, 150, and "finished Reaping the Spring Wheat in the Cow yard" on August 10, 157. He "began to plant potatoes in the old Cow yard" on May 7, 1767, 193. "The boys plowed for Cow yards" on May 21, 1771, 266. On May 23 and 24, 1780, Patten "made up the fence about our last years Cow yards and fitted them for plowing," 414. He planted the first of his corn in "two peices north from the house which were Cowyards" on May 21, 1781, 431. Patten was frequently laying up new log or rail fences about old fields on his place.

20. *Massachusetts Society for Promoting Agricultural Papers and Extracts* (hereafter *Mass. Ag. Soc. Papers*) (Boston: Adams and Rhoades, 1807), 17. Thomas Hubbard Jr. of Concord returned the pamphlet of inquiries distributed by the society about 1800. There is no reason to believe Concord manuring practices had changed greatly from the colonial period by that date, except that more manure was available for every acre, and more potatoes (which also required manure) were being grown.

21. Patten, *Diary, passim.*

22. Grains were distinguished in tax valuations beginning in 1784. Inventories and widows' portions follow the same pattern.

23. Hubbard, *Mass. Ag. Soc. Papers*, 14. Matthew Patten followed similar practices, although he also planted a good deal of spring rye as well. James Kimenker, "Concord Farmer," 153, calculated that corn yields in late colonial Concord were probably about twenty bushels, rye about eight bushels. Hubbard reported thirty-bushel corn and twelve-bushel rye in 1800, an improvement which Kimenker's calculations confirm.

24. Hubbard, *Mass. Ag. Soc. Papers*, 17.

25. Ibid., 16.

26. Kimenker, "Concord Farmer," 151. See also table 8 in Bidwell and Falconer, *Agriculture in the Northern United States*, 90. For example, Job Brooks's widow, Anna, was to be provided with two bushels of malt yearly, along with five barrels of cider, MP 2842, 1791. Thomas Brooks's widow, Hannah, was to receive a bushel of malt and four barrels of cider, MP 2909, 1791.

27. Beans and peas were included for the first time on the 1793 valuation,; they accounted for almost no acreage. But they were being grown. By the late colonial period beans had largely replaced peas at New England tables. See McMahon, "A Comfortable Subsistence," for a full discussion of New England diet. Matthew Patten often planted peas, on the other hand. The Massachusetts Society for Promoting Agriculture reported it was a common practice to drop one pumpkin seed in each corn hill, *Mass. Ag. Soc. Papers*, 19. The field was cross-cultivated until mid-July, after which the pumpkins ran, covering the ground. The polycultural yield from these fields was somewhat better than the reported corn production alone indicated.

28. Potatoes appeared in Londonderry about 1719. See Howard Russell, *Long, Deep Furrow*, 137–39, on their spread. Matthew Patten grew potatoes, of course, and dunged them. By 1800, Concord was growing about one acre per farm. *Mass. Ag. Soc. Papers*, 21.

29. This is suggested by Joshua Brooks's deed of gift from his father, Joshua, in 1745, wherein what was later called the Great Field south of the house was described as "plowland and mowing land," MRD 45-623.

30. *Mass. Ag. Soc. Papers*, 28.

31. MRD 49-259, 158-163.

32. MRD 9-78, 1684, emphasis added.

33. MacMahon, "Comfortable Subsistence."

34. CTR 2:23, 188. It is possible that Pellet was finding a way to transport his onions to Boston. The estate of Peter Bulkeley, Esq. (the Reverend Peter Bulkeley's grandson) in 1688 credits him with onions shipped to the West Indies via a Charlestown merchant. Wheeler, *Concord: Climate for Freedom*, 66.

35. According to Robert Gross, 63 percent, *Minutemen and Their World*, 214. Seventy-six percent reported cider production in 1771.

36. This orchard is described not in any Meriam deeds but in the deed to a neighboring Taylor piece of meadow and orchard, MRD 62-106, 1757. The mill is found in Ebenezer Meriam Jr.'s estate, MP 15051, 1751.

37. In 1685, the way into the Cranefield from Billerica Road was said to begin "at John Meriam's orchard." CTR 1:268-69. It was still an orchard in 1788, MRD 158-163.

38. When he passed the homestead to his son George in 1765, Deacon Minot reserved as many apples as he and his wife might need. MRD 67-189, 1765; MRD 63-222, 1765. The exact location of the orchard north of the road is not specified. The orchard south of the road may have dated from the earlier house that stood there.

39. MRD 35-619, 1729.

40. MP12859, 1726. Joseph Stow had half an acre of orchard, but the location is uncertain—it was probably behind his house.

41. MRD 253-459, 1823; MP 25661, 1803.

42. Samuel Brooks reported cider production in both Concord and Lincoln in the 1771 valuation, indicating an additional orchard either south of the road before the house or half a mile away at his Suburb lot.

43. This orchard apparently did not exist in 1726, MRD 35-193, but a 1747 road survey mentions it, CTR 2:122. Also see MP 2886, 1790; and MP 2888, 1809.

44. The location of this orchard was not specified until 1790, MP 2864. However, Joshua Brooks Jr. had an orchard in the 1749 valuation, and no other location was mentioned in any of the farm's deeds or inventories. This part of the farm was called Top of the Hill. The orchard may have been planted after Joshua's father, Lt. Joshua Brooks, acquired the land from his cousin Daniel Brooks Jr. (Gershom's son) in 1733, MRD 39-519. By 1790 Joshua Brooks had another orchard on similar soil at the "school house pasture" northwest of this one.

45. MRD 44-713, 1728; MRD 36-198, 1735.

46. MP 10587, 1745; MRD 39-161, 1736; MRD 48-605, 1743; MP 10560, 1793. The eastern orchard faded from the records as the century wore on. It may have dated from late seventeenth-century residences of the Rice family. The cider mill appears on a 1779 survey of the Hartwell Farm by Stephen Davis, Lincoln Historical Society, Lincoln, Mass. It was part of Samuel Hartwell's inventory in 1745 and was in Ephraim Hartwell's will in 1793. As an innkeeper, Hartwell had an obvious use for cider—however, he did not receive an innholder's license until 1756. The large production of cider by many of the Elm Brook Hill farmers indicates a considerable local market for it.

47. McMahon, "Comfortable Subsistence."

48. MRD 4-459, 1665. Howard Russell gives a good account of the enthusiastic propagation of apples in early New England, *Long, Deep Furrow*, 47-48, 78-80. Many Concord settlers came from Kent, a noted fruit-growing country in England, which may partly account for the early adoption of cider in Concord.

49. Orchards do exceedingly well on new grounds, especially if stony, observed Massachusetts farmers in 1800, *Mass. Ag. Soc. Papers*, 46.

50. Samuel Deane, *The New-England Farmer*, 3d ed. (Boston: Wells and Lilly, 1822), 8. See also Henry

Thoreau's "Wild Apples" for an elegiac description of the last of these ungrafted orchards and trees in Concord. Henry David Thoreau, *The Natural History Essays* (Salt Lake City: Peregrine Smith, 1980), 187, 209.

51. For example, Matthew Patten got in the last of his corn on October 9, 1781, and ground apples at his brother's cider mill October 12. The cider pressing sometimes overlapped with the corn harvest, but at least it did not interfere with the heavier work of mowing in August.

52. The Connecticut River valley appears to be different, with more English than meadow hay by 1771 — see table 2:2, J. Ritchie Garrison, *Landscape and Material Life in Franklin County, Massachusetts, 1770-1860* (Knoxville: University of Tennessee Press, 1991), 30–31, and discussion, 65–67. It also appears that some of the terraces along the Connecticut that were called meadows were high and dry enough to be tilled.

53. Concord lost several hundred acres of meadow by the setting off of Lincoln (1754) and to a lesser extent Carlisle (1780), after which reported meadow acreage fluctuated near twenty-one hundred acres until 1840. During the latter half of the nineteenth century most meadows were either converted to English hay or tillage by tile drainage or, if not improvable, abandoned. While meadow acreage remained level, English hay acreage steadily rose until it surpassed meadow hay in 1850.

54. Meriam had more meadow in Bedford, however. But so did Brooks.

55. Sixteen and three-quarters acres, to be "exact" according to deeds. Landowners' meadow acreage as derived from deeds was always a larger figure than that reported on the 1749 valuation and may somewhat overstate the acreage that was actually mowed.

56. Whether Samuel Brooks mowed all these acres of meadow is doubtful—the 1749 valuation didn't show him with much hay. He doubtless grazed part of it. Joshua Brooks Jr. had just received his inheritance from his father, and the young tanner was about to embark on a land-buying campaign that would include more meadow, from the Great Meadow by the river to Flint's Great Meadow in Lincoln, four miles apart in opposite directions, MRD 55-37, 1756; MRD59-535, 1760; MRD 62-169, 1764.

57. Russell, *Long, Deep Furrow*, 276–77. Matthew Patten recorded the creation of a new mowing lot at his Little Meadow. He had it cleared in September 1760—two men were at it for two weeks. He fenced it in the spring of 1761—cutting rail stuff, hauling it to the meadow, splitting rails, and building the fence. He burned the brush in June and sowed grass seed in October—he doesn't say what kind of grass he planted, native or English. He took the first cutting of hay from the "new cleared lot" three years later, in August 1764. In all probability he fed the grass off the two preceding falls. Patten, *Diary*, 86, 92–93, 95, 100, 140.

58. Matthew Patten lost an ox this way. The ox put his hip out and could not stand after being drawn out of the mire and died within a few weeks. Patten, *Diary*, 109.

59. Golet and Larson, *Classification of Freshwater Wetlands*. Marsh emergents had been great favorites of foraging Indian women, and the Native people may well have managed for them.

60. Many Concord inventories included this tool: Asa Brooks, MP2790, 1809; Noah Brooks, MP2888, 1809; Samuel Hartwell, MP10586, 1725; Samuel Hartwell, MP10587, 1745. David Wheeler worked on the ditches on the far side of Brickiln Island on September 18, 1792, August 14, 1793, and August 15, 1794. In this case, Wheeler was diverting the course of Elm Brook. Court of Common Pleas, September 8, 1795, in Nathan Brooks Family Papers, box 14a, folder 3, CFPL.

61. An example comes from the petition of Elisha Jones and seven of his neighbors in the nearby town of Weston, near the headwaters of Four Mile Brook. The petition stated that "s[d] meadows are unprofitable & of small benefit by reason of several banks and stoppages in the course of the brook that runs through the same" and requested a "Commission of Sewers for the cleaning and removing of the Banks and Obstructions in the passage of the waters . . . & also to cut such other sluices as shall be necessary effectively to drain them at the Proprietors Cost." Massachusetts Archives 1:366.

62. Court of Common Pleas, September 8, 1795, in Nathan Brooks Family Papers, box 14a, folder 3, CFPL. It is possible that farmers also held water (what there was) on the meadow in late summer after haying, to irrigate and bring on the aftermath.

63. Court of Common Pleas, September 8, 1795, in Nathan Brooks Family Papers, box 14a, folder 3, CFPL.

64. For example, in Weston in 1713 seven men purchased from a neighbor the liberty to make a dam on Four Mile Brook at a point where the upland almost meets "for flowing their meadow lands lying above sd necks of land," MRD 35-305, 1713. When James Jones of Weston divided land in Nonesuch Meadow between two sons, he gave each half the privilege of turning the water from the intervening brook onto their meadows, MRD 59-570, 1753. Jared Eliot remarked that winter flooding enriched lowland but killed English grass. Eliot, *Essays Upon Husbandry* 17.

65. Jared Eliot describes mixing muck with dung, Eliot, *Essay Upon Husbandry*, 89. No sources mention the use of muck in colonial Concord. However, the practice of carting muck to the barnyard or cowyard to mix with dung was universal during the first half of the nineteenth century in Concord and regarded as old-fashioned. See the Records of the Concord Farmers Club, 1852-83, CFPL. Thoreau describes the muck being "thrown up" during the summer and then carted out in winter, over the frozen ground. *Journal* 3:207.

66. Herdsgrass was named twice, first after a man named Herd who grew it in New Hampshire and again after one Timothy Hanson, who popularized the species in America. Bidwell and Falconer, *Agriculture in the Northern United States*, 103-05.

67. The figure of 15 percent is from Kimenker, *Concord Farmer*, 160. Joshua Brooks had "orchard and mowing" on the hill, MP2864, 1790. Nathaniel Whittemore had "orcharding and mowing land," MRD 36-198, 1735. The rise in English hay between the 1749 and 1771 valuations appears to have come at the expense of meadow hay, but this deduction is complicated by the loss of land with the setting off of Lincoln in 1754. Jared Eliot describes planting English grass and clover in his drained meadow, Eliot, *Essays Upon Husbandry*, 11-14.

68. Only two Middlesex County towns, Weston and Pepperell, reported more English mowing than fresh meadow in the 1784 valuation—but many Weston farmers also owned or leased river meadow lots in the Sudbury meadows just to the west. These were upland towns. Even in Weston, English hay production had still lagged well behind meadow hay as late as 1768. Massachusetts Archives 162:265. Jared Eliot discussed the greater forwardness of inland towns in growing English hay in his second essay, Eliot, *Essays Upon Husbandry*, 29. Planting of English grass was urged by colonial leaders in the seventeenth century and appears to have taken hold in the Connecticut River valley thanks to William Pynchon. See Garrison, *Landscape and Material Life in Franklin County*, 65-67; and Romani, "Our *English Clover-grass*," 33-35; citing Innes, *Labor in a New Land*, 8-9, who cites Russell, *Long, Deep Furrow*, 129-30.

69. Even the usually reliable Matthew Patten says almost nothing about the movement of his stock, except when they did something notable like giving birth, or wandering off and getting lost, or dying, or all three together.

70. The difference between a sandy outwash soil and an upland till can be dramatic in a dry summer—no grass on the former by mid-July, steady growth on the latter. The late 1740s were a period of drought in Massachusetts. A ditchbank on Joseph Stow's meadow at Blosses Island was consumed by fire in the Great Drought of 1749, Nathan Brooks Family Papers, box 141, folder 3, CFPL.

71. MP 2842, 2863, 2864, 2886, 2990.

72. Samuel Brooks, MP 2891, 1758. Inventories of more elderly farmers often included little or no livestock because most of the farm operation had been passed to the sons. The 1771 valuation in Concord lists only "*Cows* four years and older," not *cattle*, as shown in Bettye Hobbs Pruitt, ed., *The Massachusetts Tax Valuation List of 1771* (Boston: G. K. Hall, 1978).

73. MRD 51-230, 1746; MRD 63-222, 1765; MRD 67-189, 1765. This was sixty-nine acres of upland and swamp on the deed of purchase, but apparently forty acres of woodland and pasture on the deeds to Samuel's son George. George Minot's 1808 inventory included forty-five acres of pasture and woodland in Carlisle, which included the very northern part of Concord; MP 15229, 1808.

74. MRD 45-577, 1740; MP 2842, 1794; MP 2790, 1816. Job received half of this parcel from his father,

Hugh, and passed half on to his son Asa, but Asa's inventory in 1816 indicates he owned the entire property outright.

75. MRD 67-189, 1765. I do not know when Meriam and Minot purchased this land. Nathan Meriam's son Amos moved to Princeton, possibly to part of this land. Samuel Minot's son George retained forty-eight acres of pasture in Princeton, with a small barn, MP 15229, 1808.

76. MP 10560, 1793. Eighty-nine acres were given by Ephraim to his eldest son, Ephraim, mostly by deed of gift and the remainder in his will.

77. MP 2842, 1794; MP 2864, 1790; MP 2886, 1790. Noah's back pasture was acquired sometime after 1749, Joshua's in 1759, MRD 56-510. Joshua's son John moved to the Pepperell land.

78. William Jones, "A Topographical Description of the Town of Concord," in *Collections of the Massachusetts Historical Society, 1792*, 1:237–42. Jones's telling use of the phrase "in proportion" is another indication of the systematic way these farmers thought about their landholdings.

79. MP 2790, 1816. The fattening oxen were mostly picked up from other farmers. To complicate matters, in 1816 Asa also owned two cows, a steer, two heifers, and three yearlings at a farm he owned in north Concord. Job and Asa together reported a very similar herd on the 1784 valuation: eight cows, ten oxen, fourteen younger cattle.

80. CTR 2:276, 336, 390, 413; 4:139. The practice of allowing swine to run at large was hotly contested in other towns as well. In 1710, the Middlesex Court of General Sessions received a petition from those inhabitants of Charlestown who were aggrieved by a new town bylaw that ordered swine to be confined. The petition argued that this injured the poor, who could not afford to pen and feed their swine, and that ringing and yoking free-ranging swine were practices sufficient to prevent them from damaging crops. Middlesex County Court Folios, folder 37X, group 3,

81. Fire continued as part of the ecological landscape of Concord once the English arrived, at levels similar to those of the Native regime—see Winkler, "Changes at Walden Pond," 205. The better quality of forage that grows where brush piles have been burned remains visible for decades, as the thick establishment of white clover puts these spots on a self-reinforcing cycle of enrichment.

82. Russell, *Long, Deep Furrow*, 130.

83. From Eaton, *Flora of Concord*. Eaton lists Kentucky bluegrass as native; most botanists believe it was introduced. Some fescues are native. Even native species suitable for pasture would have been infrequent until grazing created favorable conditions for them to spread along with the introduced species.

84. Wood, *New England's Prospect*, 13.

85. Jake Potter's recollection was passed to Henry Thoreau by George Minott in 1858, Thoreau, *Journal* 11:190. The value of the meadows for late grazing was still being touted by farmers in their struggle against flowage brought by the Billerica dam, which they claimed rendered the river meadows "inaccessible not only to teams, but also to *grazing cattle*, which used to get plenty of fresh fall feed, a great resource for the farmer, every season, but especially when the pastures and other uplands were parched by drought," David L. Child, "Memorial of D. L. Child and others, in relation to the Concord and Sudbury meadows," in *Report of the Joint Special Committee upon the Subject of the Flowage of Meadows on Concord and Sudbury Rivers* (Boston, 1860), xci.

86. Until 1801 woodland and unimproved acreage were lumped, so it is impossible to be precise about forest cover using these data. Even in nineteenth-century valuations the *total* reported acreage in Concord barely surpassed fourteen thousand acres, whereas the actual area of Concord is over sixteen thousand acres, so there was always plenty of acreage unaccounted for.

87. MRD 56-508, 1751; 56-509, 1751; 55-36, 1753; 59-535, 1758.

88. CTR 2:103, May 15, 1731; MRD 35-193, 1726; MRD 45-577, 1740. There are no deeds of sale for any of the Cedar Swamp lots, but they did not appear in any Brooks probated estates or deeds of gift after 1740.

89. Concord pine was mostly pitch pine, a better fuel than white pine. Values from inventories. Pitch pine at three-quarters white oak closely reflects the measured heat output of these two woods—it may even slightly undervalue the pine.

90. These would have been oak timbers of various lengths. The typical load carted from Jones's yard to tidewater during the summer months ranged from thirty-five to fifty feet. Ephraim Jones's Wast Book No. 2, CFPL.

91. Samuel Jones's Account Book, CFPL. "Showing" means shoeing.

92. MRD 324-442. There is no specific mention of the bark mill until the early nineteenth century, in an 1833 deed in which Isaac Brooks, the last of the Brooks tanners and Noah's great-great grandson, reserved the right to flow the brook and build a bark mill.

93. MRD 39-517, MRD 45-623; MP 2864, Joshua Brooks, 1790. Joshua Brooks was taxed for a "tan-house and slaughterhouse" on the north, Concord side of the road in the 1784 valuation, CFPL.

94. Noah Brooks was identified as a tanner on the deed with which he purchased his inheritance from his father in 1695, MRD 12-71. This is the first mention we have of the family business, although it must have been in operation earlier. By 1713, when Noah passed on a holding to his twenty-five-year-old son Joshua, Noah called himself a yeoman and young Joshua was now the tanner, although Noah sometimes reverted to "tanner" in later deeds. The tannery itself passed to Joshua in 1725. Joshua remained a tanner in his deeds until 1745, when he passed on the trade to *his* son Joshua Brooks, tanner, and briefly identified himself as a husbandman. Later, he seemed to realize he was a wealthy man and began styling himself a gentleman near the end of his life in the 1760s. His son, Deacon Joshua Brooks Jr., was already calling himself a tanner and a gentleman by this time, MRD 45-623.

95. MRD 45-577, 57-262, 55-95, 65-338.

96. The proliferation after about 1715 of young men taking up trades to help assemble the capital for a farm is discussed in Vickers, *Farmers and Fishermen*, 252-59.

97. There was little lime to be found in Concord, a lack of great significance to the long-term prospects of farming in the town. A small vein of limestone was quarried, and a lime kiln operated in the north part of Concord in Estabrook Woods, as well as others in Lincoln, which supplied enough lime for chimney mortar and plaster.

98. MacLean, *Rich Harvest*, chap. 6, "Occupation and Improvement." See also Allen Rogers, *Practical Tanning* (New York: Henry Carey Baird, 1922); and Thomas C. Thorstensen, *Practical Leather Technology* (New York: Van Nostrand Reinhold, 1969).

99. It is also possible that some of the slaughterhouse and tannery waste was saved to be used directly as a manure on surrounding farmlands. See Deane, *New England Farmer*, 250-58.

100. CTR 3:336. The deer officers were essentially game wardens, or "Informers of the Breach of the Province law relating to Deer." Wolves were trapped in Concord at least into the 1730s, and wildcats heard in Fairhaven woods into the 1770s. Thoreau, *Journal* 8:35; *Journal* 11:190.

101. MP 15106, 1769. There are only tidbits such as this concerning wildlife and game, and how their populations were changing, in the records of Concord. For a good discussion of New England in general, see Judd, *Common Lands, Common People*, 44-45.

102. The "great and peevish" complaint is from a 1707 petition to the Great and General Court, one of many that Concord sent over the years seeking relief and assistance because of problems caused by its rivers. Petition of Selectmen of Concord, August 13, 1707, *Acts and Resolves of the Province of Massachusetts Bay*, vol. 8, *Resolves, 1703-1707*.

103. MRD 35-193; MP 2886, 1790.

104. Depending on the breed, on whether or not the animals are in milk, and, of course, on the weather, grazing stock require varying amounts of water in the course of a day. When grass is young and fresh and the weather is not too hot or if there has been rain or even heavy dew, certain stock may be able to get through

the day without any water other than that they take in while grazing. In the heat of summer, when the grass is coarse and dry, it is another matter.

105. MRD 25-173, 25-175, 32-402, 351-237.

106. MRD 45-623, MP 2888, 1809. The "Dam piece" or "mowing lot near the old dam" bounded north on the lane which ran over the old milldam, MRD 39-515.

107. This was a small dam and pond (a version of which still exists) on the stream coming down to the river meadows through John Wood's farm, west of the Sudbury River. When the estate was divided among John's children in 1747, his youngest son, Zephaniah, received part of the "stonefield" south of the dwelling, but his oldest son, John, who received the bulk of the estate, was also granted the right to dig across Zephaniah's land in order to draw one-half of the water from the pond to the "grassfield" below the house. MP 25459, 1747; MRD 65-571.

108. MRD 33-129.

109. Division of the estate of Daniel Brooks Sr., 1733. Eleazer Brooks Papers, Lincoln Public Library.

110. Nathan Brooks Family Papers, box 14a, folder 3, CFPL.

111. Ibid. (emphasis added).

112. Or perhaps it was simply easiest to rework the ditches at this dry season of the year.

113. CTR 2:96. This apparently opened a new way back to Brickiln Way from the Bay Road farther east than the old highway that had made its way through the now-consolidated Brickiln Field, and then followed the same route east to the island. Numerous small private ways connected to it, accessing meadow lots: for example, in 1735, Samuel Potter granted Job Brooks Sr. "for good causes" the right to pass with teams or drive creatures through a slip of land from the island way down to his lot, MRD 39-575.

114. That was May 11 to September 11 by today's calendar. Nathan Brooks Family Papers, box 14a, folder 3, CFPL. This agreement may well have formalized a long-standing customary practice. Jarvis, *Traditions and Reminiscences*, 15, recalls that the miller was allowed "one flush daily" from May 12 to September 12. Note the precise four feet ten inches—draining a sprawling network of wetlands which had scarcely any perceptible fall from one end to the other was a game of inches. But then, as the poet A. R. Ammons once remarked, "If anything will level with you, water will."

115. Johnson, *Wonder-Working*, 110. They said the falls in these early documents, but presumably they really meant the bar at the fordway just upstream, which stands a few feet higher than the top of the falls.

116. This is my speculation. In his *History of Concord* (1835), Lemuel Shattuck scoffed at Concord's founders for believing they could cut a canal to *Watertown* for £100. One wonders where Shattuck got the idea that such a thing was contemplated. This odd supposition of Shattuck's seems more an indication of his generation's attitude toward the practical judgment of the town's forefathers than an accurate reflection of anything that would have been seriously considered by levelheaded yeomen. Shattuck, *History of Concord*, 16.

117. See Theodore Steinberg, *Nature Incorporated: Industrialization and the Waters of New England* (New York: Cambridge University Press, 1991), 33.

118. Massachusetts Archives 113:335. Petition of Town of Sudbury, Oct. 15, 1702.

119. Massachusetts Archives 113:331. Petition of Town of Concord, Oct. 15, 1702.

120. Copy taken February 18, 1722/3, of Billerica town meeting of October 4, 1708. Middlesex County Folio 80X, Massachusetts Archives. See also Henry A. Hazen, *History of Billerica, Massachusetts* (Boston: A. Williams, 1883), 278–79. Billerica in turn voted to defend Osgood against flowing a local meadow in 1710, but it is not clear if this indemnity was to extend to all upstream meadows.

121. *Acts and Resolves of the Province of Massachusetts Bay*, vol. 10, *Resolves 1720–1725*, chap. 74, March 6, 1721, 129 (emphasis added). Concord town meeting had first empowered its selectmen to petition the General Court to remove the Billerica milldam because of the fish in 1715, CTR 2:231.

122. Middlesex County Folio 80X, Massachusetts Archives. Another possible solution might have been to allow the dam to be raised higher through the winter but to lower it from April 1 to October 30. This was the compromise accepted by the Town of Concord itself in a similar (but smaller-scale) controversy concerning a dam on the Assabet River in 1704. CTR 2:93. I can find no account of the resolution of the Billerica case after 1722.

123. Massachusetts Archives, vol. 105, "Petitions 1643–1775," 209, 221, July 15, 1742. If nineteenth-century records are any clue, the "stoppages" may have included sandbars and also water weeds which grew within the channel of the river, which the farmers diligently cut every year to keep the water moving. But in 1742 the dam was not mentioned.

124. The meadow owners gave up after 1828, when the Middlesex Canal dam in Billerica was rebuilt to a height that drowned all their efforts. For a discussion of factors affecting the flow of the river and of the nineteenth-century flowage controversy, see Donahue, "Dammed at Both Ends." In fact, today's river stewards *are* once again cutting weeds—the exotic, invasive water chestnut.

125. Massachusetts Archives 113:497, May 25, 1709. A "hedgeware" was a weir made of wooden paling and interwoven rods, in the manner of a hedge. Such weirs have been constructed in both England and New England for thousands of years.

126. Middlesex County Court of General Sessions of the Peace, Record Books, 1686–1748, 249, 392, 408. By 1727 the right to build the weir was being granted by Concord town meeting to a small group of men as a monopoly, as had first been proposed in 1724. CTR 2:308, 339.

127. Unfortunately, there is little hard evidence concerning fish populations in Concord's rivers, brooks, and ponds during the colonial period. In 1851, Henry Thoreau was told that a few generations before, alewives had been caught in Bateman's Pond, in the north part of Concord. To reach the pond, these fish would have had to pass Barrett's mill on Spencers Brook. Richard Judd provides a nice instance of eighteenth-century farmers building fishways and stocking a previously landlocked lake in Maine with alewives. Judd, *Common Lands, Common People,* 129.

128. The "tending and husbanding" phrase is from John T. Cumbler, "The Early Making of an Environmental Consciousness: Fish, Fisheries Commissions, and the Connecticut River," *Environmental History Review* 15 (1991): 74–75.

129. The inclusion of such towns as Sherborn, Medfield, and Watertown Farms (Weston) in the 1709 petition was telling. These towns lay primarily (or entirely) within the Charles River drainage, yet apparently some inhabitants were traveling west to the Sudbury River, or to ponds feeding it, to catch fish. Similarly, only a small corner of Lancaster (now within the town of Harvard) drained into the Assabet. It is possible that the Charles River towns were acting in solidarity because of similar obstructions downstream on their own river, although the wording of the petition seems to refer very clearly to the Concord River. For an extended discussion of the struggle between common and private access to fish, see Judd, *Common Lands, Common People,* 123–45.

130. *Acts and Resolves of the Province of Massachusetts Bay,* vol. 2, *1715–1741,* 426.

131. Ibid., vol. 11, *Resolves 1726–1733,* chap. 136, January 21, 1732, 639.

132. Ibid., vol. 2, *Resolves 1715–1741,* chap. 21, January 16, 1736, 786–88; chap. 16, January 15, 1742, 1087–88.

133. CTR 3:238, 241. The outcome of this entreaty is not recorded.

134. CTR 3:179, 213, 297, 372.

135. CTR 3:393. William Jones mentions shad and alewives in 1792, Jones, *Description of Concord,* 238.

136. Shattuck, *History of Concord,* 202. Shattuck says the auction began in 1732, but he seems to have missed an earlier privilege granted in 1727, CTR 2:339. Apparently the tenacious eels still made it through even the tighter canal dam and are still in the river today. Farmers believed the eels were able to dig through the gravel under the dams. An effort to restore alewives in the Concord River began in April 2000.

137. Edward Jarvis, *Traditions and Reminiscences of Concord, Massachusetts, 1779–1878* (Amherst: Univer-

sity of Massachusetts Press, 1993), 60–61. It is not clear from Jarvis's account that he is even referring to anadromous fish. See also Cumbler, "Early Making of an Environmental Consciousness," 77.

138. CTR 2:52.

139. Massachusetts Archives, Middlesex County Folios, 29-1, 35-1, 37-1, 47-5, 53-1; *Acts and Resolves of the Province of Massachusetts Bay*, vol. 8, *Resolves 1703–1707*, chap. 50, August 13, 1707, 761–66; *Middlesex County Court of General Sessions Record Book, 1686–1748*, 206, 349, 367.

140. CTR 3:113, 117; 4:32, 319.

141. During the course of the eighteenth century these holdings gradually came to be commonly referred to as farms, whether they lay in one piece or not.

CHAPTER 8. A Town of Limits

1. Concord population stood below 500 in 1650. By 1700 it was close to 1,000; by 1750, perhaps 2,000 — after the separation of Bedford, but just before the departure of Lincoln. As families were having children at a rate capable of doubling the population every twenty-five years, this tells us both that a good many off-spring were emigrating right along and that the remaining population was still growing steadily. After 1750 Concord's population stabilized, sitting at about 1,650 in 1800 (after the departure of Lincoln in 1754 and Carlisle in 1780). Birthrates had slowed somewhat, but emigration had risen even more, bringing growth to a standstill. Gross, *Minutemen and Their World*, 76, 209–10; Marc Harris, "The People of Concord: A Demographic History, 1750–1850," in Fischer, *Concord: Social History*, 77–82.

2. The idea that overcrowding led to soil exhaustion and agricultural decline in late colonial New England began with Kenneth Lockridge, "Land, Population and the Evolution of New England Society, 1630–1790; and an Afterthought," in Stanley N. Katz, ed., *Colonial America: Essays in Politics and Social Development* (Boston: Little, Brown, 1971); and James A. Henretta, *The Evolution of American Society, 1700–1815: An Interdisciplinary Analysis* (Lexington, Mass.: Heath, 1973). It was fully developed for Concord in Gross, *Minutemen and Their World*, especially in chapter 4, "A World of Scarcity." Although I now believe Gross's picture of agricultural decline in late colonial Concord was overdrawn, I have no quarrel with the overall thrust of his argument concerning the social impact of living in a world of tightening ecological constraints. Elsewhere Gross emphasized, as I do here, agricultural adjustments which averted a large decline in yields. See Robert A. Gross, "The Problem of Agricultural Crisis in Eighteenth-Century New England: Concord, Massachusetts, as a Test Case," paper presented at a meeting of the American Historical Association, 1975. C. Pam. 10 #7, CFPL.

3. Greven, *Four Generations*, is the classic work on this dilemma. Gross, *Minutemen and Their World*, develops a similar thesis for Concord. Vickers, "Working the Fields," details the organization of family labor. Merchant, *Ecological Revolutions*, 185–87, calls this dilemma a "fundamental contradiction between the requirements of production and those of reproduction" in the patriarchal New England system. But that is only strictly true if so many children were *required* to help run an established farm, which is not entirely clear.

4. This process of migration and town making in New England is dealt with at length in Jaffee, *People of the Wachusett*.

5. Land hunger in Concord is described by Gross, *Minutemen and Their World*, 76–80.

6. MRD 73-208, 69-275, 147-265.

7. MP 15106, 1769.

8. MP 15095, 1782; MRD 158-163. Like so many other Meriam boys, Ephraim also married a Brooks girl, Timothy's daughter Mary. Ephraim's sister Abigail remained near Meriam's Corner, marrying Nathan Stow just a few doors to the west.

9. MP 15090, 1809; MRD 1064-72, 249-331.

10. MRD 54-169. This later became the birthplace of Henry David Thoreau—Jonas Minot was Thoreau's grandmother's second husband, his mother's stepfather.

11. MRD 63-22, 67-189, 68-315, 101-72.

12. MRD 73-281, 73-283. Interestingly, Edward and Hepzibah were married in Litchfield, New Hampshire, just before returning home to inherit the Fletcher home place.

13. MP 2891, 1758. Enoch moved to Princeton, and Elisha died before he married.

14. MP 2842, 1794. Asa Brooks began acquiring land in his own right in 1768 and was listed with half of the farm on the Concord tax valuation of 1784.

15. MRD 55-169; MP 2909, 1791; MP 2886, 1790. Noah died in 1790, father Thomas in 1791. The second inventory of Noah Brooks's estate, taken in 1792 after his father's death, includes what had "accrued to him after his decease." The entire estate passed to Noah's only son, Noah. His sisters Rebecca and Lydia received the right to live with their mother while single, liberty of the well, the garden, and the yard to hang clothes to dry, and one good cow each.

16. MRD 45-622, 45-623, 61-254, 60-374; MP 2863. Benjamin Brooks had moved to a new farm in western Lexington which would also become part of Lincoln. Ephraim Brooks's land already had a house and barn when he received it, perhaps built by Ephraim himself in the 1750s.

17. MRD 91-157, 121-182, 121-183; MP 2863, 1790.

18. MRD 52-431; MP 14765, 1789; MP 14756, 1802.

19. MRD 52-457; unrecorded deed of January 2, 1769, in Hartwell Papers, Minute Man National Historical Park; MP 10560; Middlesex County Court of General Sessions Record Book 1748-77, 597.

20. MRD 56-193, 71-38, 76-32. The Whittemore farm became the homestead of the Dodge's daughter Catherine and her husband, Capt. William Smith.

21. MP 24776, 1735; MP 24786, 1743; MRD 55-169.

22. Ebenezer Brooks was the weaver, and he did not prosper. Laurel Uhrich argues that in the early eighteenth century weaving declined as a male craft in the face of cheap imported textiles, but that it was increasingly taken up by women instead for home consumption, substituting for some purchased cloth. Laurel Thatcher Ulrich, "Wheels, Looms, and the Gender Division of Labor in Eighteenth-Century New England," *William and Mary Quarterly* 55 (1998): 3-38.

23. Max George Schumacher, *The Northern Farmer and His Markets during the Late Colonial Period* (New York: Arno Press, 1975), was among the first to describe these "dense networks" of local exchange in his 1948 dissertation. See also Pruitt, "Self-Sufficiency and the Agricultural Economy of Eighteenth-Century Massachusetts," and Vickers, "Competency and Competition: Economic Culture in Early America."

24. Gloria Main celebrates this slow but steady economic growth and creation of wealth, primarily through farm improvement and local trade. Main, *Peoples of a Spacious Land,* 203-10; Gloria L. Main and Jackson T. Main, "The Red Queen in New England?" *William and Mary Quarterly* 56 (1999): 121-50. Her own data, however, show that after the period 1725-50 only the top 10 percent were still increasing in wealth, suggesting that this expansion was encountering severe limits. Daniel Vickers describes much the same process and is more cautious in his assessment of how much surplus colonial farmers were able to generate, beyond limited local exchange. Vickers, *Farmers and Fishermen,* 205-59.

25. Quote is from Anderson, *New England's Generation,* 123. Also see Vickers, "Competency and Competition."

26. Jarvis, *Traditions and Reminiscences,* 47-48; Main, *Peoples of a Spacious Land,* 211-24; Carole Shammas, "How Self-Sufficient Was Early America?" *Journal of Interdisciplinary History* 13 (1982): 247-42; T. H. Breen, "An Empire of Goods: The Anglicization of Colonial America, 1690-1776," in Stanley N. Katz, John N. Murrin, and Douglas Greenberg, eds., *Colonial America: Essays in Politics and Social Development* (New York: McGraw-Hill, 1993), 367-98; Uhlrich, "Wheels, Looms," 3-38.

27. As historical data, tax valuation figures are subject to ineluctable uncertainty owing to variations in seasonal timing and weather and simply the vagaries of what was and wasn't counted. The slightly more than a dozen farmers along the Bay Road constitute a small sample, and unfortunately several of the most

important among them are missing from the surviving 1771 Lincoln valuation list. I have substituted less complete data from the 1774 tax list where I could. Apparent changes from 1749 to 1771 may be simply a fluctuation resulting from any number of factors and may not constitute a trend. I have included valuation figures for the town of Concord as a whole covering an entire century beginning in 1749, in order to put the late colonial figures into a longer context. These figures should be viewed with caution and used only to gain a sense of the broad shape and general direction of husbandry in Concord.

28. Grain production in Concord declined from 19,799 bushels to 18,672. Grain yield is simply the gross amount of corn, rye, and other grains divided by the total number of tillage acres as reported. We have no information about the proportion of tillage land in various grain crops or about the production of nongrain food crops like potatoes. With only two data points and many unknowns in the equation, one is reluctant to declare a marked downward trend here—but crop yields were surely not getting any better.

29. See James Kimenker, "Concord Farmer," 151–54. Kimenker argued convincingly that the unreported acreage planted to potatoes that would have been required in 1749 and 1771 to push corn and rye yields to the levels reported by Concord observers at the turn of the century was unreasonably large, and that therefore yields must have been closer to those cited above. His analysis appears sound, except that he may have underestimated the acreage already planted in potatoes in 1749. He allowed none, but see Russell, *Long, Deep Furrow*, 137–39. If Concord farmers were already growing potatoes in 1749, then the fall in grain yields was not quite as bad as it looks. But this does not alter his calculation of where yields stood in 1771. Yields for the town of Reading, Massachusetts, in 1771 were about twenty bushels for corn, twelve bushels for rye, Merchant, *Ecological Revolutions*, 180. An average corn crop of twenty to twenty-five bushels is given by Bidwell and Falconer, *Agriculture in Northern United States*, 101.

30. Bidwell and Falconer, *Agriculture in Northern United States*, 85–86; Cronon, *Changes in the Land*, 151. Robert Gross makes a similar argument for Concord, *Minutemen and Their World*, 86–87.

31. Harry J. Carman, ed., *American Husbandry* (New York: Columbia University Press, 1939), 58.

32. The author of *American Husbandry* never set foot in New England—this is made evident when in several pages criticizing New England's crop rotations he never mentions rye, which would have been simply impossible for any husbandman to miss. The author was probably a compiler who lived in England, most likely Arthur Young, who never set foot anywhere in America. See Carman, *American Husbandry*, li.; and Carl R. Woodward, "A Discussion of Arthur Young and American Agriculture," *Agricultural History 43* (1969): 65–66. He probably combined several sources on New England, which would explain the disjointed nature of his essay. One of his informants obviously had a very low opinion of New Englanders' treatment of livestock. To provide more winter fodder he (being an English agricultural improver) wanted New England farmers to grow turnips, which would have been foolish because New Englanders already had a well-adapted, hoed fodder crop, maize. This imposter should be read for amusement only. Yet he will doubtless be cited (directly or indirectly) as an authority on colonial American husbandry for centuries to come. His place in history is secure.

33. Eliot did have an unfortunate flirtation with Jethro Tull's horse-hoeing husbandry, but that was not the most telling part of his work. A social context for Eliot's writing is given by Christopher Grasso, *A Speaking Aristocracy: Transforming Public Discourse in Eighteenth-Century Connecticut* (Chapel Hill: University of North Carolina Press, 1999), 190–229. By improving agricultural productivity, Eliot hoped to preserve the agrarian moral order of the New England town composed of yeomen farmers minding elite leaders such as himself, and to that end he addressed himself largely to the practical experience of ordinary husbandmen.

34. Eliot, *Essay Upon Husbandry*, 17.

35. Ibid., 16–17.

36. Ibid., 89. Wrote Eliot, "The Dung and Stale of Beasts is so abundantly charged and impregnated with Salts, proper to promote Vegetation; which, to preserve intire, it is proper to mix with hungry poor Stuff, which will imbibe all the Salts both fixed and volatile, otherwise the Sun will exhale much of the volatile Part, and the Rains will wash off a great deal of the fixed Salt." A more precise scientific prescription

for the preservation of nitrates and ammonia through composting could not be desired. Such routine methods for composting manure were widely reported in responses to the Massachusetts Agricultural Society's "Inquiry" of 1800, *Mass. Ag. Soc. Papers,* 38–44. I use these responses to illustrate late-colonial farming practices, which I do not believe had changed significantly in the twenty-five years since the Revolution. Other historians may argue that these descriptions instead show dramatic improvements under way in the early Republic.

37. Gross, *Minutemen and Their World,* 214; Merchant, *Ecological Revolutions,* 181–85; Pruitt, "Self-Sufficiency."

38. Kimenker, *Concord Farmer,* 160, 171–76. Concord farmers kept more cows partly by keeping fewer sheep, but the rise went beyond that. It is possible (even likely) that many of these men were forced to reduce the number of beef animals they wintered, some of which were not recorded—an economically painful adjustment.

39. Many upland towns in the region took up English hay more rapidly. In Weston, for example, English mowing surpassed meadow hay by 1781; in Hopkinton, by 1791.

40. Eliot, *Essays Upon Husbandry,* 27.

41. Ibid., 29. Reverend Eliot seems a bit harsh here. In Concord it appears that farmers had gone about improving their available meadow diligently enough—they had simply reached the limit.

42. Ibid., 37. Even at a heavy seeding of twenty pounds, that would be enough seed for half a dozen acres. That much clover seed could be gathered from the second growth on a single acre.

43. Nineteenth-century improvements in grain yield were actually even greater than it appears because more potatoes were being grown without being accounted for in tax valuations. That is, potato acreage was presumably being counted as tillage, but their yield was unfortunately not recorded. More hay was being marketed by the mid-nineteenth century, so the manure supply probably did not increase quite as much as the hay to tillage ratios suggest.

44. One would expect a few anomalies in a series like this because of any number of variations among farms. There was a wide discrepancy in the number of livestock each farmer apparently kept on a given amount of hay, for example. Farmers who were short could purchase hay from their neighbors. This was expensive, however, and most farmers tried to avoid it if at all possible. On the whole, the amount of hay a man cut gave a good indication of the amount of land he could successfully till, and those who were forced to till more than they had manure to support came up short in yields.

45. The dung from two tons of hay alone wouldn't have gone far. But this manure was not going to *all* the tilled land each year, only to the corn and potatoes; and more manure was produced from summer pasture (in the cowyard in the evening), from corn stover and rye straw, and from the hog pen. The livestock kept by the farmers along the Bay Road would have produced fifty to sixty loads of manure each year—more, if well managed. That was enough to concentrate on three to five acres of corn and potatoes every year, and no more. *Mass. Ag. Soc. Papers,* 17–19, 22, 42–44.

46. This adage was called an "old proverb" by Elijah Wood in 1866, "Records of the Concord Farmer's Club, 1864–66," CFPL. How old it really was, I cannot say. In practice, an ideal balance was seldom achieved without continual exchange of labor and goods among neighbors. See Pruitt, "Self-Sufficiency."

47. Or even more, given the disappearance of Lincoln.

48. Gross, *Minutemen and Their World,* 214; Kimenker, *Concord Farmer,* 161–62. Gross and Kimenker missed this anomaly in the valuation data because the corrective information available in deeds and wills lay beyond the scope of the studies they undertook. In "Problem of Agricultural Crisis" Gross calculates a more modest decline in pasture productivity based on farmers' own estimates of how many cows their pastures would keep, which is much closer to the picture I give here.

49. Eliot, *Essays Upon Husbandry,* 155. Carolyn Merchant calculates an average of 1.9 acres/cow across a sample of fifteen younger inland towns in 1771. Merchant, *Ecological Revolutions,* 180–81, 279.

50. Lime was hard to come by in eastern Massachusetts. A small amount of lime was dug and burned in Estabrook in Concord's North Quarter, but only enough for a little chimney mortar. Anyone who doubts this is invited to go take a look at this "quarry"—and then step across to view it from the other side. See Wheeler, *Concord: Climate For Freedom*, 66.

51. I have found only one example of land apparently falling back into this "unimproved" state by the mid-eighteenth century, on the other side of Concord. In 1747, John Wood Jr. inherited seventy-eight acres of the homestead which his father had established on the slopes overlooking the Sudbury River in 1709. Forty-eight acres were "improved by mowing, plowing, orcharding and pasturing," while thirty acres consisted of "rough brushy land." This may well have been degraded pasture. MP 25459, 1747. It is likely that by 1771 many, if not most, Concord farms contained some land in a similar condition, but I am not able to determine how much.

52. We read much in environmental history about the European weeds that flourished after American soil was disturbed by Europeans, but within these introduced European pastures it was native American species that emerged as the most persistent weeds.

53. Garrison, *Landscape and Material Life in Franklin County*, 65–79. See also Pruitt, "Self-Sufficiency." Just how many acres of backcountry pasture were owned by Concord farmers, and how many cattle were kept there, is very difficult to calculate because the land was taxed in the many towns in which it lay, not in Concord.

54. Farms in Nine Acre Corner, which were on average younger, reported even more woodland.

55. Schumacher, *Northern Farmer and His Markets*. Some responses to the 1800 inquiry suggested thirty acres would safely supply twenty cords per year; others that one acre per year cut clean would be sufficient. *Mass. Ag. Soc. Papers*, 46. Wood consumption had fallen to twelve to fourteen cords by the early decades of the next century, as fireplaces became smaller and more efficient.

56. MP 10560. Most inventories did not specify the number of cords widows were to receive, but merely "a sufficiency for her fire both winter and summer."

57. Carroll, *Timber Economy*. See also Innes, *Creating the Commonwealth*, 287–96.

58. Ephraim Jones's Wast Book, April 19, 1743, CFPL.

59. This trade went on both in Concord and in Medford and ended in a falling out among the partners that went to court. Joshua Brooks lost the case. Middlesex County Folio Collection, folio 74—group IV, *"Joshua Brooks vs. John Graves and Benjamin Chamberlain,"* 1677. Collection at Massachusetts State Archives, Boston. Gobles and Billings were still working together in 1684, when the South Quarter hired them to repair the caps on the arches of the South Bridge, instructing them to use good white oak or swamp oak (*Quercus bicolor*, whose wood is virtually identical to white oak). CTR 1:105.

60. MRD 39-515, 45-623. Evidence for this mill along Elm Brook south of the Bay Road appears in references to an "old sawmill dam" in eighteenth-century deeds. It appears in no other town records that I have seen and must have been a small, local operation.

61. MP 2790, 1816.

62. Massachusetts Society for Promoting Agriculture, *Papers on Agriculture, 1803*, 50–52.

63. Whitney and Davis, "Thoreau and Forest History," 70–81; Winkler, "Changes at Walden Pond," 205–07.

64. One respondent even criticized most farmers for reserving too much woodland, instead of building better-finished rooms in their houses and conserving fuel instead. *Mass. Ag. Soc. Papers*, 12, 46. The second half of the eighteenth century saw the introduction of the Rumford fireplace and the Franklin stove, which eased pressure on forest by burning wood more efficiently.

65. Livestock that graze in the forest are definitely hard on most trees, but there is little evidence that they cause any significant soil compaction or soil erosion, at least in Eastern forests, as is sometimes alleged. See J. H. Patric and J. D. Helvey, *Some Effects of Grazing on Soil and Water in the Eastern Forest* (USDA, U.S.

Forest Service—Northeast Forest Experimentation Station, NE-GTR-115, 1986), for a thorough review of this literature.

66. William Jones reported in 1792 that Concord produced "great quantities" of onions. Jones, "Topographical Description of Concord," 238.

CHAPTER 9. Epilogue: Beyond the Meadows

1. Thoreau, *Journal* 4:72–73. Thoreau reported a sea breeze which reaches inland as far as Concord on warm afternoons in late spring. According to the survey of the "Noah and Joshua Brooks Farm" itself, Thoreau was at the job May 26, 28, 29, and 31, 1852. Original in CFPL.

2. Thoreau, *Journal* 3:215–16.

3. J. S. Keyes, "Houses in Concord in 1885," CFPL.

4. George M. Brooks, "Nathan Brooks," *Memoirs of Members of the Social Circle in Concord*, 2d ser., 1795–1840 (Cambridge, 1888).

5. Asa Brooks Jr. inherited the old Job Brooks farm in 1816 at age sixteen along with his twin brother, Job. Upon reaching his majority, Job sold his interest to Asa and became an innkeeper in Natick. When Asa sold in 1847, he moved to another farm a mile away on Virginia Road. Leppleman was milking fourteen cows by 1850 and was growing far more corn and hay than most of his neighbors along the road. MP 2790, 1816; MRD 243–82, 267–114, 519–100; U.S. Census Agricultural Schedule, Concord, 1850.

6. George Brooks, "Nathan Brooks," MP2866, 1825.

7. MRD 246–314, 271–100, 282–460, 282–462, 288–167, 324–442, 326–15, 331–457.

8. MRD 446–429.

9. Nathan Brooks Papers, box 1, CFPL; MRD 814–226, 894–571.

10. MRD 707–170; MRD 718–138; MRD 774–373; U.S. Census Agricultural Schedule, Lincoln, 1860, 1870, 1880.

11. MP15106, 1769; MRD 79–324; MP15076, 1804; MRD 266–408, 365–108.

12. MP 15056, 1803; MP 21994, 1822; MRD 1160–21, 282–462; MP 37608. See also Grindall Reynolds, "Ephraim Meriam," *Memoirs of the Social Circle*. Meriam's worth when he died was $34,547, which, after several sizeable bequests to individual brothers and sisters, was divided among his siblings. See also Yocum, *Meriam House*, 51–52.

13. Thoreau, *Journal* 13:84. The 1860 census gives Meriam fifteen cows, though poor ones. That the cows were giving milk in midwinter was a radical shift in management and feeding. U.S. Census Agricultural Schedule, Concord, 1860.

14. MP 37626, 1870; U.S. Census Agricultural Schedule, Concord, 1860; U.S. Census Population Schedule, Concord, 1850, 1860; Keyes, "Houses in Concord"; Yocum, *Meriam House*, 60–64.

15. For a full discussion (and detailed demographic analysis) of this complex transition, see Harris, "People of Concord," 65–138. See also Daniel Scott Smith, "Population, Family, and Society in Hingham, Massachusetts, 1635–1880" (Ph.D. diss., University of California, 1973).

16. Harris, "People of Concord," presents the demographic evidence for migration. See also Gross, *Minutemen and Their World*, 177–82. James A. Henretta, "The Transition to Capitalism in America," in James A. Horvath, Michael Kammen, and Stanley N. Katz, eds., *The Transformation of Early American History: Society, Authority, and Ideology* (New York: Knopf, 1991), Clark, *Roots of Rural Capitalism*, and Merchant, *Ecological Revolutions*, among others, have explored the role of the land market in pulling farmers into commercial activity.

17. Jarvis, *Traditions and Reminiscences of Concord*, 58, 61. MacMahon, "A Comfortable Subsistence," discusses advances in diet and gardens. *Mass. Ag. Soc. Papers*, 21, reports that Massachusetts farmers generally planted one acre of potatoes. Kimenker, "Concord Farmer," 152, confirms that two hundred acres of

potatoes is consistent with 1801 valuation data. Cabbage is not technically a root crop, but like them it can be stored.

18. *Mass. Ag. Soc. Papers,* 17; confirmed by Kimenker, "Concord Farmer," 152.

19. *Mass. Ag. Soc. Papers,* 16, 28. The small quantities of oats and English hay reported on the 1801 valuation suggest that this rotation was being practiced only on part of the tillage land, although Concord farmers may have substituted winter or spring rye for oats after corn. English hay and oats increased slowly to 1830, then took off. Concord's sandy tillage soils were ill-suited to grass, and it appears that much of the spread of English hay, when it came, took place on till upland that was seldom plowed for grain crops. See "Records of Concord Farmers Club," February 23, 1865, January 11, 1866, CFPL.

20. Jarvis, *Traditions and Reminiscences,* 63–64; MacMahon, "Comfortable Subsistence." Wheat from the south and west was largely providing bread to New England's ports by this time and was slowly making its way inland.

21. Jarvis, *Traditions and Reminiscences,* 20, 49.

22. Ibid., 19–20; Gross, "Culture and Cultivation"; letter from John Tuttle, Tilly Buttrick, Nathan Barrett, and Peter Wheeler to the Directors of the Middlesex Canal, August 24, 1807, Papers of the Middlesex Canal Corporation, University of Lowell Library, box 1, folder 14. See Garrison, *Landscape and Material Life,* 36–60, on "enterprising men." The improvements in Massachusetts agriculture between 1771 and 1801 claimed by Rothenberg appear to be largely increases in the volume of production, not productivity. Rothenberg, *Market-places to a Market Economy,* 214–27. Like Vickers, *Farmers and Fishermen,* 289–94, I am skeptical about how much additional surplus farmers could squeeze from the same acreage, at least until the 1820s.

23. These ecological changes are discussed at greater length in Donahue, "Dammed at Both Ends"; Donahue, *Reclaiming the Commons;* and Brian Donahue, "Skinning the Land: Economic Growth and the Ecology of Farming in Nineteenth-Century Massachusetts," a paper presented at the American Social History Society, Chicago, November 1988.

24. Among many examples see Jarvis, *Traditions and Reminiscences;* Horace Greely, *Recollections of a Busy Life* (New York: J. B. Ford, 1868); Samuel Griswald Goodrich, *Recollections of a Lifetime* (New York: Auburn, Miller, Orton, and Mulligan, 1857).

25. The characterization of early New England towns as "Christian utopian closed corporate communities" with deep peasant as well as Puritan foundations, found in Lockridge, *A New England Town,* and other social historians of the 1970s, still provides important insights into some of the roots of this society, even if the capitalist tendencies were underplayed. The framework employed here is derived from Vickers, *Farmers and Fishermen,* 1–10; Taylor, *American Colonies,* 159–86; and much of the language in this paragraph is drawn most directly from Innes, *Creating the Commonwealth,* 308–09.

26. Thoreau, *Journal* 12:209–87. Quote is from page 249.

27. For the full story of the flowage controversy, see Donahue, "Dammed at Both Ends." Meadow haying appears to have persisted longer in the Sudbury Meadows upstream.

Index

Brooks, Asa I, 200, 215; cattle, 174

Brooks, Asa II, 222

Brooks, Benjamin, 148-49, 180, 200

Brooks, Caleb, 116, 131, 146

Brooks, Daniel I: Cedar Swamp, 114, 146, 180-82, 185; land inheritance, 146, 148; meadow drainage, 182-84

Brooks, Daniel II, 146, 148, 185

Brooks, Dorothy, 148-49

Brooks, Ebenezer, 15-17, 19, 148-49, 151, 152, 158, 201

Brooks, Eleazer, 182, 184

Brooks, Ephraim, 200

Brooks, Gershom, 116, 131, 146, 148, 222; orchard, 165

Brooks, Hannah (Mason), 131

Brooks, Hugh, 147, 148, 180

Brooks, Isaac, 178, 221-23, 224

Brooks, James, 149

Brooks, Job I, 114, 146, 147; meadow drainage, 182-84

Brooks, Job II, 148, 158, 182

Brooks, Job III, 11, 148, 158, 160, 174, 178, 200; orchard, 165; meadow, 167; pasture, 172, 173; woodland, 176

Brooks, John, 200

Brooks, Jonas, 200

Brooks, Joseph I: Cedar Swamp, 114; land inheritance, 146-49; meadow drainage, 182-84

Brooks, Joshua I, 11-12, 116, 151, 178, 215; Cedar Swamp, 114, 176; hay trip, 128-34; land inheritance, 146-47

Brooks, Joshua II, 148, 180, 184, 200, 221

Brooks, Joshua III, 18, 158, 200; garden, 162; orchard, 165; meadow, 167; pasture, 172, 173; woodland, 176

Brooks, Joshua IV, 200, 221-22

Brooks, Joshua V, 223

Books, Luke, 200

Brooks, Matthew, 200

Brooks, Nathan, 221-23, 229

Brooks, Noah I, 221; Cedar Swamp, 114, 147, 178, 180; land inheritance, 146-49; meadow drainage, 182-84

Brooks, Noah II, 200; pasture, 173

Brooks, Samuel I, 148, 158, 200; orchard, 165; meadow, 167; pasture, 172; cattle, 173; woodland, 176

Brooks, Samuel II, 200

Brooks, Sarah, 223

Brooks, Thomas I, 11, 116, 128, 131, 138; Second Division, 109, 133; land inheritance, 146

Brooks, Thomas II, 148-49, 158, 180, 200; orchard, 165; meadow, 167; pasture, 172; woodland, 176

Brooks, Timothy, 200

Brooks, William, 200

Brooks Hill, 157, 158

Brooks tannery, 12, 147, 177-79, 222-23

Brooks tavern, 223

Bulkeley, Peter, 69, 77, 80, 88, 106, 107, 131; tribulations, 99-100

Bull, Ephraim, 2, 4

Burning, Native. *See* Fire, Native use of

Buss, William, 69, 105, 109

C

Cambridge Farms. *See* Lexington

Cambridge Turnpike, 223

Carlisle, 153

Cattle, 123-26, 173-74, 208, 213

Cedar, white, 114, 216

Cedar Swamp. *See* Great Cedar Swamp.

Champney, John, 143, 198

Champney, Sarah (Meriam), 143, 198

Charles River, 187

Chestnut Field, 81, 84, 89-90, 146, 147; lots in, 136-39, 176, 177

Cider, 166, 227

Clam Shell Bank, 38, 44

Climate changes, Holocene, 24-26, 33

Coal, 72

Commission of sewers: England, 63; Concord, 94, 170, 186-89

Commons system: England, 58-60, 64-65, 69; Concord, 86-98; decline of, 99-101; persistence of, 117-27

Concord, settling of, 77-78

Concord River, 1, 29, 31, 41, 77, 93, 182, 232; flow of, 186-89; fish runs, 187-91

Concord Village. *See* Acton

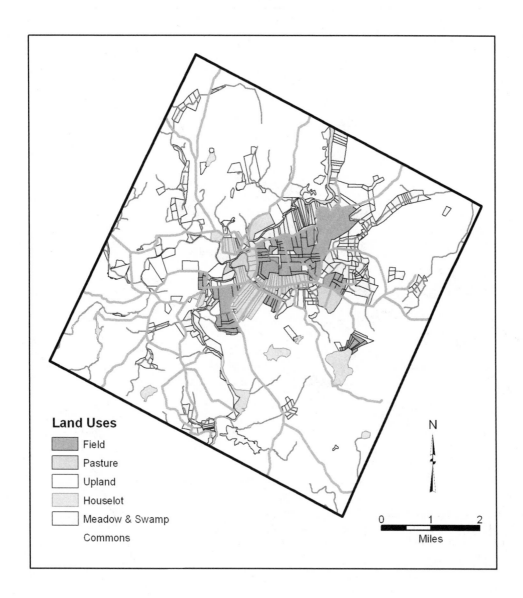

Land Uses

Field

Pasture

Upland

Houselot

Meadow & Swamp

Commons

N

0   1   2

Miles

PLATE I.

The First Division. By 1652, about fifty houselots clustered around the meetinghouse and millpond. Most of the plowlands were grouped within several general fields (some of which encompassed the backs of many houselots as well). A few parcels of detached upland were granted. Meadow lots flanked the rivers and brooks. A few small common pastures were enclosed, but about three-quarters of the town was left open as common grazing and forest.

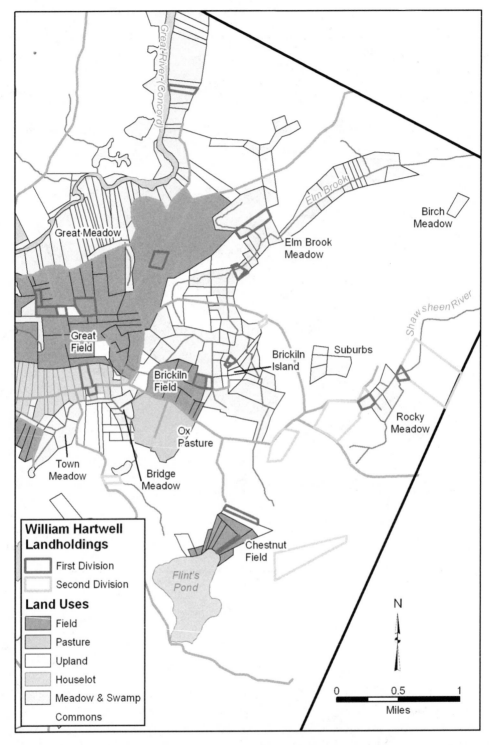

PLATE 2.

William Hartwell: First and Second Division. Hartwell's First Division included his houselot;
lots in the Great Field, the Brickiln Field, and the Chestnut Field; and mowing in the Great
Meadow, Elm Brook Meadow, and Rocky Meadow. His larger Second Division grants
were mostly on the uplands near Rocky Meadow two miles east of his house,
and were settled by his grandsons nearly half a century later.

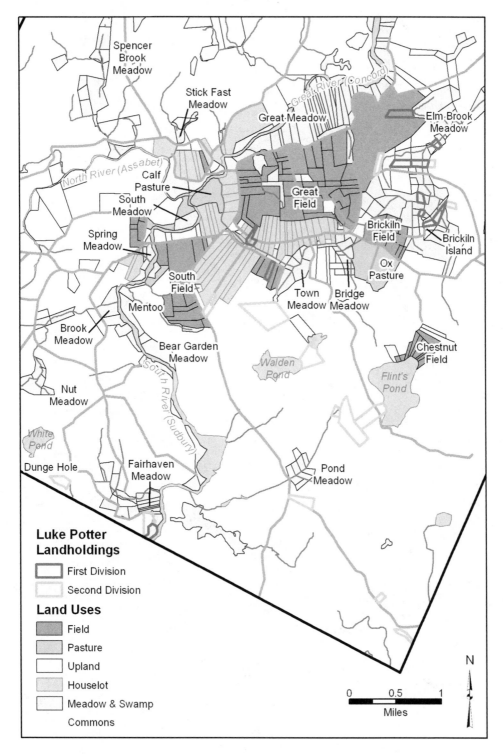

Spencer Brook Meadow

Stick Fast Meadow

Great River (Concord)

Great Meadow

Elm Brook Meadow

North River (Assabet)

Calf Pasture

South Meadow

Great Field

Spring Meadow

Brickiln Field

Brickiln Island

Ox Pasture

South Field

Mentoo

Town Meadow

Bridge Meadow

Brook Meadow

Bear Garden Meadow

Walden Pond

Chestnut Field

Flint's Pond

Nut Meadow

South River (Sudbury)

White Pond

Dunge Hole

Fairhaven Meadow

Pond Meadow

Luke Potter Landholdings

First Division
Second Division

Land Uses

Field
Pasture
Upland
Houselot
Meadow & Swamp
Commons

0   0.5   1

Miles

N

PLATE 3.

Luke Potter: First and Second Division. Potter's First Division included his houselot by the millpond; tillage lots in a small field across Town Meadow and in the Great Field; and mowing in Elm Brook Meadow and distant Fairhaven Meadow. His Second Division upland, woodland, and swamp grants were widely scattered, mostly across the South Quarter. Luke's grandson Samuel relocated the family homestead to the upland between the South Field and Walden during the early eighteenth century.

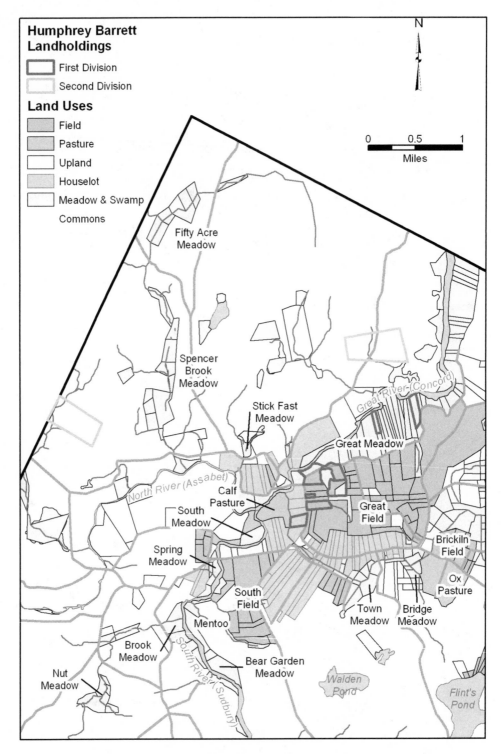

PLATE 4.

Humphrey Barrett: First and Second Division. Barrett's First Division included his houselot, and a set of tightly clustered plowlands and mowing lots in the small field across from his house, the Great Field, and the Great Meadow. His Second Division consisted of two solid 100 acre blocks, settled by his grandson and great-grandson in the eighteenth century.

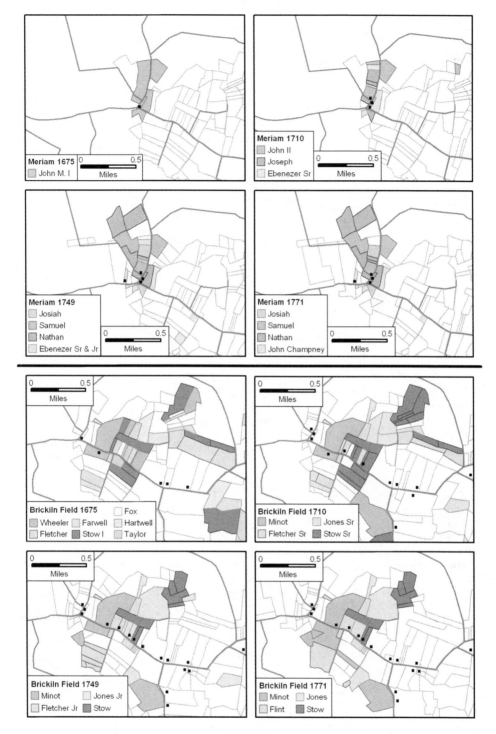

PLATE 5.

Meriam and Brickiln Land Ownership. John Meriam's homelands were split into intermixed parcels by his three sons John, Joseph, and Ebenezer. By 1749 more homeland had been acquired, and was split among Ebenezer (who owned in common with his son) and his three nephews. 1771 shows no further subdivision, or consolidation. Aside from the Minot farm, the Brickiln Field neighborhood was not aggregated into homesteads until the third generation, and did not change much through the rest of the colonial era.

PLATE 6.

Brooks and Hartwell Land Ownership. Thomas Brooks's large grants were homesteaded by his three sons Caleb, Gershom, and Joshua. Over the next two generations this land grew increasingly subdivided and crowded, but by the end of the era in 1771 fragmentation had ceased. Farms on the uplands to the east, settled by Samuel Hartwell and Benjamin Whittemore of the third generation, were each subdivided once but remained in solid blocks. By 1771, Samuel Hartwell actually occupied about half of his father Ephraim's farm (but the land was not confirmed to him until 1794), while the two Whittemore farms had been sold from the family.

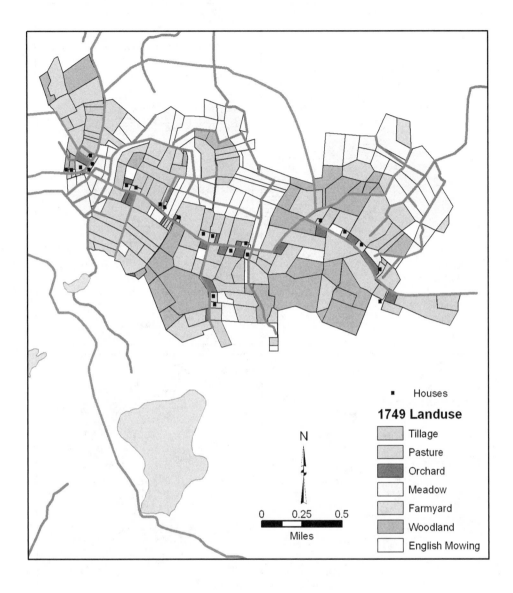

Houses

1749 Landuse

Tillage
Pasture
Orchard
Meadow
Farmyard
Woodland
English Mowing

N

0   0.25   0.5
Miles

PLATE 7.

1749 East Quarter Land Use. Clusters of homesteads (often tied by kinship) were located near patches of tillable land that were subdivided among them—except on the eastern uplands, where settlement was more dispersed. Orchards were strung along the road, and broad meadows, divided into many small lots, flanked carefully ditched brooks. Very little upland "English" hay had been planted at this date. Pastures were spread throughout the landscape on a wide variety of soils, and a good deal of woodland remained, often on rocky hills and in remote swamps.

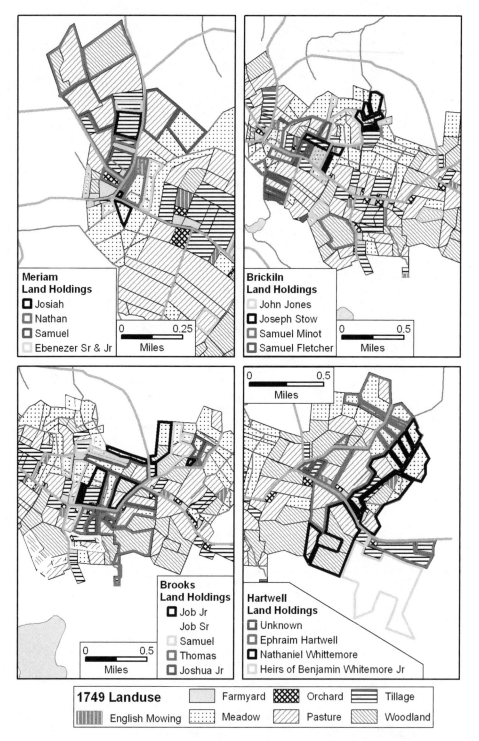

Meriam Land Holdings
- Josiah
- Nathan
- Samuel
- Ebenezer Sr & Jr

0   0.25
Miles

Brickiln Land Holdings
- John Jones
- Joseph Stow
- Samuel Minot
- Samuel Fletcher

0   0.5
Miles

0   0.5
Miles

Brooks Land Holdings
- Job Jr
- Job Sr
- Samuel
- Thomas
- Joshua Jr

0   0.5
Miles

Hartwell Land Holdings
- Unknown
- Ephraim Hartwell
- Nathaniel Whittemore
- Heirs of Benjamin Whitemore Jr

1749 Landuse
- Farmyard
- Orchard
- Tillage
- English Mowing
- Meadow
- Pasture
- Woodland

PLATE 8.

1749 Land Use: Meriam, Brickiln, Brooks, and Hartwell. These maps show details of landholdings spread over different kinds of land. Note the intricately subdivided tillage and meadow lots in the first three neighborhoods, and the more consolidated pattern found on the uplands surrounding the Hartwell Tavern.